Everything you always wanted to know about…

Physics

4th edition

STERLING
Education

STERLING
Education

4 3 2 1

ISBN-13: 978-1-9475566-2-1

Sterling Education
6 Liberty Square #11
Boston, MA 02109

info@sterling-prep.com

Published by Sterling Education

 Printed in the U.S.A.

Our guarantee – the highest quality education materials.

Our books are of the highest quality and error-free.

Be the first to report a typo or error and receive a
$10 reward for a content error or
$5 reward for a typo or grammatical mistake.

info@sterling–prep.com

*We reply to all emails – **check your spam folder***

From the foundations of Newtonian physics to atomic and nuclear theories, this clearly explained text is a perfect guide for those who want to become knowledgeable about topics in college physics. This book can be used as a refresher or for the introduction and application of essential physics. As it navigates through the material, it provides readers with the information necessary to define and understand physics concepts. Readers will also develop the ability to comprehend basic physical laws that govern our universe, as well as skills to apply the theoretical knowledge by solving the included questions that are explained in complete detail.

This book was designed for those who want to develop a better understanding of our physical universe, as well as the relationships between different laws of physics. The content is focused on an essential review of all major physics theories, principles, and experimental approaches.

Highly qualified physics instructors navigate a journey through this fascinating, complex, and sometimes incomprehensible world of physics as they share their years of experience in applied physics and decades in the classroom. This book educates and empowers the reader, regardless of whether they took college physics or are unfamiliar with the basics.

Develop and increase your understanding of how physical phenomena continuously intersect with our daily life. You will learn about kinematics and dynamics, statics and equilibrium, foundations of gravity, energy, work, sound and light, electricity and magnetism, basic principles of atomic physics, as well as heat and thermodynamics.

The editors sincerely hope that this guide is a valuable resource for learning fundamental concepts of physics.

Our Commitment to the Environment

Sterling Test Prep is committed to protecting our planet's resources by supporting environmental organizations for conservation, ecological research, education, and the preservation of vital natural resources. A portion of our profits is donated to help these organizations continue their critical missions.

Since 1992, the Ocean Conservancy advocates for a healthy ocean by supporting sustainable solutions based on science and cleanup efforts. Ocean Conservancy laid the the groundwork for an international moratorium on commercial whaling and played an instrumental role in protecting fur seals from overhunting and banning the international trade of sea turtles.

Since 1988, the Rainforest Trust saves critical lands for conservation through land purchases and protected area designations. Rainforest Trust has helped protect more than 72 protected areas in over 16 countries, such as the Falkland Islands, Costa Rica, Peru.

Since 1980, Pacific Whale Foundation saves whales from extinction and protects oceans through science and advocacy. As an international organization, with ongoing research projects in Hawaii, Australia, and Ecuador, PWF participates in global efforts for threats to whales and other marine life.

Your purchase helps support environmental causes around the world.

Table of Contents

Table of Contents *(continued)*

Table of Contents *(continued)*

Table of Contents *(continued)*

Table of Contents *(continued)*

Table of Contents *(continued)*

Table of Contents *(continued)*

Chapter 1

Kinematics and Dynamics

- **Introduction: The Nature of Science**
- **Units and Dimensions**
- **Vectors, Components**
- **Vector Addition**
- **Speed, Velocity**
- **Acceleration**
- **Freely falling bodies**
- **Chapter Summary**

Introduction: The Nature of Science

Observation is the first step toward scientific discovery. Observations recognize patterns and anomalies in unexplained phenomena. From objective observation, a hypothesis is advanced to propose the reason for the observation.

A *hypothesis* is a proposed explanation for an observation. The testing of a hypothesis involves experiments and additional observations (i.e., recording and collecting data). By manipulating discrete variables, such as how long the plant is exposed to sunlight per twenty-four hours, additional observations are made. The additional observations are the data collected about the proposed explanation (hypothesis).

Observations either 1) support the hypothesis (i.e., not the same as proves) or 2) refute the hypothesis. The hypothesis can then be modified by narrowing or expanding its proposed application to explain additional data. More experiments are performed, and the results either support (not prove) or refute the hypothesis. This is an iterative process.

Principles are the initially proposed explanations that are specific and apply to a narrow range of phenomena.

A *theory* is proposed to account for the observations if the hypothesis is valid over a range of variables (e.g., sunlight, temperature, or humidity). Theories are proposed to explain the phenomena and to predict what will happen under other conditions with different variables. Further observations determine whether the predictions were accurate.

Research (re-search; again search) continues as these many observations support hypotheses and theories. When theories are accurate over a broad scope of scenarios and experiments, Laws are developed based on these repeatedly replicated results.

Laws are brief descriptions of how nature behaves in a broad set of circumstances. A Law is the consolidation of several theories. They are the products of multiple theories that have been tested and proven, culminating in a broad claim regarding those predictions. Laws are simple explanations that are widely applicable. Laws that make broad and straightforward claims are more robust than those who make claims on a narrower, more specific set of variables and parameters.

The relative number of Laws to describe physical phenomena is minuscule compared to the number of theories and many magnitudes less than the number of supported hypotheses. The number of refuted (unsupported) hypotheses may be several magnitudes larger.

A *failed experiment* does not typically refer to the physical process, such as the researcher made a human error during the experiment, but that the data does not support the hypothesis. Therefore, after additional trials to validate the "failed" results, the hypothesis must be modified to predict the results of either future experiments or objective observations in the physical world.

Theories, Models, and Laws

Theories emerge from a hypothesis that has been repeatedly supported by experimental data and observations. Theories are detailed statements that provide testable predictions of the behavior of natural phenomena. They are established to explain observations and then tested (supported or refuted) based on their predictions. Tested theories can articulate the boundaries of the tested hypothesis. Theories are more specific than hypothesis but vaguer and more encompassing than models.

A *model* is constructed to explain in detail the observed behavior if a hypothesis accurately predicts a phenomenon. Models are far more specific in their application than theories. The data collected to develop the theories provide the conceptual foundation upon which models are constructed. The model creates mental pictures (e.g., complicated weather patterns display as visual pictures for the viewer), of the physical phenomena.

The model should be consistent with the predictions of scientific theories and are useful in the process of a comprehensive understanding of a phenomenon. Models allow for predictive outcomes under specific theories (e.g., when the wind speed reaches x, then y will occur). Multiple models are constructed to explain a portion, or the entire scope, of a theory.

Understand the limitations of a model and do not apply it dogmatically unless its applicability is evaluated. A model is an underlying representation of a part of a theory; it is not intended to provide a complete picture of every occurrence of a phenomenon under that theory. Instead, it provides a familiar and understandable way of envisioning the fundamental aspects of the theory.

Units and Dimensions

Measurement and Uncertainty: Significant Figures

No measurement is exact; there will always be some uncertainty due to limited instrument accuracy and precision.

The *accuracy* of an instrument depends on how well-calibrated that instrument is concerning the actual quantitative value of a measurement. For example, an empty scale holding no objects that are initially calibrated to 0 grams of mass is more accurate than the same empty scale calibrated to 1 gram of mass.

An instrument's *precision* is based upon the scale of the measurement, with measurements on a smaller scale being more precise than measurements on a larger scale.

A meter stick that includes millimeter markings (0.001 m) is more precise than a meter stick with only centimeter markings (0.01 m). The estimated uncertainty of a measurement is written with a ± sign. For example, 8.8 ± 0.1 cm reads as "8.8 centimeters, *give or take* 0.1 centimeters."

The *percent uncertainty* is the ratio of the uncertainty to the measured value, multiplied by 100. For example, the percent uncertainty in the measurement is:

$$\frac{0.1}{8.8} \times 100 \approx 1\%$$

The number of *significant figures* is the number of reliably known digits in an expressed value. The number is written to indicate the number of significant figures:

- 22.41 cm has four significant figures.
- 0.058 cm has two significant figures because the initial zeroes before the significant digits are ignored. If it has three significant figures, it is written as 0.0580 cm.
- 70 km has one significant figure (zeroes that serve as placeholders without a decimal point are not considered significant). If it has two significant figures, it is written as 70. km (notice that the decimal point changes the significance of the trailing zero). If it has three significant figures, the value is written as 70.0 km.

When multiplying or dividing numbers, the result should have as many significant figures as the number with the fewest significant figures used in the calculation.

> $12.5 \text{ cm} \times 5.5 \text{ cm} = 68.75 \text{ cm}$, but because 5.5 cm has two significant figures, the value 68.75 cm must be rounded to 69 cm.

When adding or subtracting numbers, the answer is no more precise than the least precise value used. In this case, round the quantity with higher precision to the same decimal place as the quantity with lower precision. For example:

> $68.75 \text{ cm} - 32.1 \text{ cm} = 36.65 \text{ cm}$, but 32.1 cm has three significant figures.

Therefore, 36.65 cm must be rounded to 36.7 cm with three significant figures.

Calculators may not display the correct number of significant figures. Calculators often displaying too many digits, depending on the setting of the instrument.

Order of Magnitude: Rapid Estimating

A quick way to estimate a calculated quantity is to round numbers to one significant figure. The calculated result should be the proper *order of magnitude*.

This estimate is expressed by rounding it to the nearest power of 10.

$$3{,}321 \times 401 = 1{,}331{,}721$$

$$1{,}331{,}721 \approx 13.3 \times 10^5$$

The result can be estimated by rounding:

> $3{,}321$ is rounded to 3×10^3
>
> 401 is rounded to 4×10^2

Calculate:

$$(3.0 \times 10^3) \times (4.0 \times 10^2) = 12 \times 10^5$$

Compare: 13×10^5 (calculated) to 12×10^5 (estimated)

This estimated value is close to the accurate solution (e.g., the difference in the value from estimating compared to the actual calculated value) and of the same order of magnitude. This application of rounding is invaluable because 1) estimating saves time when solving questions, and 2) ensures that the proposed solution is the proper approach to produce a valid answer.

The percent error introduced from estimating:

$$(1.0 \times 10^5) / (12 \times 10^5) \times 100$$

$$= 8.3 \times 10^{-5}\ \% \text{ (or } 0.000083\%)$$

Dimensions and Dimensional Analysis

The dimensions of a quantity are the base units and are expressed using square brackets around the base units (e.g., [m/t] or meters/second).

speed = distance / time

dimensions of speed: [m/t]

Quantities that added or subtracted must have the same dimensions.

Dimensional analysis is useful for checking the approach for solving calculations. A quantity calculated as the solution should have the correct dimensions, so:

- One dimension only contains the magnitude of quantities.
- Two dimensions contain quantities on a 2D plane (x- and y-coordinates).
- Three dimensions contain quantities in 3D space (x-, y-, and z-coordinates).
- Four dimensions contain quantities in 3D space at a given time (x-, y-, and z-spatial coordinates, plus a time coordinate t).

Units, Standards, and the SI System

Quantity	Unit	Standard
Length	Meter (m)	Length of the path traveled by light (in a vacuum) in 1/299,792,458 seconds
Time	Second (s)	Time required for 9,192,631,770 periods of radiation emitted by cesium atoms
Mass	Kilogram (kg)	A platinum cylinder in the International Bureau of Weights and Measures, Paris

Properties of Units in Different Dimensions

A unit is a label for a quantity; like the quantities, units have similar properties:

- unit + unit = unit
- unit − unit = unit
- unit × unit = unit^2

  unless multiplying the quantities of different units (e.g., meter × second = m·s)
- unit / unit = no unit

  unless dividing the quantities of different units (e.g. meter / second = m/s)

- Powers of units represent dimensions:
  - unit = one dimension
  - unit^2 = two dimensions
  - unit^3 = three dimensions
- The product of operations involving International System of Units (SI) units is in SI units.

In the SI system, the basic units are meters, kilograms, and seconds.

Complex units (e.g., for velocity, force, power) are derived from basic SI units.

Basic Translational Motion SI Units		
Quantity	SI Unit	Name
Length	m	meter
Mass	kg	kilogram
Time	s	second

Complex Translational Motion SI Units		
Speed or Velocity	m/s	meter per second
Acceleration	m/s^2	meters per second squared
Area	m^2	square meter
Volume	m^3	cubic meter
Density	kg/m^3	kilogram per cubic meter
Force	N ($kg \cdot m/s^2$)	Newton
Energy	J ($N \cdot m$)	Joule
Power	W (J/s)	Watt

$$Volume = \frac{kg}{1} \times \frac{m^3}{kg}$$

The kilogram terms cancel because they are on opposite sides of the division bar.

Thus, the units for volume are:

$$Volume = m^3$$

As problems become more complex, first check the units to ensure that the solution represents the proper units and that no errors have been introduced.

SI prefixes for indicating powers of 10

Metric (SI) Prefixes		
Prefix	**Abbreviation**	**Value**
peta	P	10^{15}
tera	T	10^{12}
giga	G	10^{9}
mega	M	10^{6}
kilo	k	10^{3}
deci	d	10^{-1}
centi	c	10^{-2}
milli	m	10^{-3}
micro	μ	10^{-6}
nano	n	10^{-9}
pico	p	10^{-12}

Many of these prefixes are familiar, while some are rarely used.

Converting Units

Converting between metric units involves powers of 10.

1 kg = 10^3 g

1 mm = 10^{-6} km

5 s = 5 × 10^3 ms

Converting to and from imperial (SI) units is considerably more involved.

1 m = 3.28084 ft

1 kg = 2.204 lbs.

1 km = 0.621 mile

Use dimension analysis to ensure the conversion is evaluated correctly. Write the conversion factor, and then cancel units during the calculations.

Convert 2.6 kg to grams. (use the conversion of 1 kg = 10^3 g)

2.6 kg × (10^3 g / kg)

2.6 ~~kg~~ × (10^3 g / ~~kg~~)

2.6 × (10^3 g) = 2.6 × 10^3 g

Convert 3.0 miles to km. (use the conversion of 3.28 ft = 1.00 m)

3.0 miles × (1 km / 0.621 mile)

3.0 ~~miles~~ × (1 km / 0.621 ~~miles)~~

3.0 × (1 km / 0.621) = 4.83 km

Convert 3.0 km to ft. (use the conversion of 3.28 ft = 1 m and 1 km = 10^3 m)

3 km × (1000 m / 1 km) × (3.28 ft / 1 m)

3 ~~km~~ × (1000 m / 1 ~~km~~) × (3.28 ft / 1 m)

3000 m × (3.28 ft / 1 m)

3000 ~~m~~ × (3.28 ft / 1 ~~m~~)

3000 × (3.28 ft) = 9,840 ft

Vectors and Vector Components

Kinematics in One Dimension: Reference Frames and Displacement

Any measurement of position, distance, or speed must be made with respect to a reference frame. For example, if a person walks down the aisle of a moving train, the person's speed with respect to the train is a few miles per hour. The person's speed with respect to the ground outside is much higher; it is the combined speed of both the train and the person walking on the train.

Kinematics describes how objects move with respect to a reference frame.

Within a reference frame, the displacement (movement in position) signifies an object's change in position. There is a difference between *displacement* and *distance.*

The *displacement* (a solid line as shown on the graph below) is how far the object is from its starting point, regardless of how it got there.

The *distance* traveled (a dashed line) is measured along the path (*magnitude* of displacement and the *direction* of travel) of the object.

The distance traveled is 100 meters, but the displacement is 40 meters:

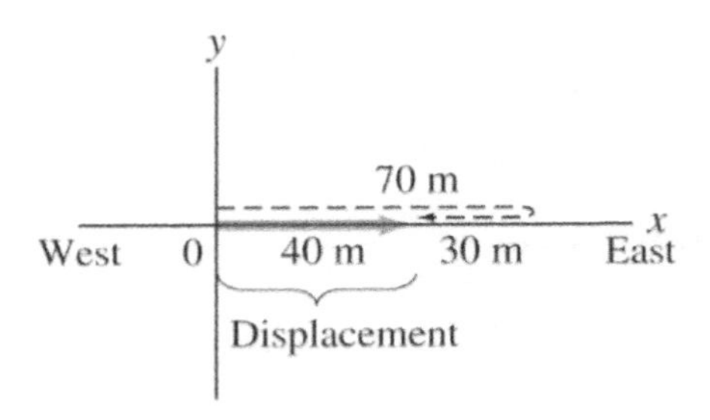

The displacement is expressed as $\Delta x = x_2 - x_1$. Since displacement involves direction, it can be positive or negative about the coordinate directions.

$\Delta x = x_2 - x_1$

$\Delta x = 30\text{ m} - 10\text{ m}$

$\Delta x = 20\text{ m}$

y
x_1 x_2 x
0 10 20 30 40
Distance (m)

An example of negative displacement:

$\Delta x = x_2 - x_1$

$\Delta x = 10 \text{ m} - 30 \text{ m}$

$\Delta x = -20 \text{ m}$

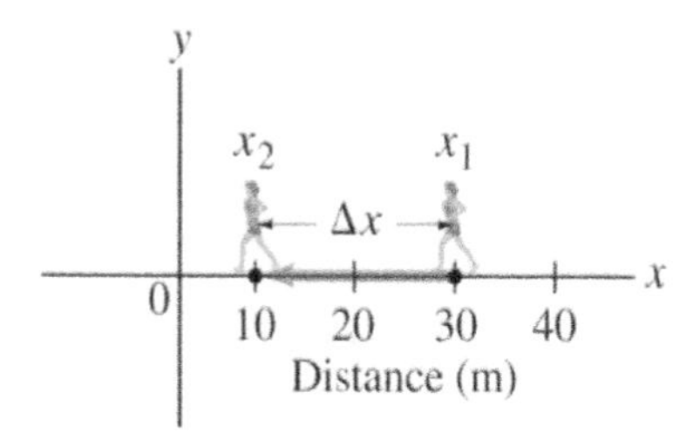

The choice of direction is arbitrary and is selected when choosing a coordinate system. The answer will be correct if the vector components are labeled and computed consistently concerning the chosen coordinate system direction.

Vectors and Vector Components

Scalar quantities represent a magnitude without a direction.

Length, time, and mass are quantities without an associated direction.

Vector quantities include both magnitude and direction.

The displacement, acceleration, and force all inherently contain a direction in which they are applied.

Three trigonometric rules apply when calculating vector components.

A mnemonic is SOH CAH TOA

- **SOH**: *sin θ* = **o**pposite / **h**ypotenuse
- **CAH**: *cos θ* = **a**djacent / **h**ypotenuse
- **TOA**: *tan θ* = **o**pposite / **a**djacent

"Opposite," "adjacent," and "hypotenuse" refer to the vectors about the angle θ.

For example (diagram below):

the vector adjacent to θ_x is v_x

the vector opposite θ_x is the vertical dotted line (equivalent to v_y)

the hypotenuse is v.

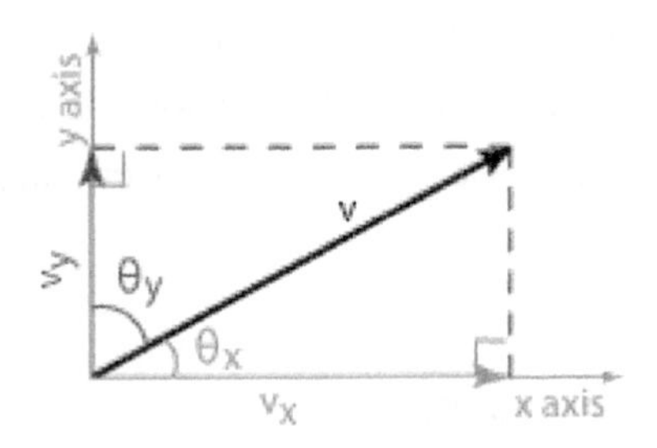

Axis-vector components

Another method to memorize the relationship expressed in "SOH CAH TOA" is from the phrase "**S**ome **O**ld **H**airy **C**amels **A**re **H**airier **T**han **O**thers **A**re."

Vector components are portions of the vector in each direction.

Typically, the components are calculated along the axis directions in a coordinate system.

A vector may have a horizontal component and a vertical component in a two-dimensional (2D) coordinate system (i.e., x-component and y-component):

$$v_x = \vec{v}\cos\theta_x = \vec{v}\sin\theta_y$$

$$v_y = \vec{v}\cos\theta_y = \vec{v}\sin\theta_x$$

The Pythagorean Theorem calculates the magnitude of the resulting vector:

$$\vec{v}^2 = v_x^2 + v_y^2$$

Examples of vectors and vector components in a 2D coordinate system:

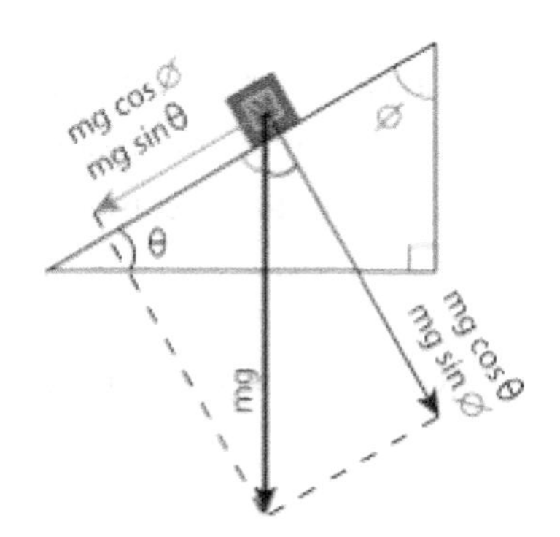

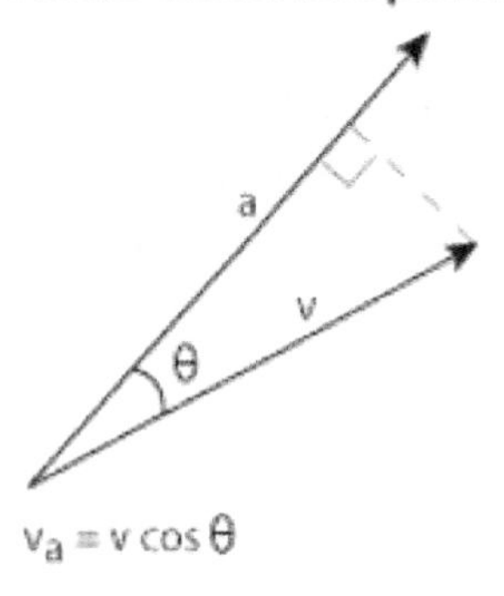

In a 3D coordinate system (i.e., *x*-, *y*- and *z*-axis) a third *z*-component of the vector is included:

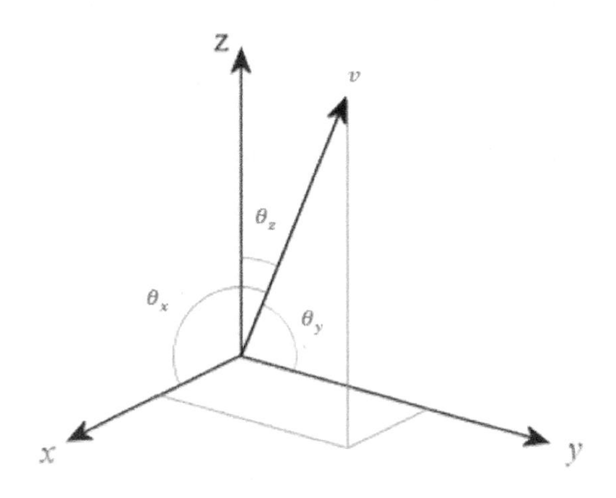

$$v_x = \vec{v}\, cos\, \theta_x$$

$$v_y = \vec{v}\, cos\, \theta_y$$

$$v_z = \vec{v}\, cos\, \theta_z$$

$$\vec{v}^2 = v_x^2 + v_y^2 + v_z^2$$

Vector Addition

Vector addition is performed either graphically or by using the vector components. Vectors can be added if they are in the same dimension (direction). Otherwise, the separate addition of each of the vector's *x*-, *y*- and *z*-components must be calculated.

The sum of all components of a vector equals the vector itself. Therefore, the added components make up the resultant vector.

Operations involving two vectors may or may not result in a vector.

For example, kinetic energy derived from the square of two velocity vectors results in a scalar quantity.

Operations involving a vector and a scalar always result in a vector.

Operations involving a scalar and a scalar always result in a scalar.

For vectors in one dimension, merely addition and subtraction are needed.

The figure below demonstrates vector addition in one dimension (same):

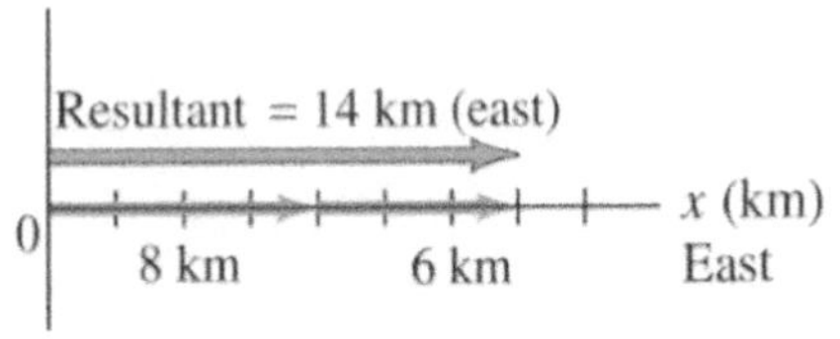

$$8\ \text{km}_{\text{East}} + 6\ \text{km}_{\text{East}} = 14\ \text{km}_{\text{East}}$$

The figure below demonstrates vector subtraction in one dimension (opposite):

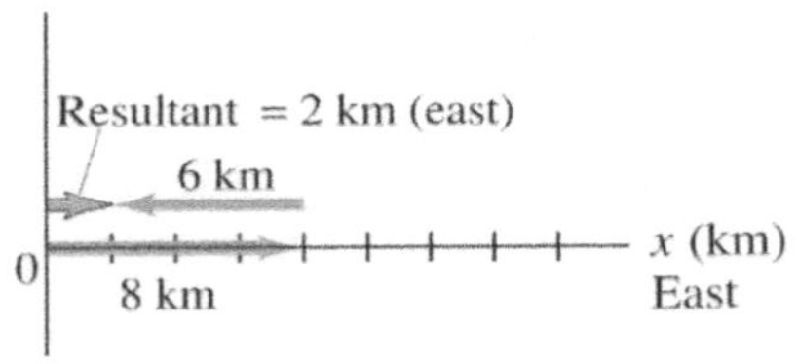

$$8\ \text{km}_{\text{East}} - 6\ \text{km}_{\text{East}} = 2\ \text{km}_{\text{East}}$$

Addition of Vectors — Graphical Methods

If there is motion in two dimensions, the analysis is more complicated.

If the travel paths are at right angles, the displacement (d) is calculated by using the Pythagorean Theorem:

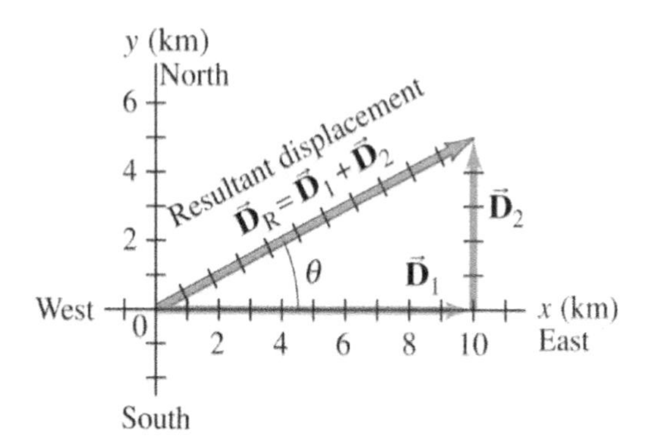

Adding the vectors in the other sequence gives the same result:

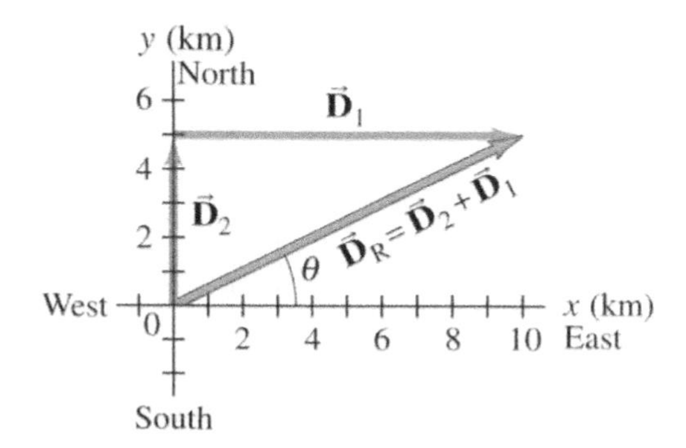

$$|\vec{D}_R| = \sqrt{\vec{D}_1^{\,2} + \vec{D}_2^{\,2}}$$

If the vectors are not at right angles, add using the "tail-to-tip" method.

To use the "tail-to-tip" method, graphically place the tail of each vector at the tip of the arrow of the vector before it.

After the vectors are arranged tail-to-tip, draw the resultant vector from the tail of the original vector to the tip of the last vector.

An example using the "tail-to-tip" method for resolving three vectors:

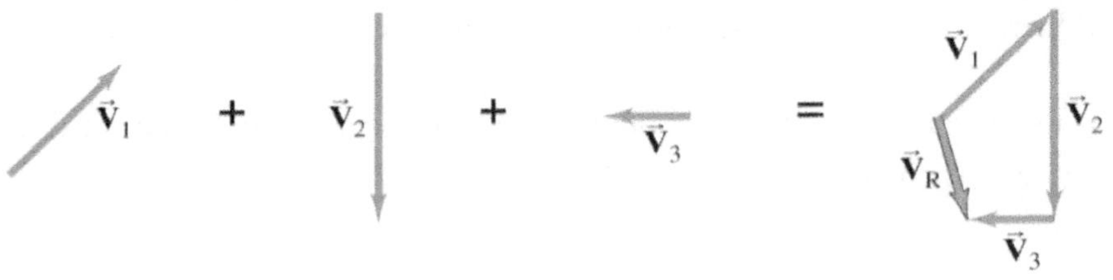

If there are two vectors, use the "parallelogram" method. For the parallelogram method, arrange the two vectors such that their tails are connected. Starting from the tips of the two vectors, draw two dashed lines that complete the parallelogram. The resultant vector is drawn from the tails of the original vectors to the corner of the drawn parallelogram.

An example using the "tail-to-tip" and parallelogram method with two vectors:

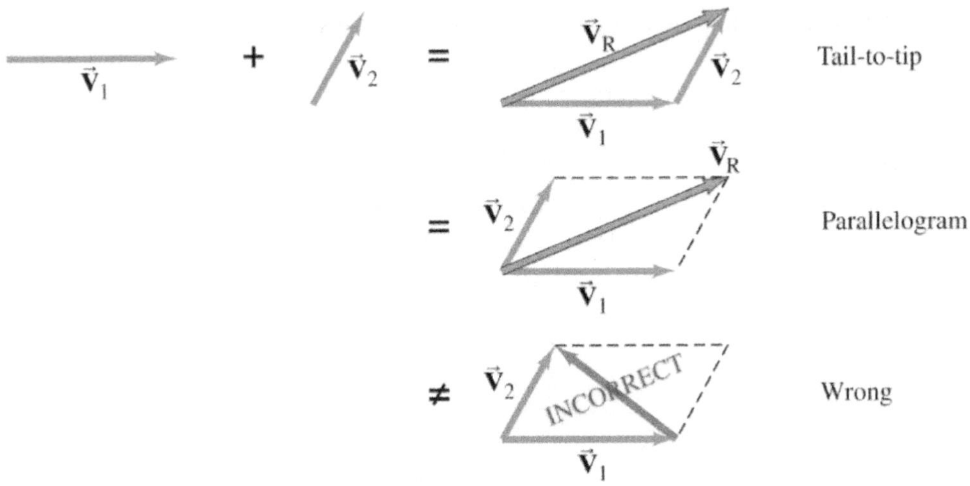

Resolving two vectors using the tail-to-tip and parallelogram method

Subtraction of Vectors and Multiplication of a Vector by a Scalar

To subtract vectors, define the negative direction of a vector that has the same magnitude, but points in the opposite direction.

Then add the negative vectors:

$\vec{V}_2 - \vec{V}_1 = \vec{V}_2 + (-\vec{V}_1) = \vec{V}_2 - \vec{V}_1$

A vector, $\vec{V}$, can be multiplied by a scalar c.

The result is a vector that has the same direction, but a magnitude of $c\vec{V}$.

If c is negative, the resultant vector points in the opposite direction.

An example of a vector, $\vec{V}$ being multiplied by different scalars:

$\vec{V}$ $\quad \vec{V}_2 = 1.5\,\vec{V} \quad \vec{V}_3 = -2.0\,\vec{V}$

Adding Vectors by Components

A vector can be expressed as the sum of its component vectors.

The component vectors are usually chosen so that they are perpendicular:

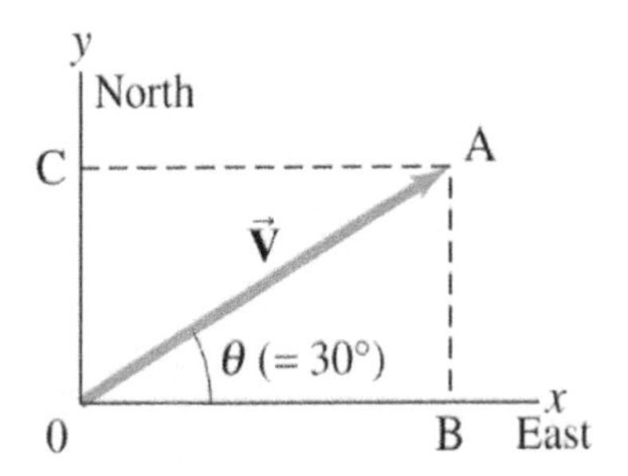

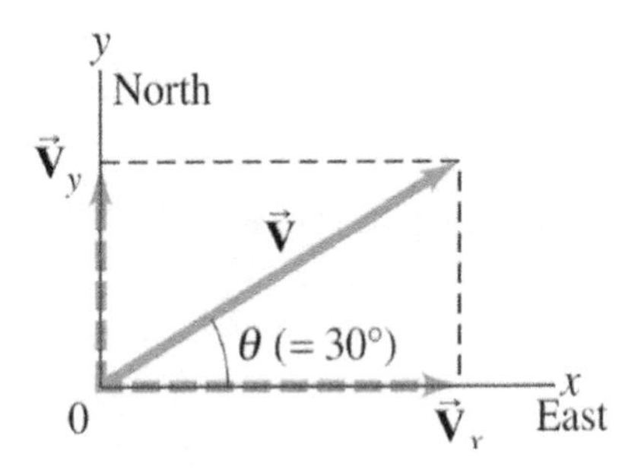

For perpendicular components, calculate the values using trigonometric functions.

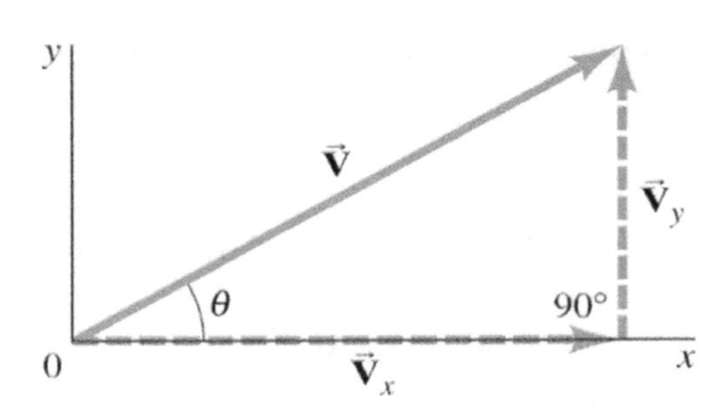

$$\sin\theta = \frac{V_y}{V}$$

$$\cos\theta = \frac{V_x}{V}$$

$$\tan\theta = \frac{V_y}{V_x}$$

$$V^2 = V_x^2 + V_y^2$$

The components are effectively one-dimensional, so add algebraically:

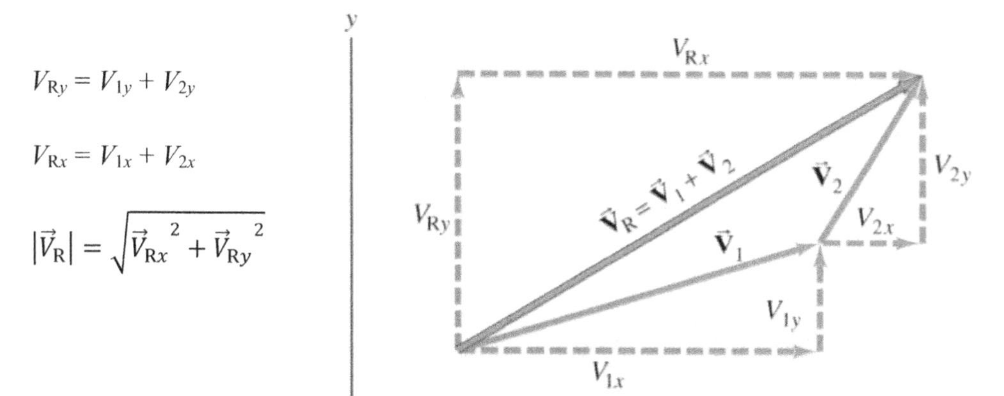

$$V_{Ry} = V_{1y} + V_{2y}$$

$$V_{Rx} = V_{1x} + V_{2x}$$

$$|\vec{V}_R| = \sqrt{\vec{V}_{Rx}^{\,2} + \vec{V}_{Ry}^{\,2}}$$

Resolving vectors from component vectors

Five rules for adding vectors:

1. Draw a diagram of the vectors on a coordinate system.

2. Establish the x and y-axes.

3. Resolve each vector into x and y components.

4. Calculate each component using sines and cosines.

5. Add the components in each direction to determine the resultant vector.

To find the magnitude and direction of the resultant vector, use:

$$\vec{v} = \sqrt{v_x^2 + v_y^2} \qquad\qquad tan\ \theta = \frac{v_y}{v_x}$$

Speed and Velocity

Speed is the quantity of how far an object travels in each time interval.

Speed is the rate of change in the *distance*; it does not include a direction and is a scalar quantity.

Average speed is the distance (scalar) traveled divided by the time elapsed:

$$\text{Average speed: } speed_{average} = v_{avg} = \frac{d}{t} = \frac{distance\ traveled}{time\ elapsed}$$

The *velocity* quantifies the speed of an object in a direction.

Velocity is the rate of change in *displacement*. Since velocity includes direction, it is a vector quantity (unlike speed, which is a scalar quantity).

Average velocity is the displacement (vector) divided by the elapsed time:

$$velocity_{average} = \vec{v}_{avg} = \frac{s}{t} = \frac{displacement}{time\ elapsed} = \frac{final\ position - initial\ position}{time\ elapsed}$$

Instantaneous speed is the speed at a point in time (infinitesimal time interval).

Instantaneous velocity is the velocity at a point in time (infinitesimal time).

Instantaneous velocity is the mathematical limit (i.e., time becomes infinitesimally small) of the displacement over this imperceptible time interval:

$$\vec{v}_{inst} = \lim_{\Delta t \to 0} \frac{\Delta x}{\Delta t}$$

The instantaneous velocity is a vector and has a direction.

The instantaneous speed is not a vector and does not indicate a direction.

The *direction of instantaneous velocity* is tangent to the path at that point. Its magnitude is equal to that of the instantaneous speed at that point.

The graphs show *constant velocity* (left) and *varying velocity* (right):

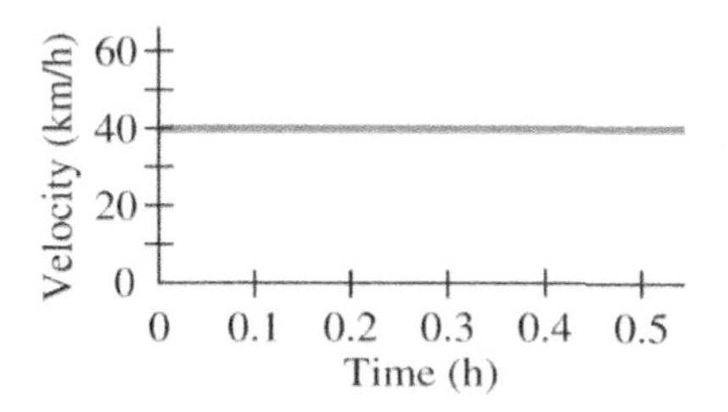

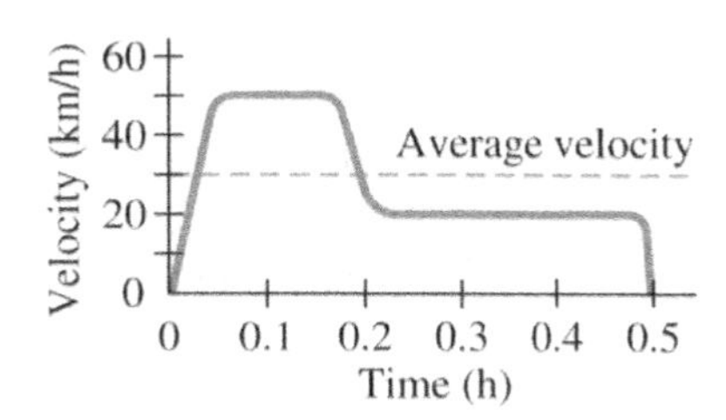

Relative Velocity

The relative speed moves in two dimensions and is calculated by adding or subtracting their magnitudes.

The velocities are vectors that must be added and subtracted accordingly.

Each component velocity is 1) labeled with the object and 2) with the reference frame in which it has this velocity.

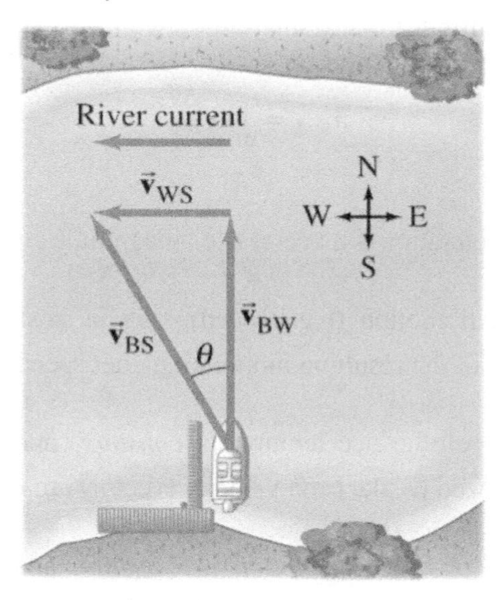

$\vec{v}_{BS}$ is the velocity of the boat relative to the shore

$\vec{v}_{BW}$ is the velocity of the boat relative to the water

$\vec{v}_{WS}$ is the velocity of the water relative to the shore

The relationship between the three velocities is:

$$\vec{v}_{BS} = \vec{v}_{BW} + \vec{v}_{WS}$$

Acceleration

Acceleration is the rate of change in the velocity of an object and is expressed as:

$$average\ acceleration = \vec{a}_{avg}$$

$$\vec{a}_{avg} = \frac{\Delta\vec{v}}{\Delta t}$$

$$\frac{\Delta\vec{v}}{\Delta t} = \frac{\vec{v}_f - \vec{v}_i}{t} = \frac{change\ in\ velocity}{time\ elapsed}$$

From the calculated average acceleration, the instantaneous acceleration of an object can be determined.

The *instantaneous acceleration* is the average velocity over the limit of the time interval as time becomes infinitesimally small:

$$\vec{a}_{inst} = \lim_{\Delta t \to 0} \frac{\Delta\vec{v}}{\Delta t}$$

Like velocity, acceleration is a vector (i.e., magnitude and direction).

In one-dimensional motion (e.g., speed), only a positive or negative sign is needed for the calculation of the resulting motion (e.g., net speed).

If the magnitude of the acceleration is *constant,* and there is *no change in direction*, the values for speed (scalar) and velocity (vector) are interchangeable.

If the magnitude of the acceleration is *constant,* and there is *no change in direction*, the values for distance (scalar) and displacement (vector) are interchangeable.

Reference the direction of the acceleration.

Negative acceleration and deceleration describe different phenomena.

- *Negative acceleration* is an acceleration in the negative direction as defined by the coordinate system (e.g., a booster rocket dislodged from the satellite). The direction of the acceleration is independent of velocity.

- *Deceleration* is relative to the object's velocity; it occurs when the acceleration is opposite in direction to the velocity (e.g., a car skidding to a stop).

If the object is traveling in a positive direction (i.e., the velocity is positive), then deceleration and negative acceleration are the same.

Positive acceleration (i.e.., car advances) is shown:

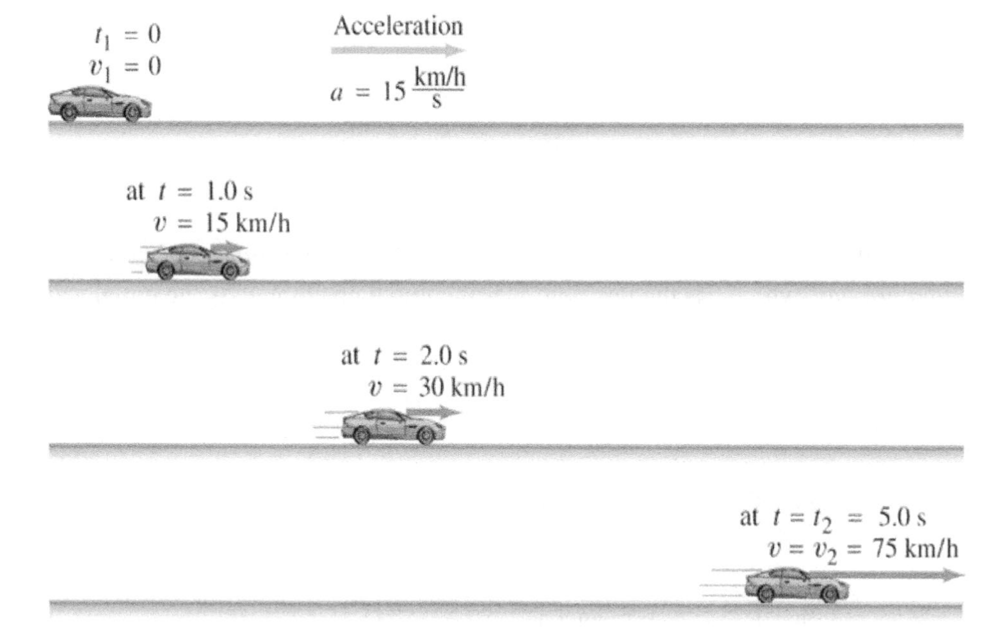

Negative acceleration (i.e., an acceleration in the negative direction) is shown:

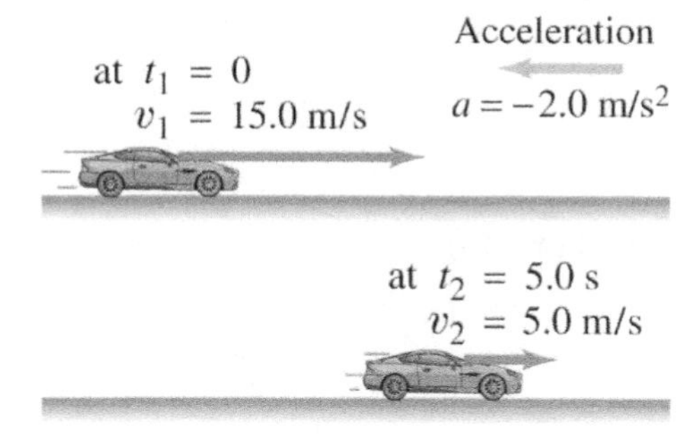

Motion at Constant Acceleration

The *average velocity* of an object, during a time interval t is:

$$\bar{v} = \frac{\Delta x}{\Delta t} = \frac{x - x_0}{t - t_0} = \frac{x - x_0}{t}$$

The *constant velocity* expression is:

$$\bar{v} = \frac{v_0 + v}{2}$$

The *constant acceleration* expression is:

$$\vec{a} = \frac{v - v_0}{t}$$

Combine the expressions for *average velocity*, *constant acceleration,* and *constant velocity*; where x is the position, x_0 is the initial position, v_0 is the initial velocity, t is the time:

$$x = x_0 + \bar{v}t$$

$$x_0 + \bar{v}t = x_0 + \left(\frac{v_0 + v}{2}\right)t$$

$$x_0 + \left(\frac{v_0 + v}{2}\right)t = x_0 + \left(\frac{v_0 + v_0 + at}{2}\right)t$$

or simplified as:

$$x = x_0 + v_0 t + \frac{1}{2}at^2$$

Solve to eliminate t:

$$v^2 = v_0^2 + 2a(x - x_0)$$

The equations needed to solve constant acceleration problems:

- $v = v_0 + at$
- $x = x_0 + v_0 t + \frac{1}{2}at^2$
- $v^2 = v_0^2 + 2a(x - x_0)$
- $\bar{v} = \frac{v_0 + v}{2}$
- $s = v_{avg}t$
- $v_{avg} = \frac{v_f + v_i}{2}$
- $a = \frac{\Delta v}{\Delta t} = \frac{v_f - v_i}{t}$
- $v_f^2 = v_i^2 + 2as$
- $s = \frac{1}{2}at^2 + v_i t$
- $l_f = l_i + s$

Learn to rearrange, combine, and apply the equations appropriately.

When solving problems with these equations, assign one direction as positive and the opposite as negative; use this orientation for all subsequent calculations.

In Cartesian coordinates (i.e., horizontal x and vertical y-axes), take upward and rightward motions as positive, and downward and leftward motions as negative.

For free falls, take downward (gravitational acceleration) as positive.

The direction for motion can be assigned arbitrarily, provided that the opposite direction has the opposite sign.

Graphical Analysis of Linear Motion

A graph of the x *vs.* t (i.e., distance *vs.* time) for an object moving with constant velocity is shown below.

Velocity is the slope (rise/run or x- / y-) of the x *vs.* t (i.e., distance *vs.* time) graph.

If the slope is positive, the velocity is positive.

If the slope is negative, the velocity is negative.

The slope of the line is linear (straight line) because the velocity is constant:

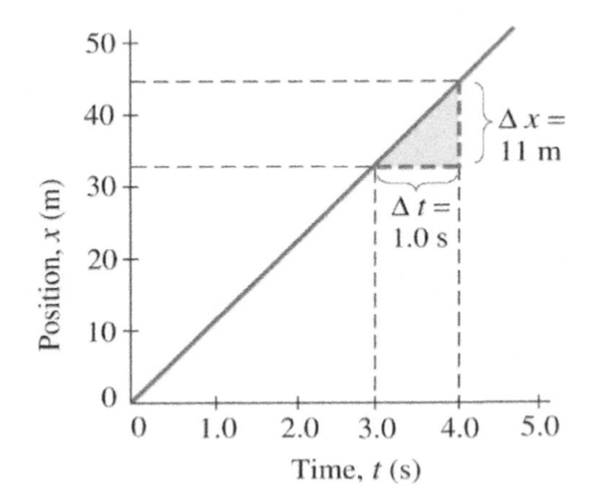

Position vs. time graph shows a linear slope for constant velocity

Freely Falling Bodies

Motion with constant acceleration is an object in *free fall* (e.g., an apple falling to the ground from the tree).

Near the Earth's surface, objects experience approximately the same acceleration due to gravity ($g = 9.8$ m/s^2 or use 10 m/s^2 for ease of calculations).

The differences in acceleration due to gravity between various objects (e.g., feather and marble) are related to air resistance. A larger surface area means higher air resistance (e.g., a flat sheet of paper compared to a crumpled sheet of the same paper).

In a vacuum (i.e., no air resistance), gravity causes all objects to fall with the same acceleration ($g = 9.8$ m/s^2). Acceleration is constant because of no air resistance.

Whenever an object is in the air, it is in free fall, regardless of whether it is tossed upwards, downward, or at an angle. For objects that are dropped, label the downward direction as positive, so g (acceleration) is positive.

For an object thrown downward, both the initial velocity and g are positive.

For an object thrown upward, initial velocity has the opposite sign to acceleration.

Choose either the upward or the downward direction as positive.

Projectiles

A projectile is an object moving in two dimensions under the influence of the Earth's gravity. The projectile's path of travel is a parabola. The object's motion along this path is *projectile motion*, is described through its vertical and horizontal components.

The speed in the horizontal (x-direction) is constant.

The speed in the vertical (y-direction) is constant acceleration g.

Projectiles are freely falling bodies, so the vertical component of the projectile motion is accelerating toward the Earth at a rate of g. Even when the projectile is traveling upward, the downward acceleration of gravity causes the projectile to decelerate when traveling upward. There is no acceleration due to gravity in the horizontal component.

The *horizontal component* of velocity is constant. The time that elapses while the projectile is in motion is dependent on only the vertical (not horizontal) component of the projectile motion.

The horizontal distance covered by the projectile is determined from the elapsed time during the projectile's path multiplied by its horizontal speed (speed × time).

For an object tossed straight up, and it comes down to where it started, and the total displacement s for the trip is zero.

The initial velocity and final velocity are equal and opposite:

$$(\vec{v}_{\text{final}} + \vec{v}_{\text{initial}} = 0)$$

The acceleration is opposite in sign to the initial velocity.

Two balls that start to fall at the same time are illustrated below. The ball on the right in each figure has an initial velocity in the x-direction, while the ball on the left was dropped straight downward (i.e., zero initial velocity in the x-direction).

The vertical positions of the two balls are identical at identical times, while the horizontal position of the ball on the right increases linearly.

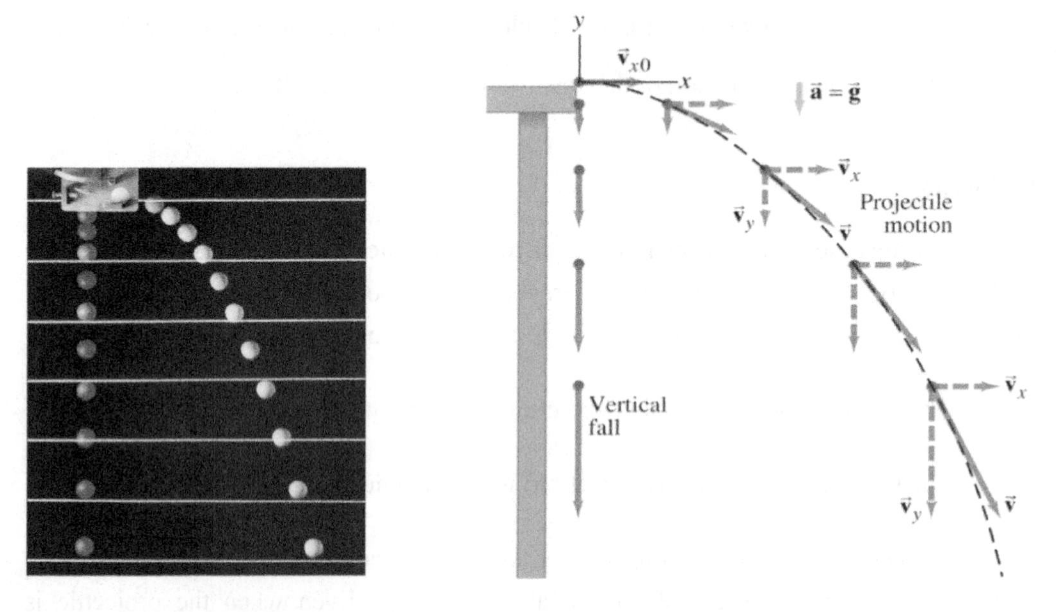

Comparison of a ball dropped vs. projected with a horizontal velocity

The object is launched at an initial angle of θ_0 with the horizontal. The analysis is similar except that the initial velocity has both a *horizontal* and *vertical* component.

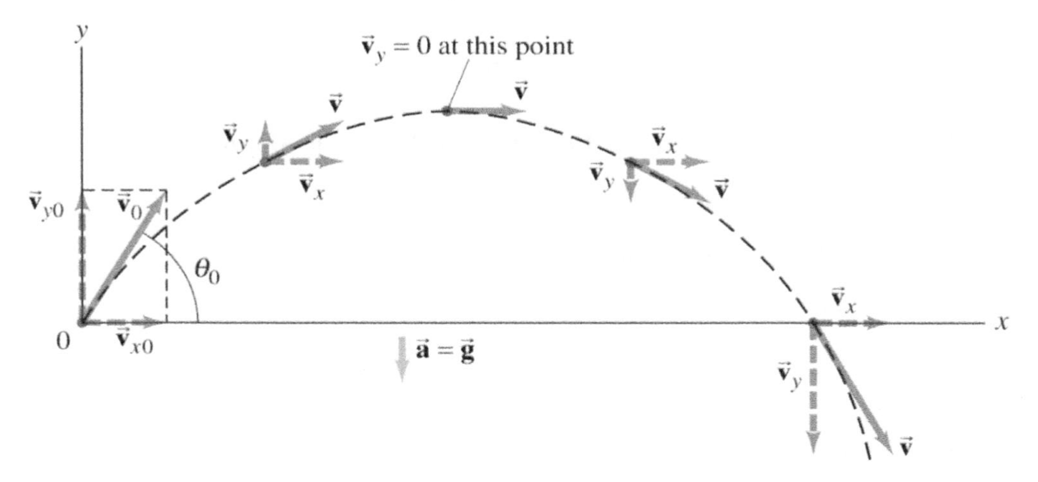

Projectile with both x- and y-component for velocity

Projectile Motion Is Parabolic

To demonstrate that projectile motion is parabolic, write y as a function of x.

A *parabola* is expressed by the equation:

$$y = Ax^2 - Bx + C$$

where A, B, and C are constants.

A is a scalar (magnitude) multiplier of the *object's acceleration.*

B as a scalar multiplier of the *velocity.*

C as a scalar multiplier of the object's *initial position.*

The *direction of the object's acceleration* is determined by the signs associated with these constants for determining the velocity and initial displacement from the origin.

Solving Projectile Motion Problems

Projectile motion is motion with a constant acceleration in one dimension.

For acceleration due to gravity, the direction is downward.

Kinematic Equations for Projectile Motion (y positive downward; $a_x = 0$, $a_y = g = 9.8$ m/s^2)	
Horizontal Motion ($a_x = 0$, $v_x =$ constant)	Vertical Motion ($a_y = g =$ constant)
$v_x = v_{x0}$	$v_y = v_{y0} - gt$
$x = x_0 + v_{x0}t$	$y = y_0 + v_{y0}t - (\frac{1}{2})gt^2$
	$v_y^2 = v_{y0}^2 - 2g(y - y_0)$

Projectile motion problems may be solved by following seven steps:

1. Read the problem carefully and choose the object(s) to be analyzed.
2. Draw a diagram of the described actions.
3. Choose a point of origin and a coordinate system.
4. Determine the time interval;

   the same in both directions, and

   includes only the time the object is moving, with constant acceleration g.
5. Evaluate the x and y-components of the motion separately.
6. List the known and unknown quantities.

   the v_x never changes, and

   the $v_y = 0$ at the highest point.
7. Use the appropriate equations; some of them may have to be combined.

Free Fall *vs.* Non-Free Fall

In free fall, the only force acting on the object is gravity.

Assuming no air resistance, the only resistance to change is the object's inertia.

All objects fall at a *constant acceleration* of 9.8 m/s^2 downward. Since Earth is not a vacuum, air resistance (force opposing the motion downward) is generally accounted for when objects are falling.

In a *non-free fall*, the upward force due to air resistance counteracts the downward force due to gravity. The net force of this interaction is the difference between the force due to gravity and air resistance.

Therefore, the total force on an object in a non-free fall is less than that of the same object in free fall, where the force from air resistance is zero.

As an object falls, the force from air resistance increases with the increased speed of the object, and the force downward decreases until the two opposing forces are equal in magnitude. The acceleration of the object is zero. The *terminal velocity* is the point at which the object is traveling at a constant velocity downward due to the maximum force of air resistance. Terminal velocity is achieved when air resistance equals the weight of the object. Weight is the mass under the influence of gravity. The mass of an object on Earth and the moon are the same. However, the weights are different due to differences in gravity. Gravity is an attractive force between two bodies. Massive objects exert a tremendous attractive force.

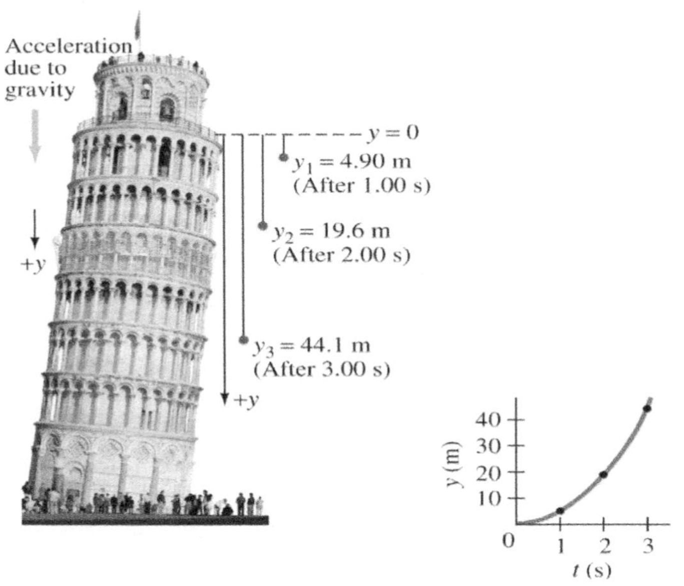

The diagram on the right graphs distance vs. time of the falling object

Satellites and *Weightlessness*

Satellites are routinely put into orbit around the Earth. The tangential speed of the satellite must be enough so that the satellite does not fall to Earth. However, the tangential speed cannot be so high that the satellite escapes Earth's gravity altogether. Satellites orbiting the Earth are in continuous free fall, and their centripetal acceleration equals the acceleration from the Earth's gravity.

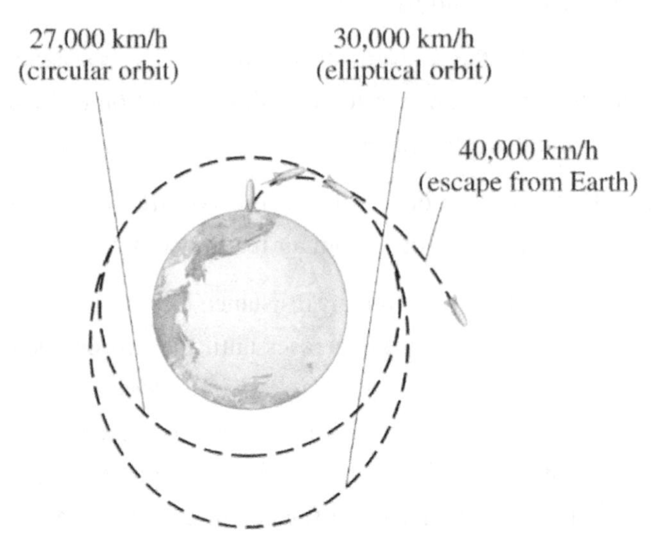

The satellite is kept in orbit around the Earth by its speed. The satellite is continually falling and accelerating toward the Earth. It never crashes into the Earth's surface because the Earth curves away underneath it at the same rate as the satellite falls toward Earth.

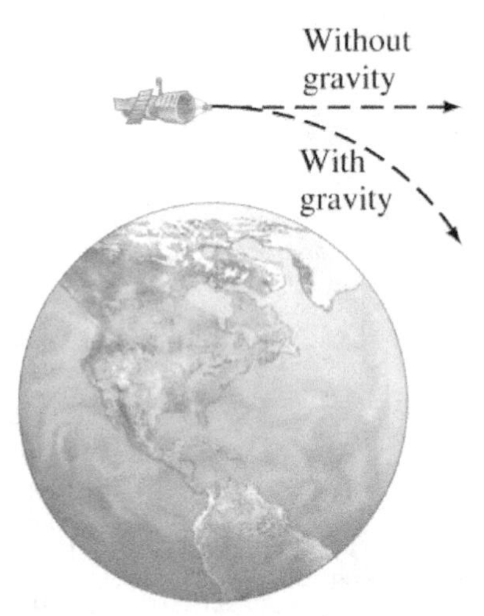

Objects in orbit are said to experience weightlessness, even though they do have a gravitational force acting on them. The satellite and all its contents are in free fall, so there is no normal force, leading to the experience of weightlessness. The term to describe this is *apparent weightlessness* because the gravitational force still exists. Apparent weightlessness can be experienced on Earth, but only briefly.

For example, when a roller coaster rushes down a hill, the normal force temporarily goes to zero, and the riders experience apparent weightlessness.

Real weightlessness occurs when there is no net gravitational force acting on an object. Either the object is in outer space:

1) void of the attractive force of gravity when other massive objects are light-years away, or

2) the object does not experience a net gravitational force because it is between two other objects with equal gravitational forces that cancel.

Chapter Summary

Graphs are useful tools to help visualize the motion of an object. They can also help solve problems once the information contained in a graph is translated to describe a physical phenomenon. When working with graphs, consider the following:

- Always look at the axes of the graph first. A common mistake for the velocity *vs.* time graph is thinking about it as a position *vs.* time graph.
- Do not assume one interval is one unit; inspect the numbers on the axis.
- Align the position *vs.* time graphs *directly above* the velocity *vs.* time graphs and *directly above* the acceleration *vs.* time graphs. This way, relationships and the key points among the graphs can be matched.

  The *slope* of a distance *vs.* time (x *vs.* t) graph gives velocity.

  The *slope* of a velocity *vs.* time (v vs. t) graph gives acceleration.

  The *area* under an acceleration *vs.* time (a *vs.* t) graph gives the change in velocity.

  The *area* under a velocity *vs.* time (v vs. t) graph gives the displacement.

- The motion of an object in one dimension can be described using Five Equations. Identify what is given, determine what must be calculated, and use the equation that has those variables. Note that sometimes there is hidden (or assumed) information in the problem (e.g., acceleration due to gravity $g = -10$ m/s^2)

		Missing variable
#1*	$\Delta x = vt$	a
#2	$v = v_0 + at$	x
#3	$x = x_0 + v_0 t + \frac{1}{2}at^2$	v
#4	$x = x_0 + vt - \frac{1}{2}at^2$	v_0
#5	$v^2 = v_0^2 + 2a(x - x_0)$	T

* Because the acceleration is uniform: $\Delta x = \frac{1}{2}(v_0 + v)t$

For projectiles, separate the horizontal and vertical components.

Horizontal Motion	Vertical Motion
$x = v_x t$	$y = y_0 + v_0 t + \frac{1}{2} g t^2$
$v_x = v_{0x} =$ constant	$v_y = v_{0y} + gt$
$a_x = 0$	$a = g = -10 \frac{m}{s^2}$

At any given moment, the relationship between v, v_x and v_y is given by:

$v^2 = {v_x}^2 + {v_y}^2$

$v_x = v \cos \theta$

$v_y = \sin \theta$

$\theta = tan^{-1}(\frac{v_y}{v_x})$

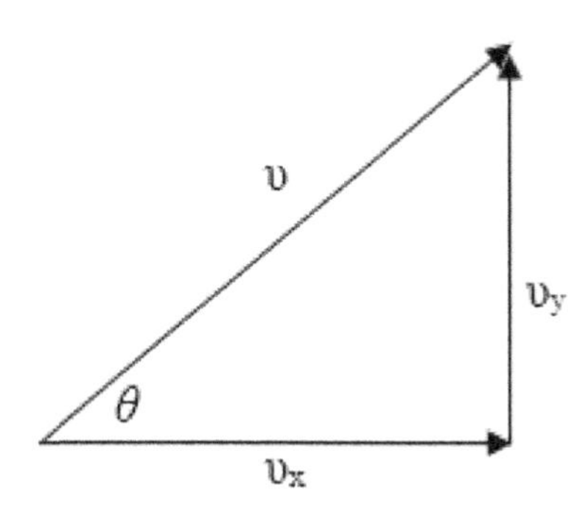

Resolving vectors from the known x- and y-component of the separate vectors

Solving Problems

General guidelines for solving problems involving motion:

1. Read the entire problem. Evaluate what is provided and what is asked to be solved for in the problem.

2. Decide which objects are under study and what the time interval is.

3. Draw a diagram and choose coordinate axes.

4. Evaluate all known quantities, then identify the unknown quantities that need to be determined.

5. Determine which equations relate to the known and unknown quantities. Solve algebraically for the unknown quantities and check that the result is sensible (e.g., magnitude and correct dimensions).

6. Calculate the solution and round to the appropriate number of significant figures.

7. Evaluate if the result is reasonable by comparing if it agrees with a rough estimate or approximations (e.g., positive or negative, the order of magnitude).

8. Check that the units on one side of the equation match the units on the other.

Practice Questions

1. Which graph represents a constant non-zero velocity?

I. 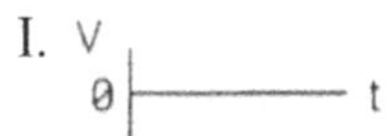

II. V 0 t

III. V 0 t

A. I only
B. II only
C. III only
D. I and II only
E. II and III only

2. A train starts from rest and accelerates uniformly until it has traveled 5.6 km and has acquired a velocity of 42 m/s. The train then moves at a constant velocity of 42 m/s for 420 s. The train then slows down uniformly at 0.065 m/s^2 until it stops moving. What is the acceleration during the first 5.6 km of travel?

A. 0.29 m/s^2
B. 0.23 m/s^2
C. 0.16 m/s^2
D. 0.12 m/s^2
E. 0.20 m/s^2

3. A cannonball is fired straight up at 50 m/s. Ignoring air resistance, what is the velocity at the highest point it reaches before starting to return towards Earth?

A. 0 m/s **B.** 25 m/s **C.** $\sqrt{50}$ m/s **D.** 50 m/s **E.** 100 m/s

4. The tendency of a moving object to remain in unchanging motion in the absence of an unbalanced force is:

A. impulse
B. acceleration
C. free fall
D. inertia
E. momentum

5. Which statement is correct when an object that is moving in the $+x$ direction undergoes an acceleration of 2 m/s^2?

A. It travels at 2 m/s
B. It travels at 2 m/s^2
C. It decreases its velocity by 2 m/s every second
D. It decreases its velocity by 2 m/s^2 every second
E. It increases its velocity by 2 m/s every second

6. What is the increase in speed each second for a freely falling object?

A. 0 m/s
B. 9.8 m/s^2
C. 9.8 m/s
D. 19.6 m/s
E. 4.8 m/s

7. A marble is initially rolling up a slight incline at 0.2 m/s and starting at $t = 0$ s, decelerates uniformly at 0.05 m/s^2. At what time does the marble come to a stop?

A. 2 s **B.** 4 s **C.** 8 s **D.** 12 s **E.** 6 s

8. An 8.7-hour trip is made at an average speed of 73 km/h. If the first third of the journey was driven at 96.5 km/h, what was the average speed for the rest of the trip?

A. 54 km/hr
B. 46 km/hr
C. 28 km/hr
D. 62 km/hr
E. 37 km/hr

9. The lightning flash and the thunder are not observed simultaneously because light travels much faster than sound. Therefore it can be assumed as instantaneous when the lightning occurs. What is the distance from the lightning bolt to the observer, if the delay between the sound and the lightning flash is 6 s? (Use speed of sound in air $v = 340$ m/s)

A. 2,040 m **B.** 880 m **C.** 2,360 m **D.** 2,820 m **E.** 1,460 m

10. If a cat jumps at a 60° angle off the ground with an initial velocity of 2.74 m/s, what is the highest point of the cat's trajectory? (Use acceleration due to gravity $g = 9.8$ m/s^2)

A. 9.46 m **B.** 0.69 m **C.** 5.75 m **D.** 1.55 m **E.** 0.29 m

11. What is the resultant vector AB when vector A = 6 m and points 30° North of East, while vector B = 4 m and points 30° South of West?

A. 8 m at an angle 45° East of North
B. 8 m at an angle 30° North of East
C. 2 m at an angle 30° North of East
D. 2 m at an angle 45° North of East
E. 8 m at an angle 45° North of East

12. How much farther would an intoxicated driver's car travel before he hits the brakes than a sober driver's car if both cars are initially traveling at 49 mi/h, and the sober driver takes 0.33 s to hit the brakes while the intoxicated driver takes 1 s to hit the brakes?

A. 38 ft **B.** 52 ft **C.** 32 ft **D.** 48 ft **E.** 60 ft

13. A projectile is fired at time $t = 0$ s from point 0 at the edge of a cliff with initial velocity components of $v_{0x} = 80$ m/s and $v_{0y} = 800$ m/s. The projectile rises, then falls into the sea at point P. The time of flight of the projectile is 200 s. What is the height of the cliff? (Use the acceleration due to gravity $g = 10$ m/s^2)

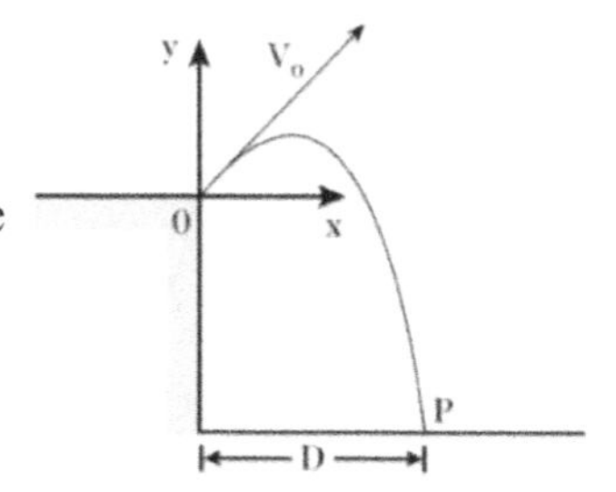

A. 30,000 m
B. 45,000 m
C. 47,500 m
D. 35,000 m
E. 40,000 m

14. Which of the following is a unit that can be used for a measure of weight?

A. kilogram
B. kg·m/s
C. Newton
D. milligram
E. liter

15. Which of the statements about the direction of the velocity and acceleration is correct for a ball that is thrown straight up, reaches a maximum height, and falls to its initial height?

A. Its velocity changes from upward to downward and its acceleration points downward
B. Its velocity points downward, and its acceleration points upward
C. Both its velocity and its acceleration point downward
D. Both its velocity and its acceleration point upward
E. Its velocity points downward and its acceleration changes from upward to downward

Solutions

1. B is correct.

Graph I depicts a constant zero velocity process.

Graph II depicts a constant non-zero velocity.

Graph III depicts non- constant velocity.

2. C is correct.

$v_f^2 = v_i^2 + 2ad$

$a = (v_f^2 - v_i^2) / 2d$

$a = [(42 \text{ m/s})^2 - (0 \text{ m/s})^2] / [2(5{,}600 \text{ m})]$

$a = (1{,}764 \text{ m}^2/\text{s}^2) / 11{,}200 \text{ m}$

$a = 0.16 \text{ m/s}^2$

3. A is correct.

At the maximum height, the velocity = 0, and the cannonball stops moving up and begins to come back.

4. D is correct.

An object's inertia is its resistance to change in motion.

5. E is correct.

$v = v_0 + at$

If $a = 2 \text{ m/s}^2$ then the object has a 2 m/s increase in velocity every second.

Example:

$v_0 = 0 \text{ m/s}, a = 2 \text{ m/s}^2, t = 1 \text{ s}$

$v = 0 \text{ m/s} + (2 \text{ m/s}^2)\cdot(1 \text{ s})$

$v = 2 \text{ m/s}$

6. **C is correct.** The acceleration due to gravity is 9.8 m/s^2.

If an object is in freefall, then every second, it increases velocity by 9.8 m/s.

$v = v_0 + at$

$v = v_0 + (9.8\ \text{m/s}^2)\cdot(1\ \text{s})$

$v = v_0 + 9.8\ \text{m/s}$

7. B is correct. To determine the time at which the marble stops:

$v_2 - v_1 = a\Delta t$

where $v_1 = 0.2$ m/s and $a = -0.05$ m/s^2

$\Delta t = (v_2 - v_1) / a$

$\Delta t = (0\ \text{m/s} - 0.2\ \text{m/s}) / (-0.05\ \text{m/s}^2)$

$\Delta t = (-0.2\ \text{m/s}) / (-0.05\ \text{m/s}^2)$

$\Delta t = 4\ \text{s}$

8. D is correct.

$v_{avg} = (1/3)v_1 + (2/3)v_2$

$73\ \text{km/h} = (1/3)\cdot(96.5\ \text{km/h}) + (2/3)v_2$

$73\ \text{km/h} = (32\ \text{km/h}) + (2/3)v_2$

$(2/3)v_2 = 41\ \text{km/h}$

$v_2 = (41\ \text{km/h})\cdot(3/2)$

$v_2 = 62\ \text{km/h}$

9. A is correct.

Distance to the lightning bolt is calculated by determining the distance traveled by sound.

Distance = velocity × time

$d = vt$

$d = (340\ \text{m/s})\cdot(6\ \text{s})$

$d = 2{,}040\ \text{m}$

10. E is correct.

Calculate the vertical component of the initial velocity:

$v_{up} = v(\sin\theta)$

$v_{up} = (2.74 \text{ m/s}) \sin 60°$

$v_{up} = 2.37 \text{ m/s}$

Then solve for the upward displacement given the initial upward velocity:

$d = (v_f^2 - v_i^2) / 2a$

$d = [(0 \text{ m/s})^2 - (2.37 \text{ m/s})^2] / 2(-9.8 \text{ m/s}^2)$

$d = (-5.62 \text{ m}^2/\text{s}^2) / (-19.6 \text{ m/s}^2)$

$d = 0.29 \text{ m}$

11. C is correct.

30° North of East is the exact opposite direction of 30° South of West, so set one vector as negative, and the direction of the resultant vector will be in the direction of the larger vector between A and B.

Since the magnitude of A is greater than the magnitude of B and vectors were added "tip to tail," the resultant vector is:

AB = A + B

AB = 6 m + (–4 m)

AB = 2 m at 30° North of East

12. D is correct.

The solution is measured in feet, so first convert the car velocity into feet per second:

$v = (49 \text{ mi/h})\cdot(5{,}280 \text{ ft/mi})\cdot(1 \text{ h}/3{,}600 \text{ s})$

$v = 72 \text{ ft/s}$

The sober driver's distance:

$d = vt$

$d_{sober} = (72 \text{ ft/s})\cdot(0.33 \text{ s})$

$d_{sober} = 24 \text{ ft}$

The intoxicated driver's distance:

$d = vt$

$d_{drunk} = (72 \text{ ft/s})\cdot(1 \text{ s})$

$d_{drunk} = 72 \text{ ft}$

The differences between the distances:

$\Delta d = 72 \text{ ft} - 24 \text{ ft}$

$\Delta d = 48 \text{ ft}$

13. E is correct.

$d = v_0 t + \frac{1}{2}gt^2$

Set g as negative because it is in the opposite direction as +800 m/s:

$d = (800 \text{ m/s})\cdot(200 \text{ s}) + \frac{1}{2}(-10 \text{ m/s}^2)\cdot(200 \text{ s})^2$

$d = (160{,}000 \text{ m}) - (200{,}000 \text{ m})$

$d = -40{,}000 \text{ m}$

The projectile traveled –40,000 m from the top of the cliff to the sea at point P, which means the cliff is 40,000 m from the base to the top.

14. C is correct.

1 Newton is the force needed to accelerate 1 kg of mass at a rate of 1 m/s^2.

$F = ma$

$\text{W} = mg$

$\text{N} = \text{kg}\cdot\text{m/s}^2$

15. A is correct.

When the ball is thrown upwards, its velocity is initially upwards, but as it falls back down, the velocity direction points downward.

The acceleration due to gravity always points down and does not change.

Chapter 2

Equilibrium and Momentum

EQUILIBRIUM

- **Concept of Equilibrium**
- **Concept of Force and Linear Acceleration, Units**
- **Translational Equilibrium ($Fi = 0$) Σ**
- **Torques, Lever Arms**
- **Rotational Equilibrium**
- **Analysis of Forces Acting on an Object**

MOMENTUM

- **Momentum**
- **Conservation of Linear Momentum**
- **Elastic Collisions**
- **Inelastic Collisions**
- **Impulse**

Concept of Equilibrium

The concept of equilibrium is related to Newton's First Law of Motion.

Newton's *First Law of Motion* (Law of Inertia) states that *an object at rest tends to stay at rest, and an object in motion will remain in motion at a constant velocity unless acted upon by an external force.*

The object is in equilibrium until an outside force is applied, upon which the object begins to accelerate. Acceleration is the change in the rate of velocity; this change may be just for the direction of motion).

The concept of equilibrium is linked to acceleration; objects undergoing an acceleration are not in equilibrium, while those with zero acceleration are in equilibrium.

There are two forms of equilibrium: translational equilibrium and rotational equilibrium. These two forms of equilibrium may exist concurrently or separately.

Objects in *translational equilibrium* have no linear acceleration.

Objects in *rotational equilibrium* have no angular acceleration.

For example, an object with zero linear acceleration and zero angular acceleration is in both translational and rotational equilibrium.

However, an object with zero linear acceleration but non-zero angular acceleration is only in translational equilibrium.

Both types of equilibrium can be separated into static and dynamic equilibrium.

Static equilibrium refers to objects with zero velocity (linear or angular).

Dynamic equilibrium refers to objects with constant non-zero velocity (linear or angular).

Concept of Force and Linear Acceleration & Units

Force causes objects to accelerate (i.e., change speed), change direction, or both.

Newton's *Second Law of Motion* states that *the acceleration of an object is directly proportional to the applied force and inversely proportional to the mass of the object*:

$$\vec{a} = \frac{F}{m} \qquad \text{or} \qquad F = m\vec{a}$$

An arrow above a symbol indicates a force with the arrowhead pointing in the direction of the force.

The *magnitude of the force* is labeled near the arrow and is measured in the units of Newtons (N).

A typically labeled force vector:

F = 20 N

The concept of force is essential when discussing equilibrium, as force imparts linear or angular acceleration to an object.

Linear acceleration is measured in units of m/s^2 and is the rate of change of the linear velocity of an object over time:

$$\vec{a} = \frac{\Delta\vec{v}}{\Delta t}$$

Angular acceleration is measured in units of rad/s^2 and is the rate of change of the angular velocity of an object over time:

$$\vec{\alpha} = \frac{\Delta\vec{\omega}}{\Delta t}$$

Translational Equilibrium

Translational equilibrium refers to an object undergoing zero linear acceleration.

In translational equilibrium, either 1) no forces are acting upon the object, and the linear acceleration is zero, or 2) all the forces acting upon the object cancel, and the object experiences zero linear acceleration in any direction.

Most typical situations refer to the latter scenario where the forces acting upon the object cancel; thus, translational equilibrium is achieved when the sum of all forces along each coordinate axis sum to zero.

Force is expressed by the equation: $\Sigma\vec{F} = 0 \text{ N}$

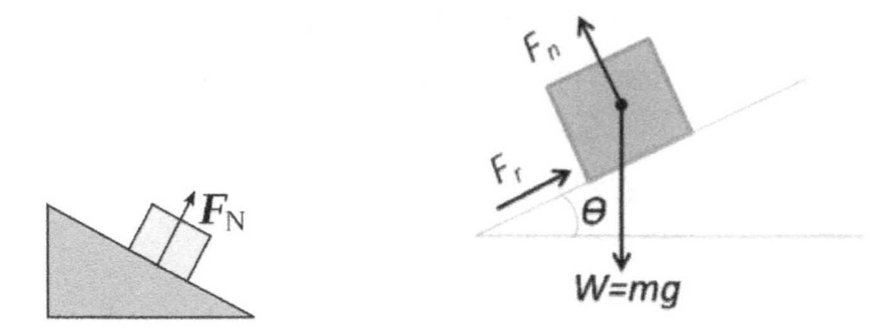

For example, a book, shown below, lies at rest on a table in translational equilibrium. The equal and opposite normal upward force of the table cancels the force due to gravity. The normal force is the component of the contact force that is perpendicular to the surface that an object contacts.

The book experiences a zero net linear acceleration and remains stationary.

The book has zero velocity and is an example of a static system.

$\Sigma F_y = F_N + F_G$

$\Sigma F_y = 0 \text{ N}$

$\vec{v} = 0 \text{ m/s}$

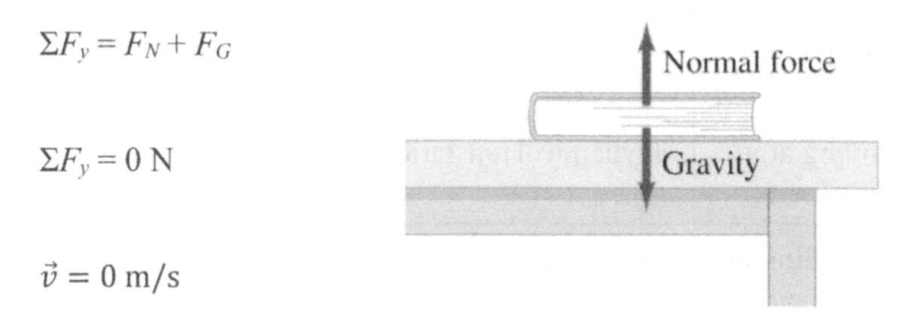

Below is another system in static translational equilibrium. A two-coordinate axis (x- and y-*axis*) is presented, and the forces along each axis sum to zero. Like the previous example, no velocity vector is labeled (no movement of the object).

Therefore, the system is *static* and has no velocity.

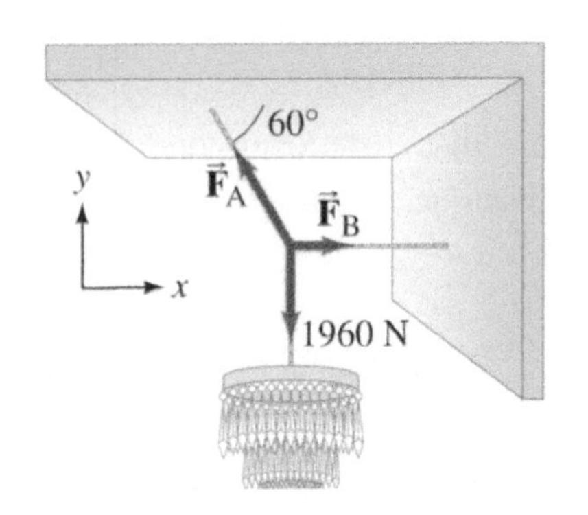

$$F_{Ax} = F_A \cos 60°$$

$$F_{Ay} = F_A \sin 60°$$

$$\Sigma F_x = F_{Ax} + F_B = 0 \text{ N}$$

$$\Sigma Fy = F_{Ay} + 1{,}960 \text{ N} = 0 \text{ N}$$

$$\vec{v} = 0 \text{ m/s}$$

The skydiver below presents an example of dynamic translational equilibrium.

A skydiver has achieved terminal velocity, and therefore the force of gravity (weight) is balanced by the equal and opposite force from air resistance.

Unlike the previous examples, the skydiver has a constant non-zero velocity and is therefore considered a *dynamic translational* system.

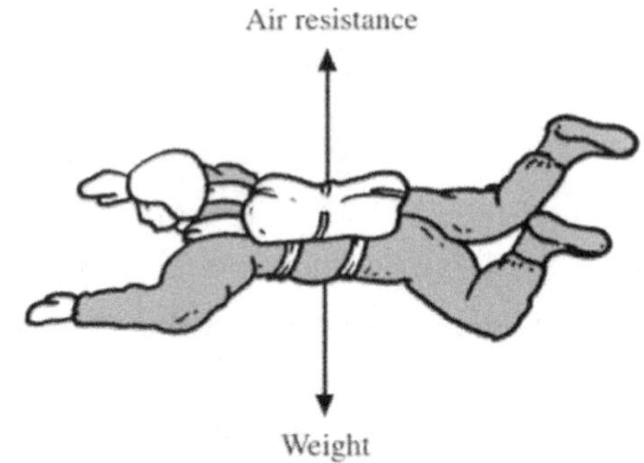

$$\Sigma F_y = F_G + F_{Air\ Resistance} = 0 \text{ N}$$

$$\vec{v} = constant$$

Translational equilibrium *vs.* non-equilibrium

Scenario	Type of Equilibrium	Translational non-equilibrium
An apple at rest	Static Translational	An apple falling toward the Earth with an acceleration of g
A car moving at a constant velocity	Dynamic Translational	A car either accelerating or decelerating
A skydiver falling at terminal velocity	Dynamic Translational	A skydiver before reaching terminal velocity

Torques, Lever Arms

Force applied to an object or system that induces rotation is torque.

Torque is the angular equivalent of force; it causes objects to rotate, have angular acceleration and change angular velocity. The *position and direction of the force* applied to the rotating body, about the axis of rotation, determines the magnitude of the torque.

The torque is the product of the applied force and the lever arm. Torque is directly proportional to the magnitude of the applied force and the length of the lever arm, expressed in the unit of the Newton-meter (N·m).

$$\vec{\tau} = r_{\perp}\vec{F} = r\ sin\ \theta\vec{F}$$

The lever arm is a distance from the axis of rotation to the line along which the perpendicular component of the force acts. The force of the lever arm from the angle between the applied force and the distance to the axis of rotation:

$r_{\perp} = r\ sin\ \theta$

$\theta \leq 90$

$r_{\perp} \leq r$

The *maximum torque* on an object occurs when the force is applied entirely perpendicular to the radius. For maximum torque, the angle θ = 90° between the force vector and the radius:

$\vec{\tau}_{max} = r\ sin\ (90)\vec{F} = r_{\perp}\vec{F})$, where sin 90° = 1

The *minimum torque* occurs when the force applied to the rotational body is directed toward the center of that body:

$\vec{\tau}_{min} = r\ sin\ (0)\vec{F} = 0$ N·m., where sin θ = 0°

For a door with two forces (A and B) applied at different points along its length:

$\vec{F}_A = \vec{F}_B$

$\theta_A = \theta_B = 90°$

$\vec{\tau} = r_{\perp}\vec{F}$

$r_{\perp A} > r_{\perp B}$

$\vec{\tau}_A > \vec{\tau}_B$

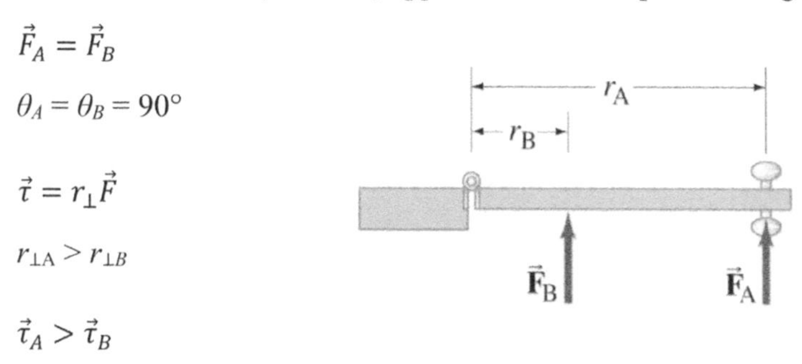

In the diagram above, both F_A and F_B are equal in magnitude, and each is applied perpendicularly to the door (i.e., $\theta = 90°$). Moving the door by F_A is easier than by using F_B because the lever arm for F_A is longer than the lever arm for F_B (torque = r $\sin\theta\, F$).

The torque produced by F_A is larger than the torque produced by F_B, and the door can be opened more easily with this larger torque.

Imagine a door with force applied at various angles concerning the doorknob:

$$\vec{F}_A = \vec{F}_C = \vec{F}_D$$

$$90° = \theta_A > \theta_C > \theta_D$$

$$\vec{\tau} = r_\perp \vec{F} = r \sin\theta \vec{F}$$

$$r_{\perp A} > r_{\perp C} > r_{\perp D}$$

$$\vec{\tau}_A > \vec{\tau}_C > \vec{\tau}_D$$

The forces A, C, and D are equal in magnitude and are applied at the same point from the axis of rotation (the length from the doorknob to the hinge).

The torques produced by forces A, C, and D are not the same because the forces are applied at different angles.

Force A produces the maximum torque because it is applied perpendicular (i.e., sin (90°) = 1) to the door, and thus has the largest lever arm.

Force C is applied at a non-perpendicular angle to the door; it has a smaller lever arm and produces a smaller torque.

Force D is applied parallel to the door and produces no torque because the force and its lever arm are equal to zero ($\sin(0°) = 0$).

Rotational Equilibrium

The Conditions for Rotational Equilibrium

When objects are in rotational equilibrium, they have zero angular acceleration.

For *rotational equilibrium*, either no torques are applied, or the sum of all torques around every rotational axis equals zero.

Rotational equilibrium is expressed as:

$$\sum \vec{\tau} = 0 \text{ N} \cdot \text{m}$$

The object either does not rotate (static rotational equilibrium), or it rotates at a constant rate (dynamic rotational equilibrium).

Positive torques act counterclockwise, while negative torques act clockwise.

Positive Convention +

Negative Convention -

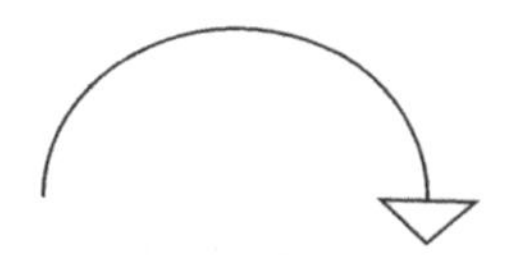

A pulley system demonstrates static rotational equilibrium. Force 1 and force 2 are equal and applied at opposite ends tangent to the pulley. The resulting torque 1 and torque 2 are equal and opposite. The sum of all torques along the pulley's rotational axis is zero. The pulley experiences no angular velocity and is a static system.

$\vec{F}_1 = \vec{F}_2$ and $\theta_1 = \theta_2 = 90°$

$r_1 = d$

$\vec{\tau}_1 = \vec{F}_1(d \sin \theta_1)$

$\vec{\tau}_2 = \vec{F}_2(d \sin \theta_2)$

$\sum \vec{\tau} = \vec{\tau}_1 + \vec{\tau}_2 = 0 \text{ N} \cdot \text{m}$

$\vec{\omega} = 0 \text{ rad/s}$

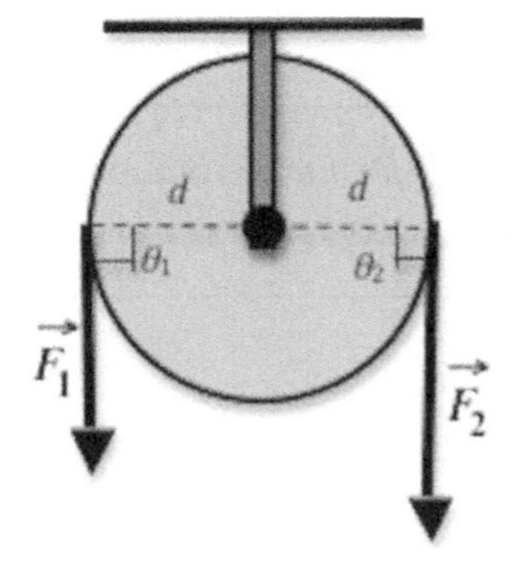

A system in static rotational equilibrium involves a mass attached to a cable-supported beam. The mass hangs perpendicularly from the beam at distance L, creating a clockwise torque from the gravitational force. The cable attached at a distance d produces its torque from the tension force and cancels the torque from the mass.

The sum of all torques about the axis of rotation is zero, and the system is static because it has no angular velocity.

$\vec{F}_G \neq \vec{F}_T$

$$\vec{\tau}_G = \vec{F}_1(L \sin 90°)$$

$$\vec{\tau}_T = \vec{F}_T(d \sin \theta)$$

$$\sum \vec{\tau} = \vec{\tau}_G + \vec{\tau}_T = 0 \text{ N} \cdot \text{m}$$

$$\vec{\omega} = 0 \text{ rad/s}$$

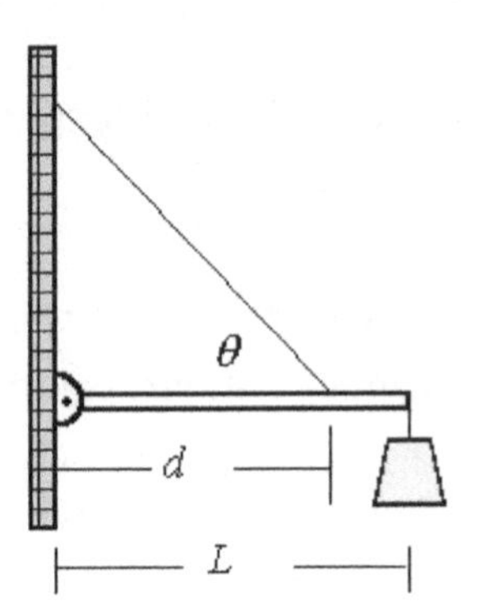

Rotational equilibrium *vs.* non-equilibrium

Scenario	Equilibrium Type	Rotational Non-Equilibrium
Equal masses on a balance	*Static* rotational	Unequal masses on a balance causing an angular acceleration
Propeller spinning at a fixed frequency	*Dynamic* rotational	Propeller spinning faster and faster
Asteroid rotating at a constant velocity	*Dynamic* rotational	Asteroid rotation slowing down

Analysis of Forces Acting on an Object

Solving Problems

1. Identify the objects and draw *free-body diagrams* of all forces acting on them and where they act.

2. Choose a coordinate system and determine the negative and positive directions for each axis.

3. Resolve forces into components (the normal and parallel components for inclined planes).

4. Add the force components; the resulting x-, y- and z-components determine the net force acting on the object.

5. Choose any axis perpendicular to the plane of the forces and select the equilibrium equations for the forces and torques. A smart choice here can simplify the problem enormously.

6. Determine the type of equilibrium (refer to the table below).

7. Use the Pythagorean Theorem to determine the magnitude of the net force from its components. Use trigonometry to solve for the angles. Angles are measured in radians, a whole circle = 2π radians.

Acceleration	Velocity	Equilibrium Type
$a = 0$ m/s^2	$v = 0$ m/s	Static Translational
	$v \neq 0$ m/s	Dynamic Translational
$\alpha = 0$ rad/s^2	$\omega = 0$ rad/s	Static Rotational
	$\omega \neq 0$ rad/s	Dynamic Rotational

Solved example 1 for a *static equilibrium* system

A 15,000 kg printing press is set upon a 1,500 kg table, as shown. Assuming the center of gravity of the table contains all its mass, what are the reaction forces from the table legs A and B?

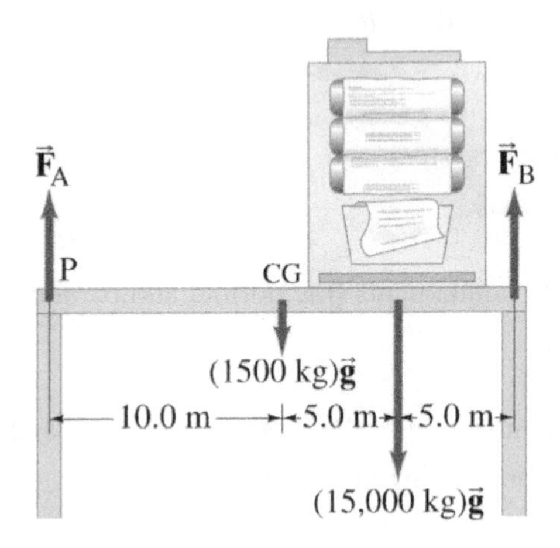

Sum all the forces in y-axis:

$$\Sigma F_y = F_A + F_B - (1{,}500\text{ kg})\cdot(9.8\text{m/s}^2) - (15{,}000\text{ kg})\cdot(9.8\text{m/s}^2) = 0$$

$$F_A + F_B = 161{,}700\text{ N}$$

Sum all the torques about point P:

$$\Sigma\tau_P = [-(1{,}500\text{ kg})\cdot(9.8\text{m/s}^2)\cdot(10\text{ m}) - (15{,}000\text{ kg})\cdot(9.8\text{ m/s}^2)\cdot(15\text{ m})] + F_B(20\text{ m}) = 0$$

$$F_B(20\text{ m}) = 2{,}352{,}000\text{ N}\cdot\text{m}$$

$$F_B = 117{,}600\text{ N}$$

Solve for the remaining forces:

$$F_A + 117{,}600\text{ N} = 161{,}700\text{ N}$$

$$F_A = 44{,}100\text{ N}$$

If the calculated force is negative, it is in the opposite direction of the direction initially chosen to be positive.

There is no "correct" way to choose which direction is positive. The requirement is that it is kept consistent throughout the problem.

Solved example 2 for a *static equilibrium* system

A beam with a mass of 500 kg located at the center of gravity is attached to two supporting legs. According to the diagram, what are the reaction forces from leg A and leg B? (Use the acceleration due to gravity, g = 9.8 m/s²)

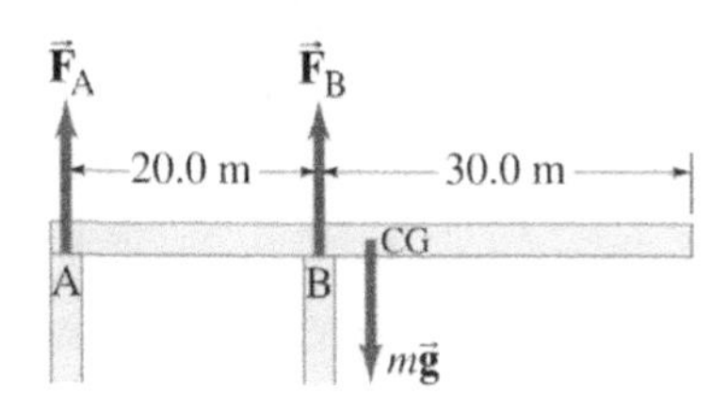

Sum the forces in the y-axis:

$$\Sigma F_y = F_A + F_B - (500 \text{ kg})\cdot(9.8 \text{ m/s}^2)$$

$$\Sigma F_y = 0$$

$$F_A + F_B = 4{,}900 \text{ N}$$

Sum the torques about point A (center of gravity is at the midpoint of the beam):

$$\Sigma\tau_A = -(500 \text{ kg})\cdot(9.8\text{m/s}^2)\cdot(25 \text{ m}) + F_B(20 \text{ m})$$

$$\Sigma\tau_A = 0$$

$$F_B(20 \text{ m}) = 122{,}500 \text{ N}\cdot\text{m}$$

$$F_B = 6{,}125 \text{ N}$$

Solve for the remaining forces:

$$F_A + 6{,}125 \text{ N} = 4{,}900 \text{ N}$$

$$F_A = -1{,}225 \text{ N}$$

For a cable attached, it can support tension forces (F_T) only along its length.

Force applied perpendicular to F_T would cause it to bend.

Solved example 3 for a *static equilibrium* system

A sign support consists of a beam attached to a hinge with a sign attached at the opposite end, and a cable supports the load. The beam has a length of 10 m, a mass of 500 kg and the center of gravity is at the 5 m point. The sign has a mass of 1,000 kg and is located 10 m from the hinge. If the support cable forms an angle of 45° with the beam, what are the reaction forces labeled in the diagram?

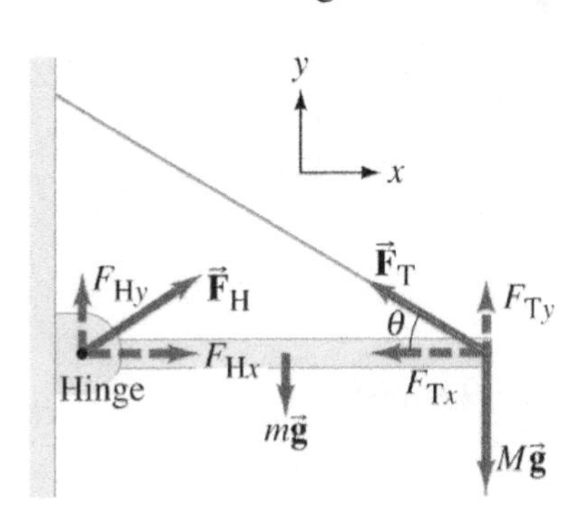

Sum the forces in the *y*-axis:

$$\Sigma F_y = F_{Hy} + F_{Ty} - (500\text{ kg})\cdot(9.8\text{ m/s}^2) - (1{,}000\text{ kg})\cdot(9.8\text{ m/s}^2)$$

$$\Sigma F_y = 0$$

$$F_{Hy} + F_{Ty} = 14{,}700\text{ N}$$

Sum the forces in the *x*-axis:

$$\Sigma F_x = F_{Hx} - F_{Tx} = 0$$

$$F_{Hx} = F_{Tx}$$

Sum the torques about hinge H (center of gravity is at the beam's midpoint):

$$\Sigma\tau_H = -(1{,}000\text{ kg})\cdot(9.8\text{ m/s}^2)\cdot(10\text{ m}) - (500\text{ kg})\cdot(9.8\text{ m/s}^2)\cdot(5\text{ m}) + F_T(10\text{ m})\, sin\,(45°)$$

$$\Sigma\tau_H = 0$$

$$F_T(10\text{ m})\, sin\,(45°) = 122{,}500\text{ N}\cdot\text{m}$$

$$F_T = 17{,}324\text{ N}$$

Solve for the remaining forces:

$F_{Ty} = F_T \sin(45°)$ N and $F_{Hx} = F_{Tx} = F_T \cos(45°)$ N

$F_{Ty} = (17{,}327\text{ N}) \sin(45°)$ and $F_{Hx} = F_{Tx} = (17{,}327\text{ N}) \cos(45°)$

$F_{Ty} = 12{,}250$ N and $F_{Hx} = F_{Tx} = 12{,}250$ N

$F_{Hy} + 12{,}250\text{ N} = 14{,}700\text{ N}$

$F_{Hy} = 2{,}450$ N

These principles of force and torque can be used to understand the human body. In the diagram below of an arm, the elbow is acting as a fulcrum, and the forearm is acting as the lever arm. The biceps can be modeled as a cable containing tension.

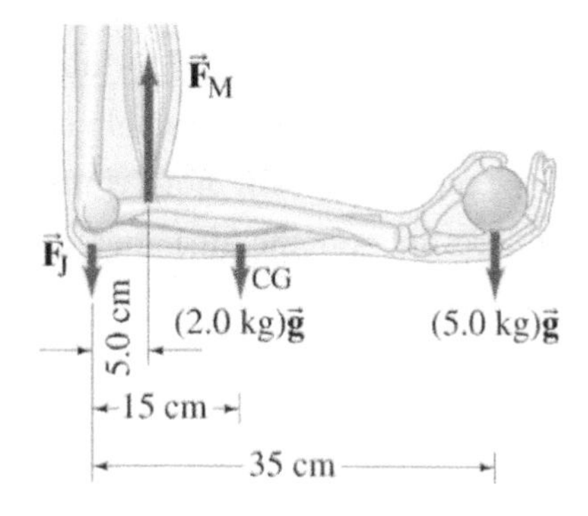

Sum the forces in the y-axis:

$\Sigma F_y = F_M - F_J - (2\text{ kg})\cdot(9.8\text{ m/s}^2) - (5\text{ kg})\cdot(9.8\text{ m/s}^2)$

$\Sigma F_y = 0$

$F_M - F_J = 69$ N

Sum the torques about point J:

$\Sigma \tau_J = F_M(0.05\text{ m}) - (2\text{ kg})\cdot(9.8\text{ m/s}^2)\cdot(0.15\text{ m}) - (5\text{ kg})\cdot(9.8\text{ m/s}^2)\cdot(0.35\text{ m})$

$\Sigma \tau_J = 0$

$F_M = 402$ N

Solve for the remaining forces:

$402\text{ N} - F_J = 69$ N

$F_J = 333$ N

Stability and Balance

Stable equilibrium is when the forces on an object return the object to its equilibrium position. A hanging ball connected to a string is an example of stable equilibrium. If the ball were to be momentarily disturbed, the net force returns the ball to its original equilibrium position.

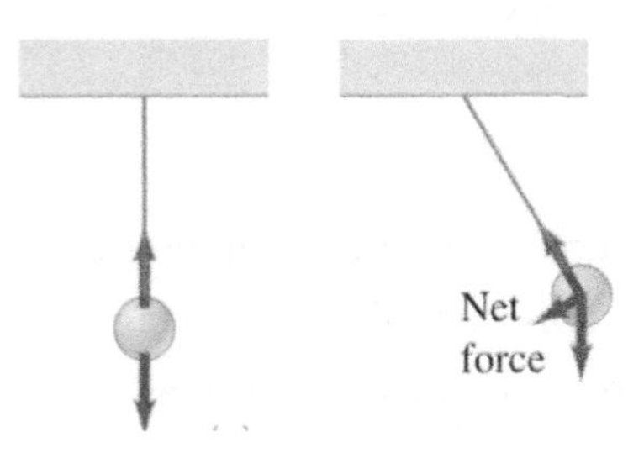

Unstable equilibrium is when the forces tend to move an object away from its equilibrium point. A pencil balanced on its tip is in unstable equilibrium. When disturbed from its equilibrium position, the net force will not return the pencil to its original equilibrium position (i.e., pencil resting on the tip), but cause it to settle in a new equilibrium position (e.g., lying on its side).

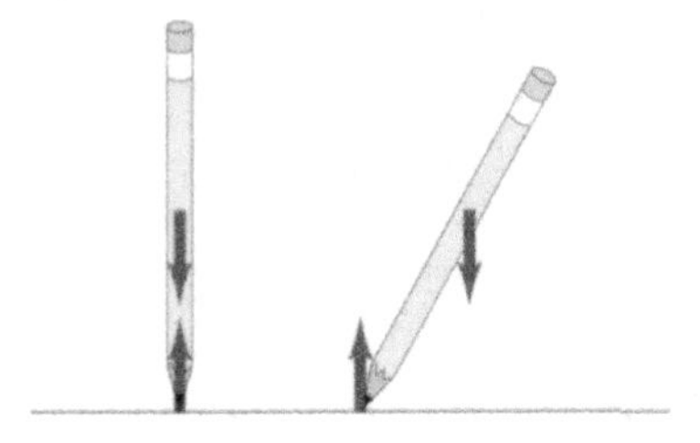

An object in stable equilibrium may transition to an unstable equilibrium if it is disturbed, so its center of gravity (CG) is outside the pivot point.

A resting refrigerator is in stable equilibrium, while its center of gravity (CG) is vertically positioned over the pivot point. If the refrigerator's center of gravity deviates from being vertically above the pivot point, it transitions to an unstable equilibrium, and the net force causes it to settle in a new equilibrium position (e.g., on its side).

Momentum

Newton's *First Law of Motion* (Law of Inertia) states that *an object at rest tends to stay at rest, and an object in motion will remain in motion at a constant velocity unless acted upon by an external force.*

This is due to the object's *inertia* (resistance to change). Inertia is directly related to an object's mass, so more massive objects have more resistant to a change in motion.

When an object is in motion (i.e., it has a non-zero velocity), this inertia translates into the object's momentum.

Momentum (p) is the vector product of mass and velocity:

$$\vec{p} = m\vec{v}$$

where m is mass measured in kg, $\vec{v}$ is the velocity measured in m/s and the momentum $\vec{p}$ is measured in units of kg·m/s or N·s (Newton·seconds)

Total momentum is the vector sum of individual momenta (note the velocity directions of individual momenta).

The *Law of Conservation of Momentum* states that *in an isolated system of objects, the total momentum of the objects is conserved.* The conservation of momentum has important implications in understanding and predicting the behavior of natural phenomena. A classic example for the Law of Conservation of Momentum is the predictable behavior of colliding objects (e.g., bumper cars or pool balls).

Conservation of Linear Momentum

In an isolated system, the total momentum for any collision is conserved.

An *isolated system* is a collection of two (or more) objects free from any net external force. The only forces present within the isolated system are those supplied by the objects.

If two balls collide on a frictionless surface, it is considered an isolated system since no external forces are present, and the conservation of momentum applies.

If non-negligible friction is present, then the balls experience an external force (force of friction), and their momentum changes according to the force supplied by friction (the force of friction is in the direction opposite to the motion).

The Law of Conservation of Momentum does not apply if:

1) an external force is supplied, and

2) the motion is not within an isolated system.

The conservation of momentum in a collision (within isolated systems), is when the total momentum (p) before a collision equals the total momentum after the collision.

$$\vec{p}_{initial} = \vec{p}_{final}$$

Momentum (p) is a vector, so assign one direction as positive and the opposite direction as negative when adding the component momenta vectors.

The momentum of a bomb at rest equals the vector sum of the momenta of all shrapnel from the explosion. Because shrapnel tends to travel in all directions, the momenta of individual pieces cancel, and the total momentum is conserved.

The conservation of momentum applies to a rocket in flight if the rocket and its fuel are considered as one system to accounts for the mass loss of the rocket's fuel.

When a rocket fires in space, it does not need air to push against, because the momentum of the combustion gasses must be conserved. Thus, the rocket moves in the opposite direction.

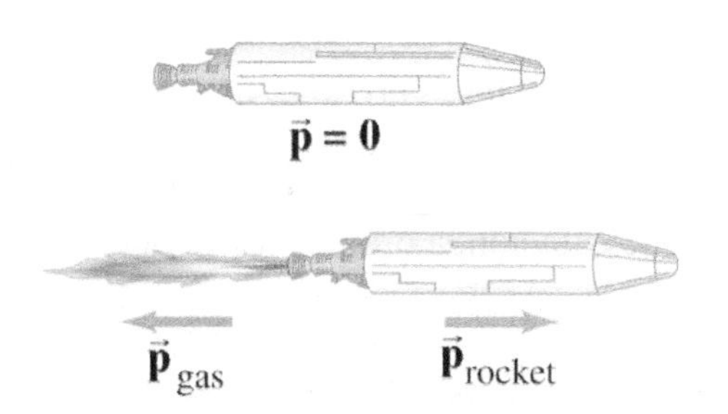

Momentum before = Momentum after = 0 N·s

$$m_{Gas}\vec{v}_{Gas} = m_{Rocket}\vec{v}_{Rocket}$$

For elastic collisions and inelastic collisions, *momentum is conserved.*

For elastic collisions, the kinetic energy (KE) *is conserved* (e.g., pool balls bounce after collision)

For inelastic collisions, the kinetic energy (KE) *is not conserved* (e.g., bullet embeds into the wooden block).

Elastic Collisions

Perfectly elastic collisions conserve both kinetic energy and momentum so that the initial total kinetic energy equals the final total kinetic energy ($KE_{Before} = KE_{After}$).

Kinetic energy is a scalar quantity (not a vector):

$$KE = \frac{1}{2}mv^2$$

The unit for energy is the Joule (J), and since energy is scalar (no direction is specified), there are no associated positive or negative signs for KE.

In a perfectly elastic collision, a dropped ball bounces back to its original height (e.g., ignoring gravity and air resistance).

A ball strikes a wall and bounces at about the same speed it struck the wall.

For elastic collisions, use the conservation of kinetic energy and the conservation of momentum.

The example below has two objects, with known masses and initial speeds, colliding elastically. The top diagram below is before impact, and the bottom diagram is after the collision. Since both *momentum and kinetic energy* are conserved, both equations can be written to solve for the two unknown final speeds via substitution:

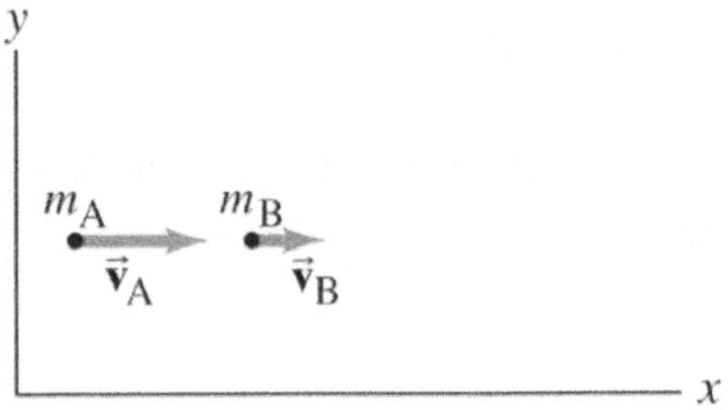

$$\vec{p}_A + \vec{p}_B = \vec{p}'_B + \vec{p}'_A$$

$$m_A\vec{v}_A + m_B\vec{v}_B = m_A\vec{v}'_A + m_B\vec{v}'_B$$

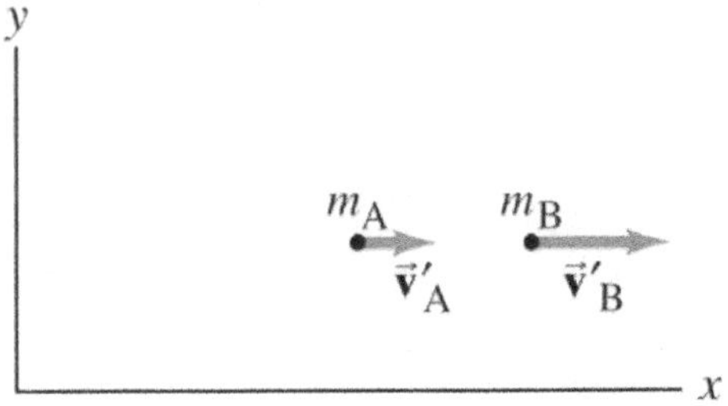

$KE_{Before} = KE_{After}$

$KE = ½mv^2$

$$\frac{1}{2}m_A\vec{v}_A^{\,2} + \frac{1}{2}m_B\vec{v}_B^{\,2} = \frac{1}{2}m_A\vec{v}_A'^{\,2} + \frac{1}{2}m_B\vec{v}_B'^{\,2}$$

For the *elastic collision* of objects traveling in opposite directions:

initial momentum equals the final momentum.

initial kinetic energy equals the final kinetic energy (the signs of the velocity changes, depending upon the coordinate axis).

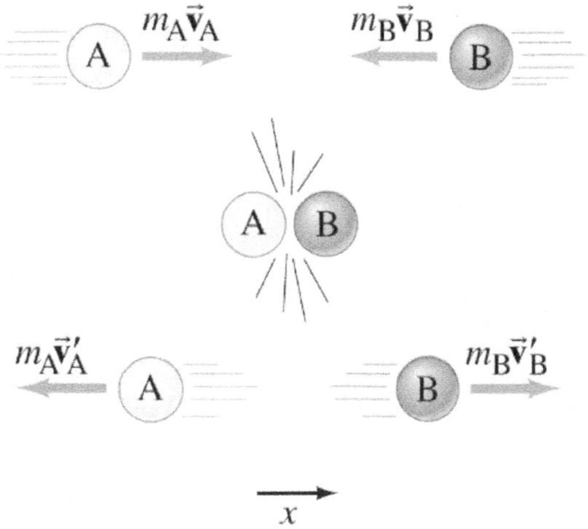

$$\vec{p}_A + \vec{p}_B = \vec{p}_B' + \vec{p}_A'$$

$$m_A\vec{v}_A + m_B\vec{v}_B = m_A\vec{v}_A' + m_B\vec{v}_B'$$

$$KE_{Before} = KE_{After}$$

$$KE = \frac{1}{2}mv^2$$

$$\frac{1}{2}m_A\vec{v}_A^{\,2} + \frac{1}{2}m_B\vec{v}_B^{\,2} = \frac{1}{2}m_A\vec{v}_A'^{\,2} + \frac{1}{2}m_B\vec{v}_B'^{\,2}$$

Inelastic Collisions

Inelastic collisions *conserve momentum* but do not conserve kinetic energy ($KE_{initial} > KE_{final}$). This does not violate the Law of Conservation of Energy because the lost kinetic energy is converted into another form of energy during the collision.

Collisions in everyday life are often, to varying extents, inelastic. If a ball is dropped from a person's hand, the ball does not reach the same height as it was released from after bouncing off the ground. This is because the collision was not elastic, but inelastic. The momentum was conserved, but the kinetic energy was not. Some of the ball's kinetic energy was converted to other forms (e.g., acoustic, thermal, vibrational).

This diagram illustrates two objects that experience elastic collision (conserves both momentum and kinetic energy) and inelastic collision (conserves momentum only; not kinetic energy).

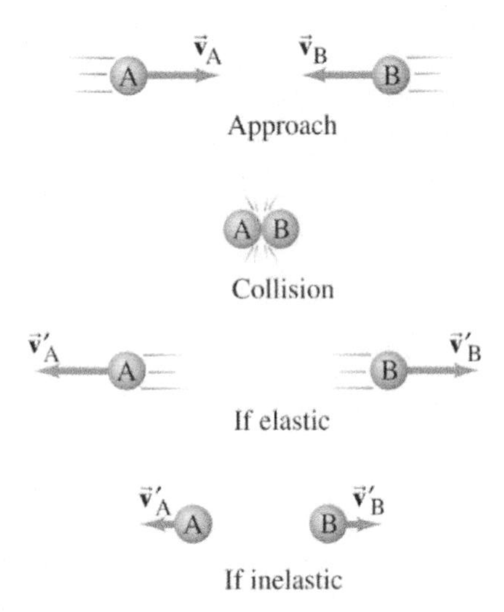

Objects that stick after a collision are typically inelastic collision.

Below, a bullet with a known velocity and mass is shot into a hanging block of known mass. To solve for the height attained by the block + bullet (inelastic collision), apply the conservation of momentum to find the velocity after the collision.

After collision, the kinetic energy of the system is converted to potential energy.

The conservation of energy is applied, and the height can be calculated.

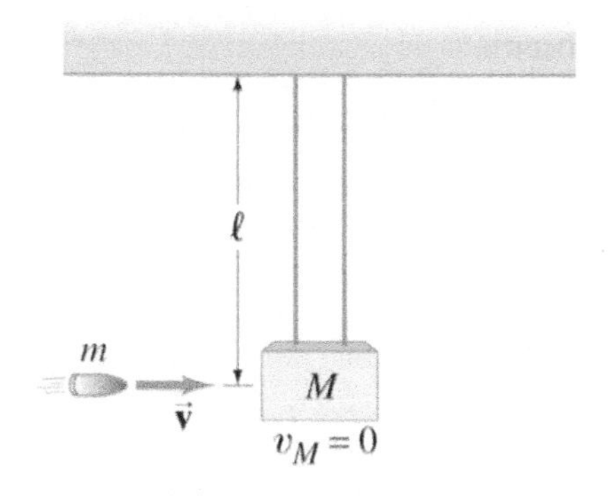

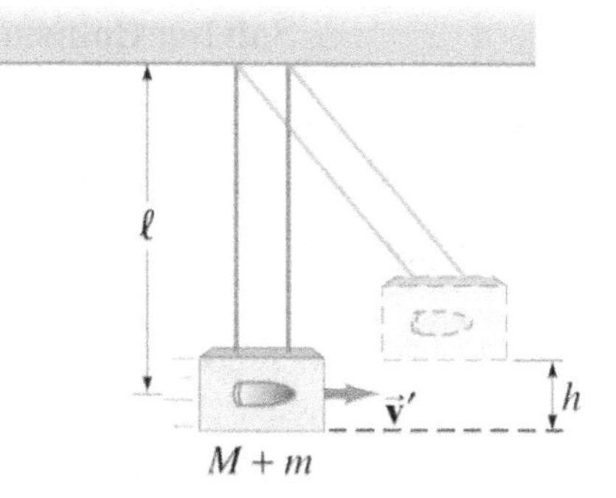

$$\vec{p}_{Bullet} + \vec{p}_{Block} = \vec{p}'_{Bullet+Block}$$

$$m\vec{v} + M\vec{v}_M = (M + m)\vec{v}'$$

$$\text{KE}_{\text{Bullet Before}} \neq \text{KE}_{\text{System After}}$$

$$\text{KE}_{\text{System After}} = \text{PE}_{\text{System After}}$$

$$KE = \frac{1}{2}mv^2$$

$$PE = mgh$$

$$\frac{1}{2}(M + m)\cdot(\vec{v}')^2 = (M + m)gh$$

Collisions in Two or Three Dimensions

Conservation of kinetic energy and momentum can also be used to analyze collisions in two or three dimensions. A moving object m_A collides with an object m_B initially at rest. Determining the final velocities requires 1) the masses, 2) the initial velocities, and 3) the angles of the object's post-collision trajectory.

$$\Sigma\vec{p}_x = \vec{p}_{Ax} + \vec{p}_{Bx} = \vec{p}'_{Bx} + \vec{p}'_{Ax}$$

$$m_A\vec{v}_A = m_A\vec{v}'_A \cos\theta'_A + m_B\vec{v}'_B \cos\theta'_B$$

$$\Sigma\vec{p}_y = \vec{p}_{Ay} + \vec{p}_{By} = \vec{p}'_{By} + \vec{p}'_{Ay}$$

$$0 = m_A\vec{v}'_A \sin\theta'_A + m_B\vec{v}'_B \sin\theta'_A$$

Solving Collision Problems

1. Choose the system. If it is a complex system (e.g., involves both elastic and inelastic collisions), subsystems should be chosen where one or more conservation laws apply.

2. Choose a system so that no external forces are present.

3. Draw diagrams of the initial and final situations, with the momentum vectors labeled.

4. Choose a coordinate system. If an object's momentum is at an angle to the axis, separate the motion into the *x*- and *y*- components:

   $p_x = \vec{p}\cos\theta$ and $p_y = \vec{p}\sin\theta$

5. Apply conservation of momentum with a separate equation for motion in each dimension.

   The total initial momentum in the *x*-direction must equal the total final momentum in the *x*-direction ($\Sigma p_{xi} = \Sigma p_{xf}$).

   The total initial momentum in the *y*-direction must equal the total final momentum in the *y*-direction ($\Sigma p_{yi} = \Sigma p_{yf}$).

6. Do not assume the type of collision (elastic or inelastic) if unknown. The type may be given or determine if kinetic energy is conserved.

   If the KE is conserved, the collision is elastic.

   If the KE is not conserved, the collision is inelastic.

7. For elastic collisions, use substitution between the conservation of kinetic energy and the conservation of momentum; for inelastic collisions, only use momentum conservation.

   $\vec{p} = \sqrt{p_x^2 + p_y^2}$, and the angle is given by $\theta = tan^{-1}\left(\frac{p_y}{p_x}\right)$

8. Check the units and the magnitudes to evaluate if they make sense.

Example of an Inelastic Collision

Train car A (m =1,000 kg) collides inelastically with train car B of equal mass. After the collision, the train cars stick (inelastic collision that conserves momentum, but not kinetic energy). What is the final velocity of the attached train cars?

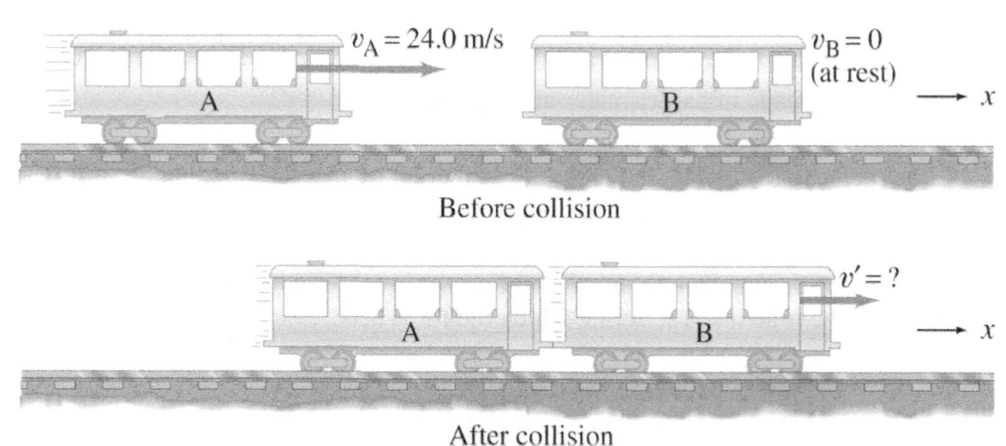

Apply the Law of Conservation of Momentum:

$$\vec{p}_A + \vec{p}_B = \vec{p}'_{A+B}$$

$$m_A\vec{v}_A + m_B\vec{v}_B = (m_A + m_B)\vec{v}'$$

Solve for the final velocity:

$$(1{,}000 \text{ kg}) \cdot (24 \text{ m/s}) + (1{,}000 \text{ kg}) \cdot (0 \text{ m/s}) = (1{,}000 \text{ kg} + 1{,}000 \text{ kg})\vec{v}'$$

$$\vec{v}' = 12 \text{ m/s}$$

Determine the magnitude of the difference in kinetic energy:

$$KE = \frac{1}{2}mv^2$$

$$KE_{\text{Before}} = \frac{1}{2}(1{,}000 \text{ kg}) \cdot (24 \text{ m/s})^2 = 288 \text{ kJ}$$

$$\text{KE}_{\text{After}} = \frac{1}{2}(2{,}000 \text{ kg}) \cdot (12 \text{ m/s})^2 = 144 \text{ kJ}$$

$$\text{KE}_{\text{Before}} \neq \text{KE}_{\text{After}}$$

Impulse

During a collision, objects are deformed to varying degrees due to the large forces involved.

Softer objects (e.g., pillows) are deformed more dramatically over a larger time frame than objects made of harder materials (e.g., metal).

The *impulse* (J) relates the extent to which an object deforms due to an impact, regarding the time the object spends in contact with another mass.

Impulse (i.e., the change in momentum, p), is represented as $F\Delta t$, where F is a force, and Δt is the time interval during which the force acts:

$$\Sigma\vec{F} = \frac{\Delta\vec{p}}{\Delta t}$$

$$\vec{F}\Delta t = \Delta\vec{p}$$

The total force equals the change in momentum divided by the time interval:

$$\Sigma\vec{F} = \frac{\Delta\vec{p}}{\Delta t}$$

Impulse ($\vec{J} = \vec{F}\Delta t$) can be derived using the equation $\vec{F} = m\vec{a}$.

$$\vec{F} = m\vec{a} = m\frac{\Delta\vec{v}}{\Delta t} \quad \rightarrow \quad \vec{F}\Delta t = m\Delta\vec{v} = \Delta\vec{p} = \vec{J}$$

The impulse determines if a change in momentum occurs by either:

1) a large force acting for a short time, or

2) a small force acting for a long time.

Automobile airbags are effective because they increase the length of time between the initial impact of the two objects (the airbag and the person) and the moment both objects come to rest. Without an airbag, this time interval would be shorter (smaller impulse), and therefore the force of impact is greater and more harmful.

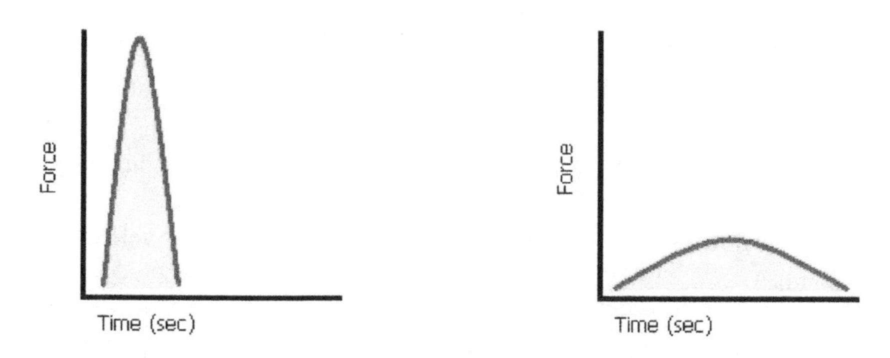

For the force *vs*. time diagrams, the magnitude of the force changes with time.

The area under the curve represents the impulse. For the above diagram on the right, the magnitude of the force is less because the impulse is extended over a greater range of time.

In practice, the time of the collision is often short, and the exact time dependence of the force is less relevant.

Since the time interval is relatively short, the average force, rather than the time-dependent force, can be used when calculating impulse (diagrams below).

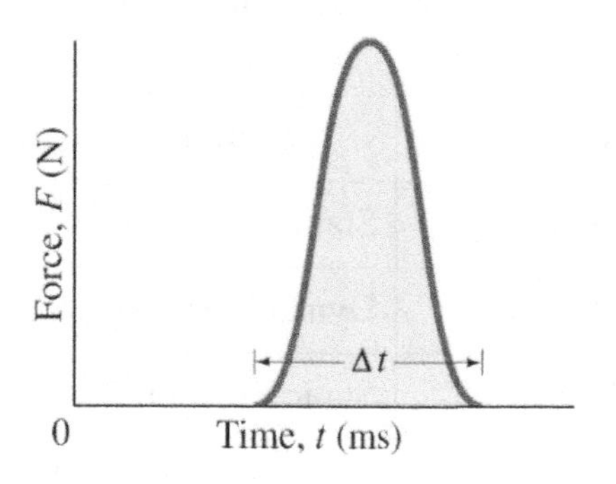

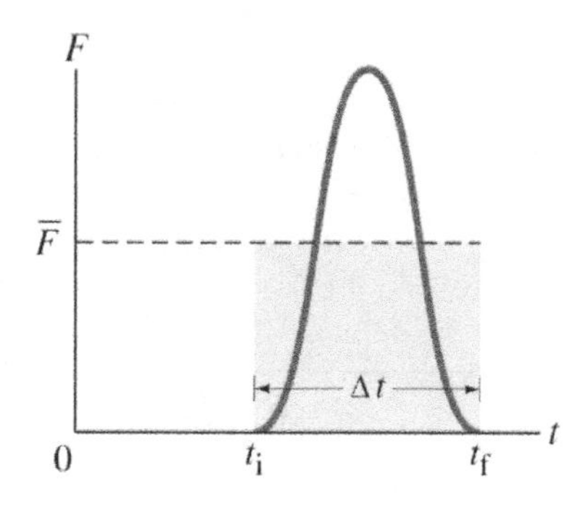

Chapter Summary

- Objects in *translational equilibrium* have no linear acceleration

- Objects in *rotational equilibrium* have no angular acceleration.

  These two forms of equilibrium may exist concurrently or separately.

- *Static equilibrium* refers to objects with zero velocity (linear or angular)

- *Dynamic equilibrium* refers to objects with constant non-zero velocity (linear or angular).

- In translational equilibrium, either 1) no forces are acting upon the object, and linear acceleration is zero, or 2) all the forces acting upon the object cancel, and the object experiences no linear acceleration in any direction.

- Force applied to an object or system that induces rotation is *torque.*

- Torque can be expressed as:

$$\vec{\tau} = r_{\perp}\vec{F} = r\, sin\, \theta \vec{F}$$

- When objects are in rotational equilibrium, the angular acceleration is zero.

  Therefore, either

  1) no torques are applied, or

  2) the sum of all torques around every rotational axis equals zero.

Acceleration	Velocity	Equilibrium Type
$a = 0$ m/s^2	$v = 0$ m/s	Static Translational
	$v \neq 0$ m/s	Dynamic Translational
$\alpha = 0$ rad/s^2	$\omega = 0$ rad/s	Static Rotational
	$\omega \neq 0$ rad/s	Dynamic Rotational

- Momentum (p) is the vector product of mass × velocity, represented by the equation:

$$\vec{p} = m\vec{v}$$

- According to the Law of Conservation of Momentum, in an isolated system of objects, the total momentum is conserved.

$$\vec{p}_{initial} = \vec{p}_{final}$$

- For an elastic collision, the momentum and total kinetic energy are conserved:

$$\frac{1}{2}m_A\vec{v}_A^{\,2} + \frac{1}{2}m_B\vec{v}_B^{\,2} = \frac{1}{2}m_A\vec{v}_A'^{\,2} + \frac{1}{2}m_B\vec{v}_B'^{\,2}$$

- For an inelastic collision, the momentum is conserved (not the kinetic energy.

  ($KE_{initial} > KE_{final}$)

  This does not violate the Law of Conservation of Energy because the lost kinetic energy is converted into another form of energy (e.g., light, heat, sound) during the collision.

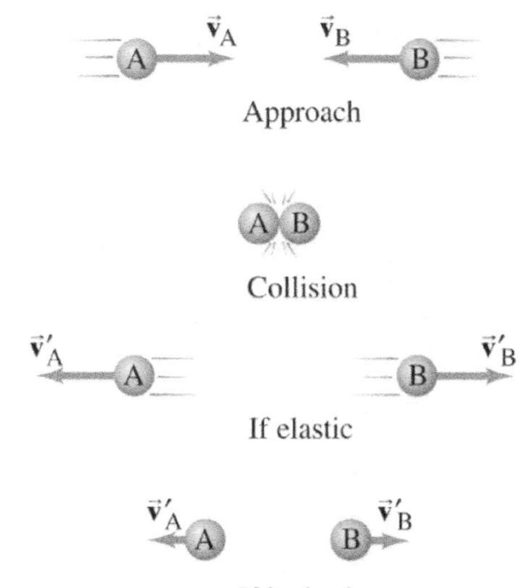

- The object's *impulse* (J), the time the object spends in contact with another mass, determines the extent to which an object deforms (e.g., airbags) due to an impact.

- Impulse, the change in momentum, is represented as $F\Delta t$, where F is a force, and Δt is the time interval during which the force acts:

$$\Sigma\vec{F} = \frac{\Delta\vec{p}}{\Delta t}$$

Practice Questions

1. A 1.2 kg asteroid is traveling toward the Orion Nebula at a speed of 2.8 m/s. Another 4.1 kg asteroid is traveling at 2.3 m/s in a perpendicular direction. The two asteroids collide and stick. What is the change in momentum from before to after the collision?

A. 0 kg·m/s
B. 2.1 kg·m/s
C. 9.4 kg·m/s
D. 12 kg·m/s
E. 19.7 kg·m/s

2. A 1,120 kg car experiences an impulse of 30,000 N·s during a collision with a wall. If the collision takes 0.43 s, what was the speed of the car just before the collision?

A. 12 m/s
B. 64 m/s
C. 42 m/s
D. 27 m/s
E. 18 m/s

3. Does the centripetal force acting on an object do work on the object?

A. No, because the force and the displacement of the object are perpendiculars
B. Yes, since a force acts and the object moves, and work is force times distance
C. Yes, since it takes energy to turn an object
D. No, because the object has a constant speed
E. Yes, because the force and the displacement of the object are perpendiculars

4. Cars with padded dashboards are safer in an accident than cars without padded dashboards, because a passenger hitting the dashboard has:

I. increased the time of impact
II. decreased impulse
III. decreased impact force

A. I only
B. II only
C. III only
D. I and III only
E. I and II only

5. The acceleration due to gravity on the Moon is only one-sixth of that on Earth, and the Moon has no atmosphere. If a person hit a baseball on the Moon with the same effort (and therefore at the same speed and angle) as on Earth, how far would the ball travel on the Moon compared to on Earth? (Ignore air resistance on Earth)

A. The same distance as on Earth
B. 6 times as far as on Earth
C. 1/6 as far as on Earth
D. 36 times as far as on Earth
E. $\sqrt{6}$ times as far as on Earth

6. A uniform meter stick weighing 20 N has a weight of 50 N attached to its left end and a weight of 30 N attached to its right end. The meter stick is hung from a rope. What is the tension in the rope, and how far from the left end of the meter stick should the rope be attached so that the meter stick remains level?

A. 100 N placed 37.5 cm from the left end of the meter stick
B. 50 N placed 40 cm from the left end of the meter stick
C. 80 N placed 37.5 cm from the left end of the meter stick
D. 80 N placed 40 cm from the left end of the meter stick
E. 100 N placed 40 cm from the left end of the meter stick

7. To catch a ball, a baseball player extends her hand forward before impact with the ball and then lets it ride backward in the direction of the ball's motion upon impact. Doing this reduces the force of impact on the player's hand principally because the:

A. time of impact is decreased
B. time of impact is increased
C. relative velocity is less
D. force of impact is reduced by $\sqrt{2}$
E. none of the above

8. A 1,200 kg car, moving at 15.6 m/s, collides with a stationary 1,500 kg car. If the two vehicles lock together, what is their combined velocity immediately after the collision?

A. 12.4 m/s **B.** 5.4 m/s **C.** 6.9 m/s **D.** 7.6 m/s **E.** 11.8 m/s

9. A 0.05 kg golf ball, initially at rest, has a velocity of 100 m/s immediately after being struck by a golf club. If the club and ball were in contact for 0.8 ms, what is the average force exerted on the ball?

A. 5.5 kN **B.** 4.9 kN **C.** 11.8 kN **D.** 7.1 kN **E.** 6.3 kN

10. Which of the following is an accurate statement for a rigid body that is rotating?

A. All points on the body are moving with the same angular velocity
B. Its center of rotation is its center of gravity
C. Its center of rotation is at rest and therefore not moving
D. Its center of rotation must be moving with a constant velocity
E. All points on the body are moving with the same linear velocity

11. Cart 1 (2 kg) and Cart 2 (2.5 kg) run along a frictionless, level, one-dimensional track. Cart 2 is initially at rest, and Cart 1 is traveling 0.6 m/s toward the right when it encounters Cart 2. After the collision, Cart 1 is at rest. Which of the following is true concerning the collision?

A. The collision is completely elastic
B. Kinetic energy is conserved
C. Momentum is conserved
D. Total momentum is decreased
E. Potential energy is conserved

12. What is the reason for using a long barrel in a gun?

A. Allows the force of the expanding gases from the gunpowder to act for a longer time
B. Increases the force exerted on the bullet due to the expanding gases from the gunpowder
C. Exerts a larger force on the shells
D. Reduces frictional losses
E. Reduces the force exerted on the bullet due to the expanding gases from the gunpowder

Questions **13-15** are based on the following:

A 5-gram bullet is fired horizontally into a 2 kg block of wood suspended from the ceiling by 1.5 m strings. The bullet becomes embedded within the block of wood, and they move together at a speed of 1.5 m/s. Gravity can be ignored during the time of the bullet's impact with the block. Then, the woodblock with the bullet swings upward by height h.

13. What best describes the energy flow as the bullet impacts the block of wood?

A. Kinetic to heat and kinetic
B. Potential and kinetic to heat
C. Kinetic to potential
D. Potential to heat
E. Kinetic to heat and potential

14. What is the velocity of the bullet just before it enters the block?

A. 30 m/s **B.** 60 m/s **C.** 600 m/s **D.** 90 m/s **E.** 180 m/s

15. Immediately after the bullet embeds itself in the wood, what is the kinetic energy of the block and bullet?

A. 2.26 J **B.** 4.5 J **C.** 9 J **D.** 18 J **E.** 36 J

Solutions

1. **A is correct.** Since the momentum is conserved during the collision, the change is $p = 0$.

$p_{initial} = p_{final}$

$m_1v_1 + m_2v_2 = (m_1 + m_2)v_3$

2. **D is correct.**

$J = \Delta p = m\Delta v$

$\Delta v = J / m$

$\Delta v = (30{,}000 \text{ N·s}) / (1{,}120 \text{ kg})$

$\Delta v = 27 \text{ m/s}$

$\Delta v = v_f - v_i$

$27 \text{ m/s} = 0 - v_i$

$v_i = 27 \text{ m/s}$

3. **A is correct.** In a uniform circular motion, the centripetal force does no work because the force and displacement vectors are at right angles.

$W = Fd$

The work equation is only applicable if force and displacement direction are the same.

4. **D is correct.**

$J = F\Delta t$

Impulse remains constant, so the time of impact increases, and the impact force decreases.

5. **B is correct.** Consider a dropped baseball on Earth:

$d_E = -½gt^2$

On the Moon:

$d_M = -½(g / 6)t^2$

$d_E / d_M = (-½gt^2) / [(-½(g / 6)t^2)]$

$d_E / d_M = 1/6$, so the distance is 6 times greater on the Moon

6. E is correct.

The total downward force on the meter stick is:

$$20\text{ N} + 50\text{ N} + 30\text{ N} = 100\text{ N}$$

The total upward force on the meter stick – which is provided by the tension in the supporting rope – must also be 100 N to keep the meter stick in static equilibrium.

Let x be the distance from the left end of the meter stick to the suspension point.

From the pivot point, balance the torques.

The counterclockwise (CCW) torque due to the 50 N weight at the left end is $50\,x$.

The total clockwise (CW) torque due to the weight of the meter stick and the 30 N weight at the right end is:

$$(50\text{ N})x = (20\text{ N})\cdot(50\text{ cm} - x) + (30\text{ N})\cdot(100\text{ cm} - x)$$

$$(50\text{ N})x = [1{,}000\text{ cm} - (20\text{ N})x] + [3{,}000\text{ cm} - (30\text{ N})x]$$

$$(50\text{ N})x = 4{,}000\text{ cm} - (50\text{ N})x$$

$$(100\text{ N})x = 4{,}000\text{ cm}$$

$$x = 40\text{ cm}$$

7. B is correct.

$J = F\Delta t$ is constant

increased t = decreased F

8. C is correct.

$$m_1v_1 = (m_1 + m_2)v_2$$

$$v_2 = (m_1v_1) / (m_1 + m_2)$$

$$v_2 = (1{,}200\text{ kg})\cdot(15.6\text{ m/s}) / (1{,}200\text{ kg} + 1{,}500\text{ kg})$$

$$v_2 = (18{,}720\text{ kg}\cdot\text{m/s}) / (2{,}700\text{ kg})$$

$$v_2 = 6.9\text{ m/s}$$

9. E is correct.

$F\Delta t = m\Delta v$

$F = (m\Delta v) / \Delta t$

$F = [(0.05 \text{ kg})\cdot(100 \text{ m/s} - 0 \text{ m/s})] / (0.0008 \text{ s})$

$F = (5 \text{ kg}\cdot\text{m/s}) / (0.0008 \text{ s})$

$F = 6{,}250 \text{ N} = 6.3 \text{ kN}$

10. A is correct. In a circular motion, all points along the rotational body have the same angular velocity regardless of their radial distance from the center of rotation.

11. C is correct.

In both elastic and inelastic collisions, momentum is conserved.

Compare the KE before and after the collision to determine if the collision is elastic (KE is conserved) or inelastic (KE is not conserved).

Initial KE:

$KE_i = \frac{1}{2}m_1v_1^2$

$KE_i = \frac{1}{2}(2 \text{ kg})\cdot(0.6 \text{ m/s})^2$

$KE_i = 0.36 \text{ J}$

Conservation of momentum:

$m_1v_1 + m_2v_2 = m_1u_1 + m_2u_2$

$m_1v_1 = m_2u_2$

$(m_1 / m_2)v_1 = u_2$

$u_2 = (2 \text{ kg} / 2.5 \text{ kg})\cdot(0.6 \text{ m/s})$

$u_2 = 0.48 \text{ m/s}$

Final KE:

$KE_f = \frac{1}{2}m_2u_2^2$

$KE_f = \frac{1}{2}(2.5 \text{ kg})\cdot(0.48 \text{ m/s})^2$

$KE_f = 0.29 \text{ J}$

KE_i (0.36 J) does not equal KE_f (0.29 J) and thus KE is not conserved.

Therefore, the collision is inelastic, and the only the momentum is conserved.

12. A is correct.

A longer barrel gives the expanding gas more time to impart a force upon the bullet and thus increase the impulse upon the bullet.

$$J = F\Delta t$$

13. A is correct.

The energy starts as kinetic. Initially, there is no change in potential energy.

Upon impact, most of the bullet's kinetic energy is converted into heat, some are transferred into kinetic energy of the block, and a small amount remains as kinetic energy for the bullet.

Since gravity is being ignored, the increase in the height of the block does not lead to potential energy.

14. C is correct.

Conservation of momentum:

$$m_1v_1 = (m_1 + m_2)v_2$$

$$(0.005 \text{ kg})v_1 = (0.005 \text{ kg} + 2 \text{ kg})\cdot(1.5 \text{ m/s})$$

$$v_1 = 600 \text{ m/s}$$

15. A is correct.

$$\text{KE} = \tfrac{1}{2}mv^2$$

$$\text{KE} = \tfrac{1}{2}(m_1 + m_2)v^2$$

$$\text{KE} = \tfrac{1}{2}(2.005 \text{ kg})\cdot(1.5 \text{ m/s})^2$$

$$\text{KE} = 2.2556 \text{ J} \approx 2.26 \text{ J}$$

Chapter 3

Work and Energy

WORK

- **The Concept of Work**
- **Derived Units, Sign Conventions**
- **Work Done by a Constant Force**
- **Mechanical Advantage**
- **Conservative Forces**
- **Path Independence of Work Done in a Gravitational Field**

ENERGY

- **The Concept of Energy**
- **Kinetic Energy**
- **Work-Kinetic Energy Theorem**
- **Potential Energy**
- **Conservation of Energy**
- **Power, Units**

The Concept of Work

Work is the relationship between a force and the distance traveled by an object acted upon by force, in a direction parallel to the force.

Work — Applied force

$$W = \vec{F}d$$

Distance parallel to applied force

The above equation is for an applied force and a displacement parallel to the force.

If the distance traveled is not parallel to the force, but at an angle, the angle must be accounted for by the formula:

$$W = \vec{F}d\cos\theta$$

Where $\vec{F}$ is force, d is the distance over which the force is applied, and θ is the angle between the applied force and the distance.

Essentially, this formula divides the forces into components, and the only value used in the work calculation is the force parallel ($\cos\theta$) to displacement.

There are some instances in which the work done is zero:

- When the displacement is zero ($\boldsymbol{W = \vec{F} \cdot 0 = 0}$)
- When the displacement is perpendicular to the applied force (the object moves at 90° to the applied force)

$$W = \vec{F}d\cos(90°) = \vec{F} \cdot 0 = 0$$

- When an object is moving at a constant velocity, with no forces acting in the direction of movement ($W = 0 \cdot d = 0$)

Derived Units & Sign Conventions

The units of work are:

$$W = \vec{F}\vec{d}$$

$$W = m\vec{a}\vec{d} = (\mathrm{kg}) \cdot \left(\frac{\mathrm{m}}{\mathrm{s}^2}\right) \cdot (\mathrm{m})$$

$$W = \frac{\mathrm{kg} \times \mathrm{m}^2}{s^2} = \mathrm{N} \cdot \mathrm{m} = \mathrm{J}$$

The Newton-meter represents the combined units of force and distance.

The magnitude of work (W) is generally expressed in *Joules* (J) since the Newton-meter is the unit of measurement for torque.

The Joule is named after English physicist James Prescott Joule (1818-1898), is also the unit for the measurement for energy.

If the applied force and the displacement are in the same direction, the work is positive. For example, pushing a crate across a rough terrain involves doing positive work (pushing forward as the crate moves forward).

If the force opposes the direction of motion (e.g., friction), the work is negative.

Work Done by a Constant Force

As noted earlier, the work done by a constant force is defined as *the distance moved multiplied by the component of the force in the direction of displacement*:

$$W = \vec{F}\vec{d}\cos\theta$$

For example, a crate on the ground has a rope attached to the side, which a woman pulls to move the crate a distance d. If she pulls with a constant force, and the rope forms an angle θ with the crate, how much work did she perform?

$$W = \overrightarrow{F_{\parallel}}\vec{d}$$

$$\overrightarrow{F_{\parallel}} = \vec{F}\cos\theta$$

$$W = \vec{F}\vec{d}\cos\theta$$

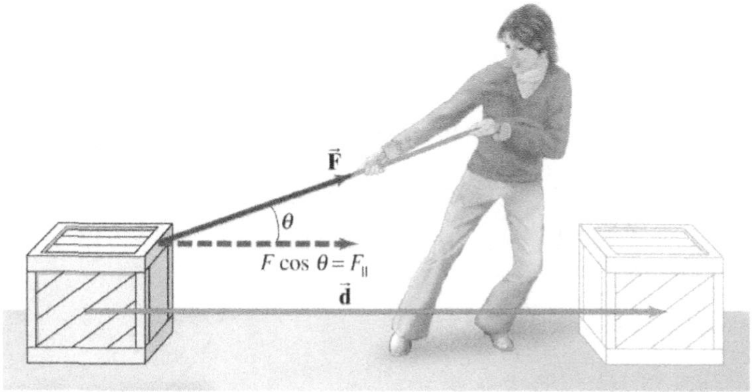

The following five steps can be useful in solving work problems:

1. Draw a free-body diagram.
2. Choose a coordinate system.
3. Apply Newton's Laws to determine any unknown forces.
4. Find the work done by a specific force.
5. To find the net work done:

   find the net force and then find the work it does, or

   find the work done by each force and add them together.

If the force and displacement are perpendiculars, the work done is zero.

The centripetal force does no work.

In a uniform circular motion, the centripetal force is constant when an object moves in a circle at a constant speed.

The centripetal force is directed toward the center of a circular path and is perpendicular to the linear movement of the object.

For a circular motion at a constant speed, the velocity is tangent to the direction of motion.

If a ball attached to a string is spun in a circle, how much work is performed?

$$W = \vec{F}_{\parallel}\vec{d}$$

$$\vec{F}_{\parallel} = \vec{F}\cos(90°) = 0\ \text{N}$$

$$W = 0\ \text{J}$$

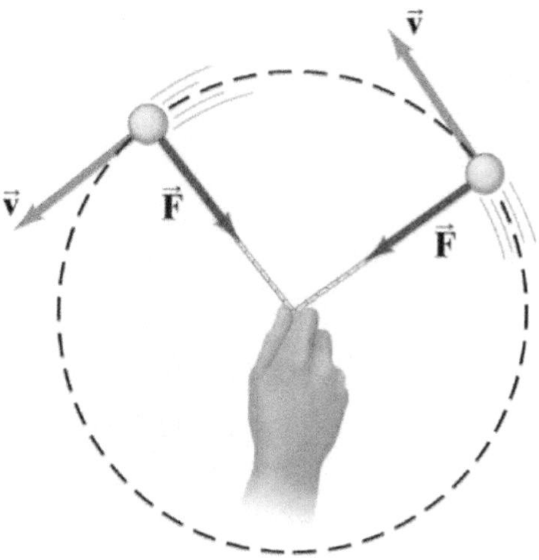

The centripetal force points toward the center and velocity is tangent to the circle

Mechanical Advantage

A mechanical advantage takes an input force of little effort and results in an output of a much larger force.

A *lever arm* or a *pulley* achieves such a mechanical advantage. Mechanical advantages are useful in the construction and movement of large objects, which are otherwise too heavy to manipulate.

Simple machines are any mechanical device that is used to apply a force. The premise of a simple machine is that the work put in is equal to the work put out. This refers to the amount of force multiplied by a ratio of distances.

Examples of simple machines:

- Pulleys
- Inclined planes
- Wedges
- Screws
- Levers
- Wheels and axles

Pulley systems are conventional simple machines. Like all simple machines, a pulley system does not reduce the amount of work done on an object, but it does reduce the amount of the force needed to manipulate it; the distance over which the force acts is increased.

The work equation shows that the force and distance are inversely related and that the distance of pulling increases by the same factor the force decreases.

$$\frac{W}{\vec{d}} = \vec{F}$$

A method for solving pulley problems is to realize that the ropes on either side of a moving pulley contribute to pulling the load. A stationary pulley, however, does not contribute to the load. The above rules apply to problems with simple pulleys. Complicated pulleys have additional ropes that contribute to pulling the load.

In the example below, a pulley is attached to the ceiling and is used to lift a box. The pulley is stationary, so no mechanical advantage is imparted to the system. Therefore, if the weight of the box is 100 N, the force needed to pull the box up is 100 N. For every 1 meter pulled, the box moves up 1 meter.

$$W = \vec{F}\vec{d}$$

$$W = (100\text{ N})\cdot(1\text{ m})$$

$$W = 100\text{ Joules}$$

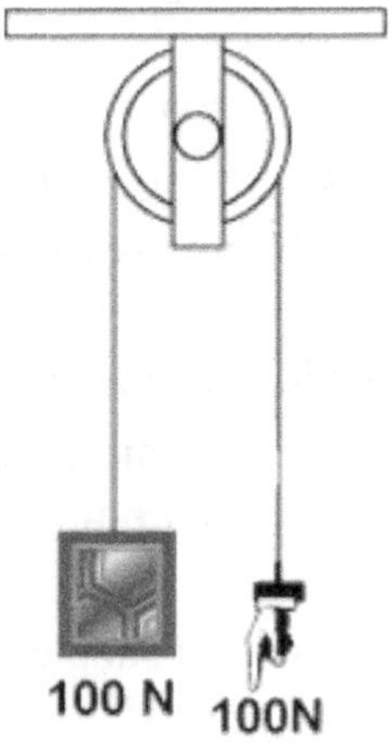

As stated, moving pulleys impart a mechanical advantage to the system.

In the diagrams below, there is one moving pulley in each system. When there is one moving pulley, the force needed to pull is halved because the rope on each side of the pulley contributes equally. The 50 N force is transmitted to the right-hand rope, while the left-hand rope contributes the other 50 N.

However, for a moving pulley system, the distance required to pull the rope is not the same as for a simple pulley system.

If the box is to be pulled up one meter (e.g., previous example), the distance of the rope required to pull the box is doubled because the force is halved.

In the diagram below, the pulley system on the right is an equivalent system to the pulley system on the left.

Although there are two pulleys on the right, only the mobile pulley halves the force needed to pull the box.

For any system, the work required to lift the box a specific distance is the same regardless of the reduction in force.

$W = \vec{F}\vec{d}$

$100 \text{ Joules} = (50 \text{ N}) \cdot (\vec{d})$

$\vec{d} = 2$ meters

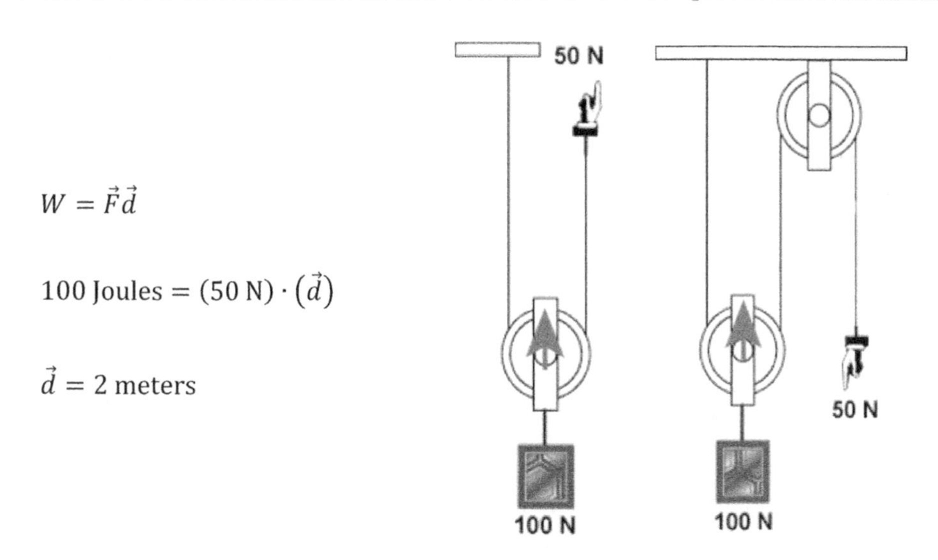

In the diagrams below, two equivalent systems are shown. Although the pulley arrangement is different, each system has two mobile pulleys, which reduce the force needed to lift the box. If each mobile pulley reduces the force by half, then the total force needed to lift the box will be reduced to a quarter of the original value.

Thus, the 100 N box can be lifted with only 25 N. Again, the work required to lift the box one meter must always be the same, so for every 1 m the box moves, 4 m of the rope must be pulled.

$W = \vec{F}\vec{d}$

$100 \text{ Joules} = (25 \text{ N}) \cdot (\vec{d})$

$\vec{d} = 4$ meters

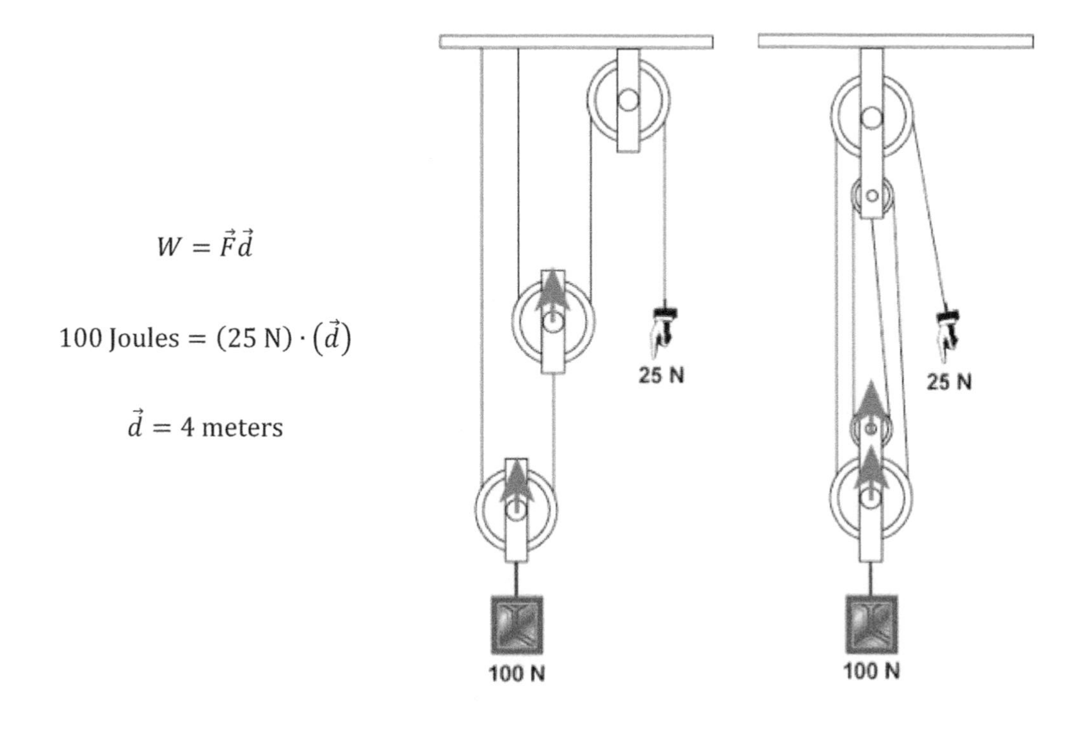

Below is an example of a complex pulley system. Like the simple pulleys, the ropes on both sides of the moving pulley contribute. However, the left-most rope also contributes. This makes three contributing ropes, which reduces the effort required by a factor of three.

The distance needed to pull is three times the distance the box travels.

$$W = \vec{F}\vec{d}$$

$$100 \text{ Joules} = (33 \text{ N}) \cdot (\vec{d})$$

$$\vec{d} \cong 3 \text{ meters}$$

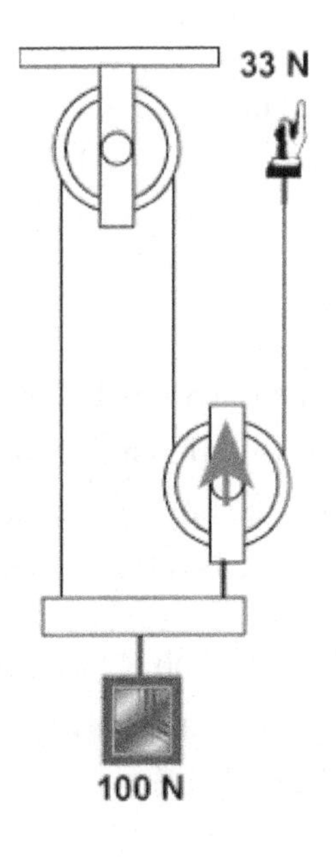

Conservative Forces

A force can be either conservative or non-conservative. Determining how to label a specific force can be slightly confusing. However, there are a few qualifications that can be checked to determine what type of conservation a force follows.

Conservative force does not dissipate heat, sound, or light. Furthermore, if the work done by a force is path independent, then that force is conservative.

Conservative forces are associated with potential energy as potential energy can only be defined for conservative forces.

The force from a spring can be stored as spring potential energy, and gravitational force can be stored as gravitational potential energy. Electromagnetic forces are also conservative. If only conservative forces are acting, mechanical energy is conserved.

A few things to consider when a particle is moving along a path:

- If the work done to move a particle in any round-trip path is zero, the force is conservative.
- If the work needed to move a particle between two points is the same regardless of the path taken, then the force is conservative.

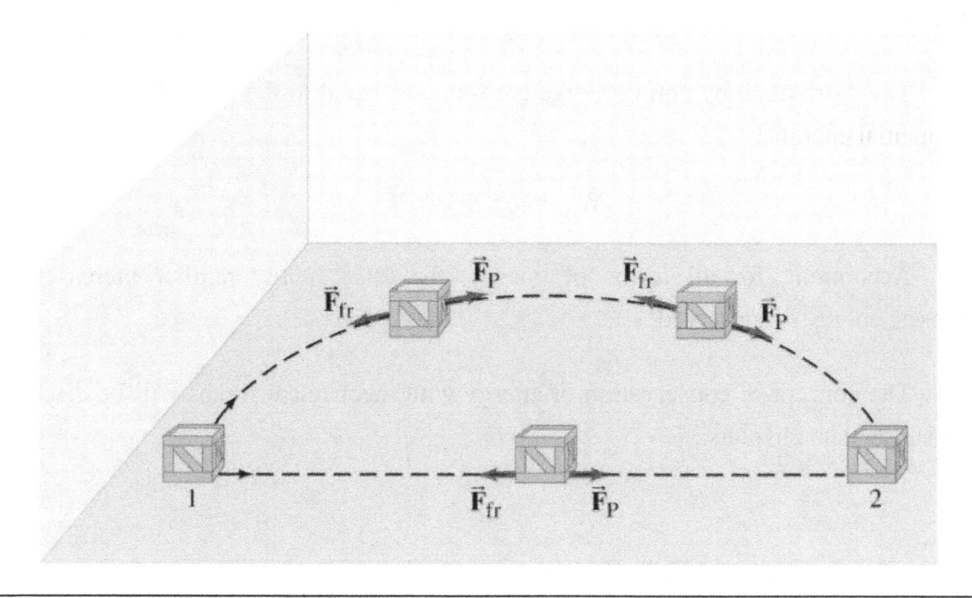

Most forces are conservative, but some non-conservative forces include friction and human exertion. If friction is present, the work done depends not only on the starting and ending points but also on the path taken, as frictional forces directly oppose the course of motion of an object.

As an object encounters friction, heat, and sound energy are released, and the total energy is not conserved. Heat energy is also lost in human exertion.

When muscles are flexed during exercise, the heat from the muscles rises off the skin and cannot be recovered or reabsorbed.

Conservative Forces	Non-conservative Forces
Gravitational	Friction
	Air resistance
Elastic	Tension in a cord
	Motor or rocket propulsion
Electric	Push or pull by a person

Potential energy is defined for conservative forces.

The work done by conservative forces must be distinguished from the work done by non-conservative forces.

The work done by non-conservative forces is equal to the total change in kinetic and potential energies:

$$W_{NC} = \Delta KE + \Delta PE$$

Accounting for all forms of energy, the total energy neither increases nor decreases; energy is conserved.

The concept of conservation of energy with mechanical forces will be discussed in the subsequent chapters.

Path Independence of Work Done in Gravitational Field

The amount of work done in a gravitational field is path-independent since gravitational forces always act downward. Sideward motion, which is perpendicular to the gravitational force, involves no work. For example, a man lifting a bag of groceries is a form of positive work in a gravitational field. The act of lifting a bag of groceries is positive work because a force is exerted (to overcome the force of gravity), and the distance the bag is lifted is in the direction of the force.

However, once the bag is being held up, it is at rest (as in, no longer traveling), and no work is being done on the bag. If the man does not lift or lower the bag, he is doing no work on it because there is no vertical displacement. Walking home with the bag of groceries also does no work on the bag. This is because the displacement (horizontal) is perpendicular to the force of gravity, and the resulting work must be zero.

$$\vec{F}_P = m\vec{g}$$

Work to lift the bag: $W = \vec{F}_P \vec{d}_y > 0\ \text{J}$

Work to walk home: $W = \vec{F}_P \vec{d}_x = 0\ \text{J}$

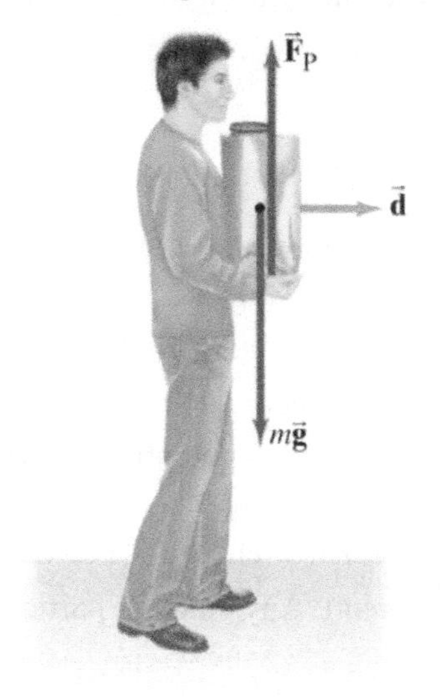

Another example of work in a gravitational field is pushing an object at constant speed up a frictionless, inclined plane. This uses the same amount of work as directly lifting the object to the same height at a constant speed because the displacement perpendicular to the force of gravity contributes no work.

Similarly, sliding an object down a frictionless, inclined plane involves the same gravitational work as the object undergoing a free fall from the same height.

Since the only forces involved in these motions are acting on a vertical plane, the work done is only dependent on the height of the ramp.

The Concept of Energy

Work and energy are closely related concepts. *Energy* is traditionally defined as the ability to do work, and, like work, the unit of energy is the Joule. However, although work and energy have the same units, the two concepts are not always interchangeable. Energy takes many different forms, while mechanical work is limited to the definition above.

For example, energy may be exerted to move a heavy box, but if the box does not budge, then no work has been done.

Another important concept of energy is that the sum of the energy of the universe in its current state is the same amount of energy that was present at the inception of the universe. There is an infinite amount of space for it to go, but it never disappears. This is the Law of Conservation of Energy and is discussed in this chapter.

Energy forms

All the energy that was created from the beginning of the universe is divided into different types of energy.

Mechanical energy—The energy associated with motion and position. It is equal to kinetic plus potential energy.

Electrical energy—The energy made up of the current and potential provided by a circuit. Any charged particle within an electrical field contains electrical energy.

Chemical energy—The energy involved in all chemical reactions. This is the potential of a substance to undergo a transformation or reaction.

Radiant energy—Consists of all the energy from electromagnetic waves. It can be viewed as the energy stored in a photon, or the motion of an electromagnetic wave.

Visible light is a small portion of the full spectrum of electromagnetic waves—there are many other sources of radiant energy that cannot be seen with the human eye.

Nuclear energy—The energy that is released during reactions involving the fusion or fission of the nucleus. Examples are nuclear bombs and nuclear power plants.

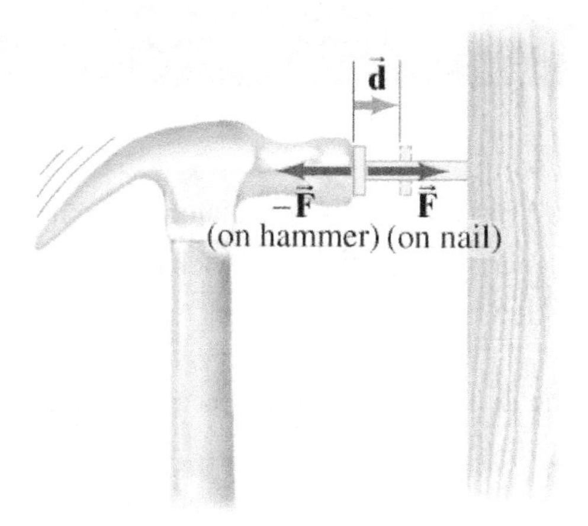

Although energy can take different forms, there are two main types of energy within mechanical physics problems:

- *Kinetic energy*: the energy of motion (energy that is being consumed or released at a given time)

- *Potential energy*: stored energy (energy that could be released for a body)

Kinetic Energy

Kinetic energy (*KE*) is the energy associated with the motion of an object and results from work or a change in potential energy. Kinetic energy is only dependent on the mass of an object as well as its speed.

As with energy, the unit of kinetic energy is a Joule ($kg \cdot m^2/s^2$).

The equation for kinetic energy is:

$$KE = \frac{1}{2}mv^2$$

An important feature of kinetic energy is its relation to the mass and speed of an object. In the equation above, the *KE* is directly proportional to mass, and *KE* is proportional to the square of the velocity.

If the mass (m) of an object is doubled, *KE* is also doubled.

If the velocity (v) is doubled, the *KE* is quadrupled (because velocity is squared).

For example, two cars are on the road, but one is twice as heavy as the other.

If both cars are traveling at the same speed, the car with twice the mass has twice as much *KE* as that of the other car.

If two cars of equal mass are on the road, and one is traveling twice as fast as the other, the faster car has a *KE* four times that of the slower car.

The importance of this is that a change in velocity is more significant, regarding kinetic energy than a change in mass because velocity is squared.

For objects of the same mass, the object with the higher velocity always has the greater *KE*, and for objects with the same velocity, the object with a greater mass always has the greater *KE*.

Work-Kinetic Energy Theorem

Work on an object can be transformed into kinetic energy. If the net work on the object is positive, the kinetic energy increases; if the net work is negative, the kinetic energy decreases. Thus, work (W) and kinetic energy (*KE*) are directly related by the equation:

$$|\,W\,| = |\,\Delta KE\,| \quad \text{(change in kinetic energy)}$$

Given that the definition of kinetic energy is:

$$KE = \frac{1}{2}mv^2$$

Work can be written, regarding a change in the kinetic energy, as:

$$W = \Delta KE = \frac{1}{2}mv_2^2 - \frac{1}{2}mv_1^2$$

For example, the force *F* provided by a car's engine moves the car a distance *d*, and thus performs positive work on the car. This work is transformed into kinetic energy and increases the total kinetic energy of the car.

The car has a higher final velocity than before the work was performed.

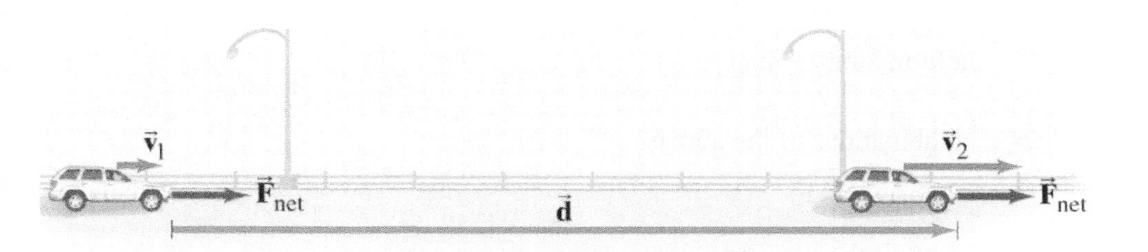

$$+W = +\Delta KE$$

$$v_2 > v_1$$

If the same car applied its brakes at a stop sign, the force provided by friction in the car's brakes does negative work on the car.

In this case, the negative work on the car would reduce its kinetic energy, and the change in kinetic energy would be negative.

As a result, the final velocity of the car would be less than its initial velocity.

$$-\mathrm{W} = -\Delta KE$$

$$v_2 < v_1$$

The kinetic energy of an object can also do work when forces are involved.

For example, a moving object on a frictionless surface can convert all of its kinetic energy to work if it slides up an inclined plane before coming to a complete stop. The kinetic energy of the object is converted to work to lift the object a distance against the force of gravity.

$$-\Delta KE = +\mathrm{W}$$

In summary, energy can perform work against:

- inertia
- gravity
- friction
- deformation of shape
- combinations of the above

Potential Energy

Potential energy (*PE*) is the energy stored in an object. Regardless if an object is in motion, an object always retains energy. This energy can be stored in many different forms, such as the chemical potential energy of food. The reason living things eat is because food is full of chemically bonded substances. Organisms' bodies store these substances in cells until they are needed. Then the cells will break the bonded substances down into individual parts. This process of breaking bonds releases potential energy.

Unlike chemical potential energy, potential energy in mechanical physics problems depends solely upon the position or configuration of objects. By this definition, an object has potential energy by its surroundings.

Some examples of potential energy:

- An object at some height above the ground
- A wound-up spring
- A stretched elastic band

There are a few types of potential energy: gravitational potential energy (local and general) and spring (elastic) potential energy.

Gravitational, local ($PE = mgh$)

Local gravitational potential energy is the energy stored in an object due to its position (height) above the ground. It only depends upon the mass of the object, its height above the ground, and the acceleration of gravity:

$$PE = mgh$$

As can be seen, by the equation, potential energy is directly proportional to mass and height. If two objects of equal mass are at different heights above the ground, then the higher object has more potential energy. Likewise, if two objects of unequal mass are at the same height, then the object with a larger mass will have higher potential energy.

On Earth, $g = 9.8$ m/s^2 and is assumed constant regardless of the location or height relative to the Earth (unless stated otherwise).

Another important feature of gravitational potential energy is that measuring h requires a reference height. Usually, the reference height is evident in a problem (i.e., the ground). However, if not, a reference height can be established at any arbitrary location if the reference height $h = 0$, and thus the potential energy is zero.

For example, if a rock of mass m is held a height h_1 above a table, its potential energy concerning the table (reference height) is:

$$PE = mgh_1$$

If the surface of the table is a height h_2 above the ground, then the rock's potential energy (PE) concerning the ground (reference height) is:

$$PE = mg(h_2 + h_1)$$

Like kinetic energy, work can be performed to increase or decrease potential energy. If all the work (W) is transformed into potential energy (or vice versa), then work, and the equation can relate potential energy:

$$|\,W\,| = |\,\Delta PE\,|$$

Work can be written, regarding a change in the potential energy, as:

$$W = \Delta PE = mgh_2 - mgh_1$$

For example, the diagram below shows a block of mass m that is raised a height h from position y_1 to position y_2. The work performed results in a positive change in potential energy because the new position of the block is at a greater height than its original position.

$$+W = +\Delta PE$$

$$y_2 > y_1$$

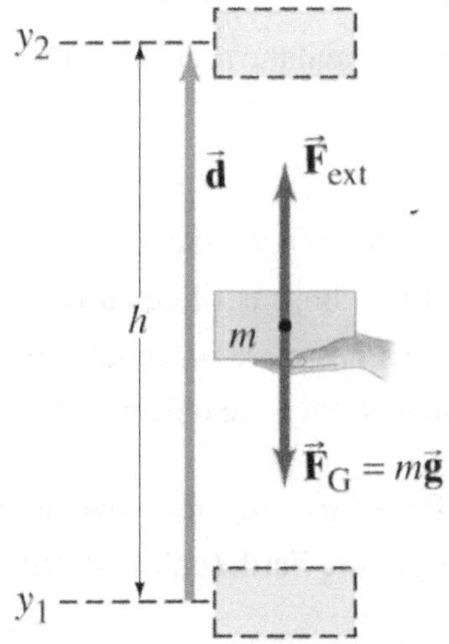

Gravitational, general ($PE = -GmM/r$)

The general formula for gravitational potential energy is found by using Newton's *Law of Universal Gravitation.* This formula was discussed in depth in the previous chapter, but it defines g as being:

$$g = \frac{GM}{r^2}$$

where G is the universal gravitation constant, M is the mass of the attracting object, and r is the distance between the two objects in question. The gravitational constant G has a value of 6.67×10^{-11} N·m²/kg².

To find the general form of gravitational potential energy g is substituted into the equation for local gravitational energy, and the equation results:

$$PE = m \times \frac{GM}{r^2} \times h$$

The height h is essentially another measurement of the distance between two objects and is equal to the radius r. This cancels an r in the denominator, and the equation for general gravitational potential energy is:

$$PE = \frac{GmM}{r}$$

Spring ($PE = \frac{1}{2}kx^2$)

Springs can also store potential energy. The potential energy of a spring is dependent upon the spring material and the stretch or compression of the spring:

$$PE = \frac{1}{2}kx^2$$

The spring constant k is a measure of the stiffness of the spring; stiffer springs have a larger k because they require more energy to stretch or compress.

The compression of the spring from its original equilibrium position is represented as x.

The PE of a spring is the elastic potential energy (PE) and is sometimes notated as PE_{el}.

The force required to compress a spring is described by the equation:

$$F_s = -kx$$

Notice the negative sign in front of the constant; this is to indicate that the spring force is a restoring force and acts opposite to the direction of the stretch.

This can be observed in the diagram below:

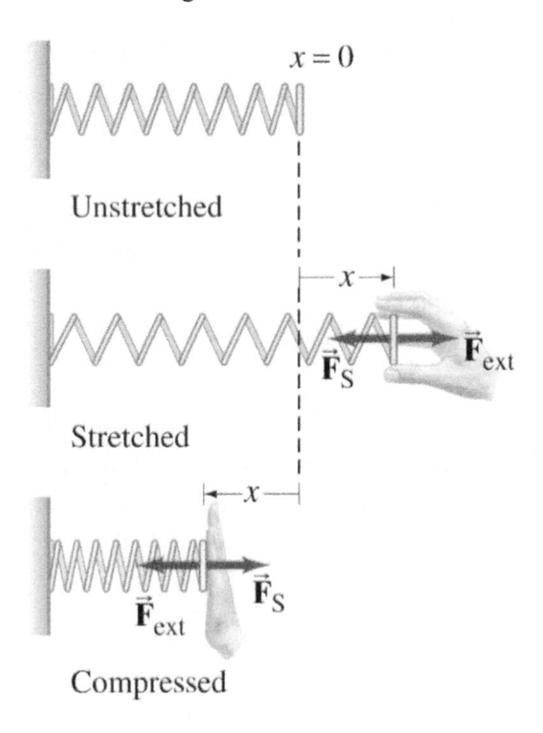

For the figure labeled unstretched, the spring is at equilibrium.

For the figure labeled stretched, if the spring is stretched a distance x from the equilibrium length, then the spring force acts opposite to the direction of the stretch.

For the figure labeled compressed, if the spring is compressed a distance x from the equilibrium length, then the spring force acts opposite to the compression.

Conservation of Energy

The conservative forces include gravity, spring forces, and the electrostatic force. Non-conservative forces include friction and human exertion. Mechanical energy, like all energy, must be conserved. If there are no non-conservative forces, the sum of the changes in the kinetic energy and the potential energy is zero. The kinetic and potential energy changes are equal but opposite in sign.

The total mechanical energy can then be defined as:

$$\Sigma E = KE + PE$$

Moreover, its conservation is:

$$E_2 = E_1 = \text{constant}$$

The total amount of initial energy equals the total amount of final energy. Gravitational potential energy is converted to kinetic energy as an object falls, but the total amount of energy stays the same. When a crate slides to a stop on a rough surface, its kinetic energy is converted into heat and sound energy.

Take the example of a man dropping a rock some distance y. By conservation of energy, the total mechanical energy of the rock at any given instant is:

$$\Sigma E = KE + PE$$

$$KE + PE = ½\, mv^2 + mgh$$

$$\Sigma E = ½\, mv^2 + mgh$$

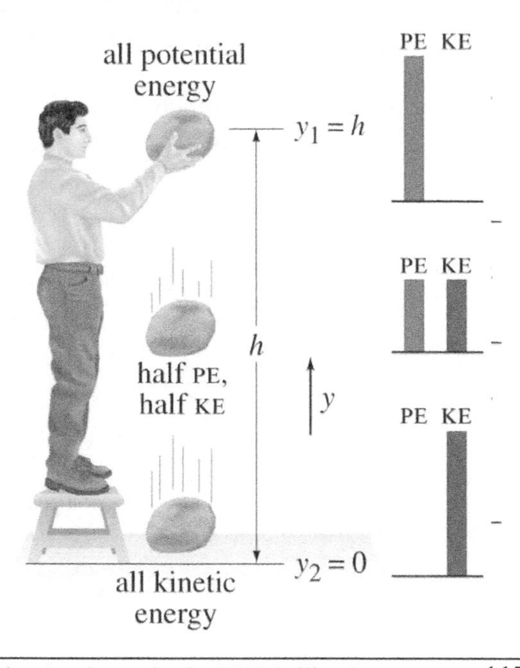

Initially, the rock is held stationary above the ground a distance *y*, and all its energy is potential energy. When the rock is released, it drops down the ground; at precisely the halfway point of the rock's descent, half of its *PE* has been converted to *KE*.

Finally, right before impact with the ground, all the initial potential energy has been converted to kinetic energy. The energy bar graphs next to the figure show how the energy moves from all potential to all kinetic.

Another example of energy conservation is a rollercoaster. The speed of a roller coaster only depends on its current height compared to its starting height.

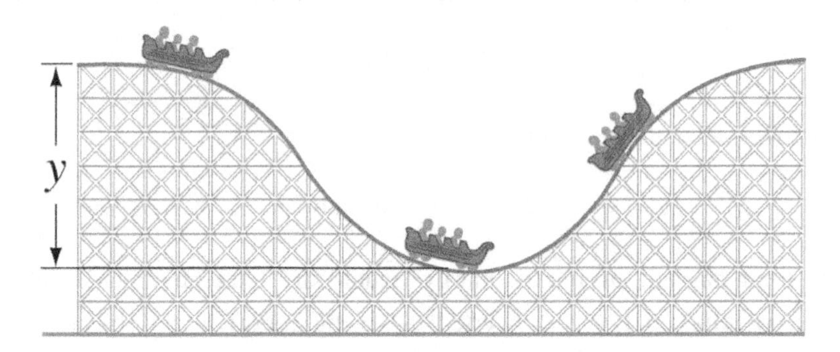

$$PE_{1st\ hill} = KE = PE_{2nd\ hill}$$

At the top of the drop, all the rollercoaster's energy will exist as potential energy. At the bottom, the rollercoaster converted all its energy into kinetic energy. If there is no friction, then all the kinetic energy at the bottom is converted back to potential energy to bring the roller coaster up the second hill of equal height to the first.

Another example of conservation of energy can be seen in the spring. The diagram below depicts a ball being shot using a spring:

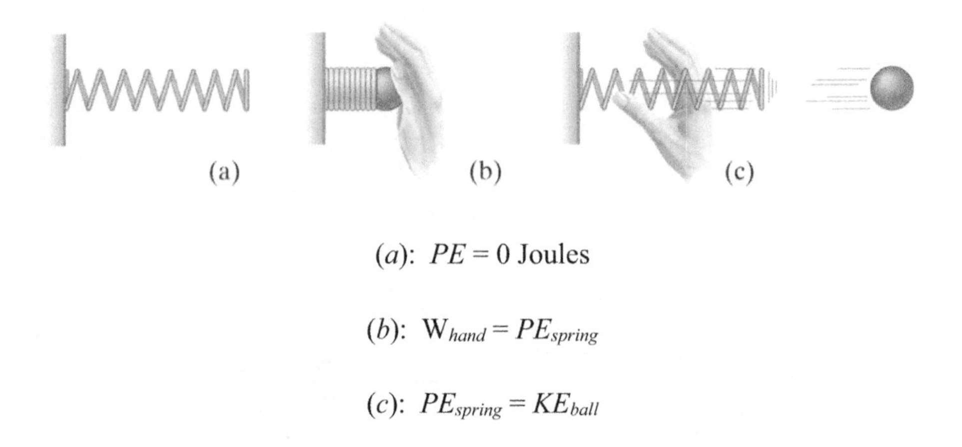

(a): $PE = 0$ Joules

(b): $W_{hand} = PE_{spring}$

(c): $PE_{spring} = KE_{ball}$

In figure (a), the spring is at its equilibrium length and has no potential energy.

Figure (b) shows that work has been performed against the restoring force of the spring. For an ideal system, the work performed is converted entirely into spring potential energy.

In figure (c), the ball is released, and all the spring's potential energy is converted into the kinetic energy of the ball.

However, if there is a non-conservative force, such as friction, where does the kinetic and potential energy go?

Energy cannot be destroyed, only converted into different forms; in this case, the friction performs work, which transforms the initial energy into heat.

Remember, in the presence of friction:

$$PE_{initial} > PE_{final}$$

$$KE_{initial} > KE_{final}$$

Tips for Problem Solving:

1. Draw a picture
2. Determine the system for which energy will be conserved
3. Determine the initial and final positions
4. Choose a logical reference frame
5. Apply the conservation of energy
6. Solve

Power, Units

Often, problems do not give the total energy or input work, but instead, give the power a system consumes.

Power is the work performed over time, and can be written as:

Power can be derived as the product of the force on an object times its velocity:

$$P = \vec{F}\vec{v}$$

The SI unit for power is the *watt* (Joules/second).

Another unit for power, more commonly used for high powered objects such as engines, is *horsepower* (hp). One hp is equal to ~745.7 Watts. A key concept of power is the time at which work is performed.

Lifting an object of mass *m* one meter off the ground always requires the same amount of work. However, lifting the object in one minute requires more power than lifting the object in an hour.

The longer the time interval, the less power is required. The difference between walking and running upstairs is power—the change in gravitational *PE* is the same.

$$P = average\ power = \frac{work}{time} = \frac{energy\ transformed}{time}$$

Power is useful for predicting the amount of work that can be performed over a given time. For example, if there are two engines, the more powerful engine can do the same amount of work in less time than the less powerful engine.

Sports cars are described by their engine's horsepower and are compared by how they accelerate from 0 mph to 60 mph. Sports cars have powerful engines that can perform work, which is, in turn, converted into kinetic energy. Electric companies also use power because it describes energy use (work done) over time.

Chapter Summary

- Work is a force applied across a displacement. Work can cause a change in energy. Positive work puts energy into a system, while negative work takes energy out of a system.

  Basic equations for work include:

  $W = \vec{F}\vec{d}\cos\theta$ $W = \Delta KE$ W = area under an F vs. d graph

Conservative Forces	**Non-conservative Forces**
Gravitational	Friction
	Air resistance
Elastic	Tension in a cord
	Motor or rocket propulsion
Electric	Push or pull by a person

- Energy is the ability to do work and is a conserved quantity.

  The total initial energy is equal to the total final energy.

  Basic equations for energy include:

  $KE = \frac{1}{2}mv^2$ $PE_g = mgh$ $PE_s = \frac{1}{2}kx^2$

- *Kinetic energy*: the energy of motion (energy that is being consumed or released at a given time)

- *Potential energy*: stored energy (energy that could be released for a certain body)

- Types of energy:

    Mechanical energy—The energy associated with motion and position. It is equal to kinetic plus potential energy.

    Electrical energy—The energy made up of the current and potential provided by a circuit. Any charged particle within an electrical field contains electrical energy.

    Chemical energy—The energy involved in all chemical reactions. This is the potential of a certain substance to undergo a transformation or reaction.

    Radiant energy—Consists of all the energy from electromagnetic waves. It can be viewed as the energy stored in a photon, or the motion of an electromagnetic wave. Visible light is only a small part of the full spectrum of electromagnetic waves—there are many other sources of radiant energy that cannot be seen with the naked eye.

    Nuclear energy—The energy that is released during reactions involving the fusion or fission of the nucleus. Examples include nuclear bombs and nuclear power plants.

- Gravitational energy: $PE = GmM / r$

- Spring energy: $PE = \frac{1}{2}kx^2$

- The Law of Conservation of Energy states that the total amount of initial energy equals the total amount of final energy,

- Often, interactions are limited to mechanical energy with no heat lost or gained.

    In this situation:

$$KE_i + PE_i \pm W = KE_f + PE_f$$

- Power is the rate at which work is performed, and is given by:

$$P = \frac{W}{t} \quad \text{or} \quad P = \vec{F}\vec{v}$$

Practice Questions

1. When a pebble is dropped from height h, it reaches the ground with kinetic energy. Ignoring air resistance, from what height should the pebble be dropped to reach the ground with twice the KE?

A. $\sqrt{2}h$ **B.** $2h$ **C.** $4h$ **D.** $8h$ **E.** $16h$

2. Which of the following situations requires the greatest power?

A. 50 J of work in 20 minutes
B. 200 J of work in 30 minutes
C. 10 J of work in 5 minutes
D. 100 J of work in 20 minutes
E. 100 J of work in 10 minutes

3. A kilowatt-hour is a unit of:

I. work II. force III. power

A. I only
B. II only
C. III only
D. I and II only
E. I and III only

4. A hydraulic press (like a simple lever), properly arranged, is capable of:

I. multiplying the energy input
II. multiplying the output force
III. exerting force only vertically

A. I only
B. II only
C. III only
D. I and II only
E. I and III only

5. 4.5×10^5 J of work is done on a 1,150 kg car while it accelerates from 10 m/s to some final velocity. What is this final velocity? (Use the acceleration due to gravity $g = 10$ m/s^2)

A. 30 m/s **B.** 37 m/s **C.** 12 m/s **D.** 19 m/s **E.** 43 m/s

6. Which of the following is not a unit of work?

A. N·m **B.** kw·h **C.** J **D.** kg·m/s **E.** W·s

7. The law of conservation of energy states that:

I. the energy of an isolated system is constant
II. energy cannot be used faster than it is created
III. energy cannot change forms

A. I only
B. II only
C. III only
D. I and II only
E. I and III only

8. A crane lifts a 300 kg steel beam vertically upward a distance of 110 m. Ignoring frictional forces, how much work does the crane do on the beam if the beam accelerates upward at 1.4 m/s²? (Use the acceleration due to gravity $g = 9.8$ m/s²)

A. 2.4×10^3 J
B. 4.6×10^4 J
C. 3.7×10^5 J
D. 6.2×10^5 J
E. 8.8×10^2 J

9. Steve pushes twice as hard against a stationary brick wall as Charles. Which of the following statements is correct?

A. Both do the same amount of positive work
B. Both do positive work, but Steve does one-half the work of Charles
C. Both do positive work, but Steve does four times the work of Charles
D. Both do positive work, but Steve does twice the work of Charles
E. Both do zero work

10. What is the change in the gravitational potential energy of an object if the height of the object above the Earth is doubled? (Assume that the object remains near the surface)

A. Quadruple
B. Doubled
C. Unchanged
D. Halved
E. Tripled

11. If 1 N is exerted for a distance of 1 m in 1 s, the amount of power delivered is:

A. 3 W
B. 1/3 W
C. 2 W
D. 1 W
E. ½ W

12. A brick is dropped from a roof and falls a distance h to the ground. If h were doubled, how does the maximal KE of the brick, just before it hits the ground, change?

A. It doubles
B. It increases by $\sqrt{2}$
C. It remains the same
D. It increases by 200
E. Requires more information

13. A helicopter with single landing gear descends vertically to land with a speed of 4.5 m/s. The helicopter's shock absorbers have an initial length of 0.6 m. They compress to 77% of their original length, and the air in the tires absorbs 23% of the initial energy as heat. What is the ratio of the spring constant to the helicopter's mass?

A. 0.11 kN/kg·m
B. 1.1 N/kg·m
C. 0.8 kN/kg·m
D. 11 N/kg·m
E. 0.11 N/kg·m

14. A 21 metric ton airplane is observed to be a vertical distance of 2.6 km from its takeoff point. What is the gravitational potential energy of the plane with respect to the ground? (Use the acceleration due to gravity $g = 9.8$ m/s^2, the metric ton = 1,000 kg)

A. 582 J
B. 384 J
C. 535 MJ
D. 414 MJ
E. 773 J

15. A tennis ball bounces on the floor. During each bounce, it loses 31% of its energy due to heating. How high does the ball reach after the third bounce, if it is initially released 4 m from the floor?

A. 55 cm
B. 171 mm
C. 106 cm
D. 131 cm
E. 87 cm

Solutions

1. B is correct.

When the pebble falls, KE at impact = PE before impact.

$PE = KE$

$mgh = KE$

$mg(2h) = 2KE$

$2PE = 2KE$

If the mass and gravity are constant, then the height must be doubled.

2. E is correct.

Power = Work / time

A: Power = 50 J / 20 min = 2.5 J/min

B: power = 200 J / 30 min = 6.67 J/min

C: power = 10 J / 5 min = 2 J/min

D: power = 100 J / 20 min = 5 J/min

E: power = 100 J / 10 min = 10 J/min

Typically, power is measured in watts or J/s.

3. A is correct.

kilowatt = unit of power

hour = unit of time

kW·h = power × time

power = work / time

kW·h = (work / time) × time

kW·h = work

4. B is correct.

Mechanical advantage:

d_1 / d_2

where d_1 and d_2 are the effort arm and load arm, respectively.

If d_1 is greater than d_2, force output is increased.

5. A is correct.

Find the final speed using the conservation of energy.

Let the energy added as work be represented by W. Then, conservation of energy requires:

$E_f = E_i + W$

$½mv^2_f = ½mv^2_i + W$

Solving for v_f:

$v_f = \sqrt{v^2_i} + (2W) / m$

$v_f = \sqrt{[(10 \text{ m/s})^2 + (2)\cdot(4.5 \times 10^5 \text{ J}) / (1150 \text{ kg})]}$

$v_f = 29.7 \text{ m/s} \approx 30 \text{ m/s}$

6. D is correct.

Work = Force × distance

W = Fd

The unit kg·m/s cannot be manipulated to achieve this.

7. A is correct.

The Law of Conservation of Energy states that for an isolated system (no heat or work transferred), the energy of the system is constant.

Energy is neither created nor destroyed; it can only transform from one form to another.

8. C is correct. The work done by this force is:

$W = Fd$

where F = force applied by the crane and d = distance over which the force is active.

Solve for F using Newton's Second Law.

There are two forces on the beam – the applied force due to tension in the crane's cable and gravity.

$F_{net} = ma$

$F - mg = ma$

Therefore:

$W = m(a + g)d$

$W = (300 \text{ kg})\cdot(1.4 \text{ m/s}^2 + 9.8 \text{ m/s}^2)\cdot(110 \text{ m})$

$W = 3.7 \times 10^5 \text{ J}$

9. E is correct.

Work = Force × displacement × $\cos \theta$

$W = Fd \cos \theta$

If $d = 0$, then work = 0

10. B is correct.

Relative to the ground, an object's gravitational $PE = mgh$, where h is the altitude.

PE is proportional to h, doubling h doubles PE.

11. D is correct.

$\text{W} = Fd$

$P = \text{W} / t$

$P = Fd / t$

$P = (1 \text{ N})\cdot(1 \text{ m}) / 1 \text{ s}$

$P = 1 \text{ W}$

12. A is correct.

$KE = PE$

$KE = mgh$

h is directly proportional to KE.

If h doubles, then the KE doubles.

13. C is correct.

The initial kinetic energy of the helicopter is entirely converted to the potential energy of the landing gear and heat. By conversation of energy:

$\Delta KE = \Delta PE + Q$

Let the equilibrium length of the landing gear's spring be L_0 and the compressed length be L. The change in potential energy of the landing gear is:

$\Delta PE = ½k(\Delta x)^2$

$\Delta PE = ½k(L_0 - L)^2$

$\Delta PE = ½k(L_0 - (0.23)L_0)^2$

$\Delta PE = ½k(0.77)L_0^2$

The energy lost to heat is 23% of the initial kinetic energy. The final kinetic energy is zero, therefore:

$Q = (0.23)\ \Delta KE$

$Q = (0.23)\ ½mv^2$

Conservation of energy becomes:

$½mv^2 = (0.77)½kL_0^2 + (0.23)½mv^2$

Solving for k / m:

$k / m = [(0.77)\ v^2] / [(0.23)^2L_0^2]$

$k / m = 819\ \text{s}^{-2}$

$k / m = 0.8\ \text{kN m}^{-1}\ \text{s}^{-1}$

14. C is correct.

$PE = mgh$

$PE = (21 \times 10^3 \text{ kg})\cdot(9.8 \text{ m/s}^2)\cdot(2.6 \times 10^3 \text{ m})$

$PE = 535 \text{ MJ}$

15. D is correct.

$h_0 = 3.5 \text{ m}$

$PE_0 = mgh$

$PE_1 = mgh(0.69)$ → after first bounce

$PE_2 = mgh(0.69)^2$ → after second bounce

$PE_3 = mgh(0.69)^3$ → after third bounce

Because mass and gravity are constant, the final height is:

final height $= h(0.69)^3$

final height $= (4 \text{ m})\cdot(0.69)^3$

final height $= 1.31 \text{ m} = 131 \text{ cm}$

Chapter 4

Force, Motion, Gravitation

- **Newton's First Law: Inertia**
- **Newton's Second Law: $F = ma$**
- **Newton's Third Law: Forces Equal and Opposite**
- **Weight**
- **Center of Mass**
- **Friction: Static and Kinetic**
- **Motion on an Inclined Plane**
- **Uniform Circular Motion and Centripetal Force**
- **Law of Gravitation**
- **Concept of a field**

Newton's First Law: Inertia

Newton's First Law of Motion states that *an object at rest remains at rest and an object in motion (with constant velocity) remains in motion (with that same constant velocity) unless acted on by an external force.*

In other terms, nothing changes about an object's motion regarding speed and direction, unless it is acted on by an outside force.

All objects resist changes in their state of motion (even if that motion is zero) until an outside force manipulates them into a new state of motion. This is because all objects have *inertia*.

Inertia is an inherent property of all objects and is the tendency for objects to resist changes in their current state motion.

Inertia cannot be calculated, but it can be measured by an object's *mass*; the more massive the object, the more inertia it has. In the SI system, mass is measured in kilograms (kg).

Mass is not the same value as weight (i.e., $w = mg$); weight (w) depends on gravity (and can change depending on location), the mass (m) of an object is always the same and gravity (g) depends on the attractive force between two objects.

Newton's First Law of Motion is only valid for an inertial reference frame, which is any frame of reference that moves with constant velocity (i.e., no acceleration) relative to the inertial system under observation.

Newton's Second Law: *F = ma*

Newton's Second Law states that force is the vector product of mass and acceleration, and it is expressed as:

$$\Sigma\vec{F} = m\vec{a}$$

(Net force equals mass times acceleration)

The unit for force in the SI system is the Newton, which is the product of the units of mass and acceleration ($N = kg{\cdot}m/s^2$).

When considering force, it is essential to note the units. A pound is a unit of force, not of mass. Therefore, it can be equated to the Newton, but not to kilograms.

Conventionally, kilograms are used as a measurement of weight (e.g., a person weighs 186 lbs. or 80 kg). This usage can be misleading, as it is not technically correct; a kilogram is a unit of mass, and the weight of an object is a measurement of force.

Remember that this everyday use of kilogram is incorrect and do not mistake kilograms for weight when solving the problems.

Units for Mass and Force		
System	**Mass**	**Force**
SI	kilogram (kg)	Newton ($N = kg{\cdot}m/s^2$)
CGS	gram (g)	dyne ($= g{\cdot}cm/s^2$)
Imperial	slug	pound (lb)
Conversion factors: 1 dyne = 10^{-5} N; 1 lb ≈ 4.45 N; 1 slug ≈ 14.6 kg		

Newton's Second Law of Motion can be separated into *x*- and *y*- and *z*-components because the force is a vector:

$$\Sigma F_x = ma_x$$

$$\Sigma F_y = ma_y$$

$$\Sigma F_z = ma_z$$

From the above equations, the force is directly proportional to the mass of the object. This further explains Newton's First Law of Motion (Law of Inertia). Much more force is required to push a boulder up a hill than to push a pebble up a hill because the mass of the boulder is much higher than that of the pebble. If this situation is applied to Newton's First Law, the boulder has higher inertia than the pebble and will, therefore, be more resistant to changes in motion.

By applying Newton's Second Law, the force needed to move (accelerate) the boulder is directly proportional to its mass and thus greater than the force needed to move (accelerate) the pebble.

Rearranging Newton's Second Law ($F = ma$) shows that acceleration is directly proportional to force and inversely proportional to mass:

$$a = F / m$$

The force on an object with a mass of 2 kg is 4 N, therefore:

$$a = \frac{F}{m} \qquad a = \frac{4\text{ N}}{2\text{ kg}} \qquad a = 2\text{ m/s}^2$$

If the force is 8 N and the mass of the object is 2 kg, then:

$$a = \frac{F}{m} \qquad a = \frac{8\text{ N}}{2\text{ kg}} \qquad a = 4\text{ m/s}^2$$

Acceleration is directly proportional to force. By doubling the force, the acceleration of the object is also doubled.

The opposite holds when doubling the mass. Since mass is inversely proportional to acceleration, doubling the mass will halve the acceleration. This intuitively makes sense considering the mass of the boulder and pebble in the previous example. Exerting the same force on both will not give the same acceleration because the mass (which represents inertia, or the resistance to change) of each object is vastly different.

If the pebble has a mass of 5 kg and its acceleration is 20 m/s^2, what force was exerted on it?

$$a = \frac{F}{m} \qquad 20\text{ m/s}^2 = \frac{F}{5\text{ kg}} \qquad F = 100\text{ N}$$

If the boulder has a mass of 20 kg and the same force of 100 N is exerted, what is the acceleration? The boulder has four times the mass of the pebble, so the acceleration is one-fourth that of the pebble: 5 m/s^2 as compared to 20 m/s^2 for the pebble.

Both force and acceleration are vectors because they have a direction. Many physics questions omit the directional attribute because it is usually assumed. For example, when an apple falls to the ground, the force of gravity acts downwards, and the apple, of course, falls downwards. Many questions involve substituting values into the appropriate formula.

However, more difficult questions have directional attributes associated with them. For example, when a bar of soap slides down an inclined plane, the force of gravity acts downwards, but the acceleration is not entirely downwards; instead, it is slanted. Therefore, vector analysis is needed.

Solving problems involving Newton's Second Law

- Visualize, draw a sketch, and always use a free body diagram.
- Clearly define an appropriate coordinate system. This is particularly important when dealing with inclined planes because the angle between the normal force and the gravitational force acting on the object is altered.
- Always apply Newton's Second Law to each mass separately, and when there are forces in two or more directions, apply Newton's Second Law to each direction separately.
- "Translational equilibrium" implies two equations: $\Sigma F_x = 0$ and $\Sigma F_y = 0$
- When $\Sigma F = 0$, $a = 0$ and vice versa.
- Solve the component equations. If the acceleration of x or y is known, or the object is moving at constant velocity ($a = 0$), then the component equations can be used to solve for unknown forces.
- Use the unit circle to calculate trigonometric functions and make sure to use degrees if the angles are given in degrees or radians if the angles are given in radians.

Newton's Third Law: Forces are Equal and Opposite

For Newton's Third Law, it is essential to understand how forces work. A force must have a source object (i.e., any force exerted on one object that is caused by another object). When a person interacts with an object by exerting a force on it, that object exerts the same amount, or magnitude, of force back on them. This is often simplified into the dogma:

Newton's Third Law: *Every action has an equal and opposite reaction.*

If body A exerts a force F_A on body B, then B exerts a force F_B on body A:

$$F_B = -F_A$$

To correctly apply Newton's Third Law, it is essential to understand that the forces are exerted on different objects, and they should not be treated as if they were acting on the same object. These matched forces are *action and reaction pairs.*

As a helpful notation, the first subscript is the object that the force is being exerted on. The second subscript is the source object. For example, a person (P) walking forward exerts a force on the ground (G), which exerts an equal and opposite force to move them forward. This can be expressed as:

$$\vec{F}_{GP} = -\vec{F}_{PG}$$

$\vec{F}_{GP}$ is the horizontal force exerted *on* the ground by the woman's foot and $\vec{F}_{PG}$ is the horizontal force exerted on the woman's foot *by* the ground.

Using Newton's Third Law for rocket propulsion, when a rocket is in flight, hot gases from combustion spew out of the tail of the rocket at high speeds (*action force*). The *reaction force* is what propels the rocket forward.

It is important to note that the rocket does not need anything to "push" against. Regardless if it is traveling within Earth's atmosphere or in the vacuum of space, the rocket can propel itself forward because of the action-reaction forces from combustion.

To summarize, all three of Newton's Laws of Motion can be extrapolated and observed in almost any situation. A cannon and cannonball provide an example:

- A cannon is initially at rest. It has a large mass and, therefore, large inertia. It remains at rest until a force is applied (Newton's First Law).

- When the cannon is fired, a large force is applied by the cannon to the cannonball to overcome the inertia of the cannonball and cause it to accelerate (Newton's Second Law).

- The tremendous force applied by the cannon is reciprocated by the cannonball back onto the cannon, the cannon recoils (Newton's Third Law).

Why does the cannonball accelerate faster than the recoiling cannon?

Consider all three laws, but the most important component to consider is inertia.

Third Law: the forces exerted on the cannon and cannonball must be the same.

The Second Law states that force and mass (inertia) determine acceleration; since the forces are equal, but the masses are considerably different, the accelerations are considerably different.

The cannonball accelerates faster because of its mass (i.e., inertia) is less than the mass of the cannon: $a = \frac{F}{m}$

Once the cannonball is airborne, it will remain in flight (according to the First Law) unless a force acts upon it to change its motion. In this case, the forces of air resistance and gravity eventually cause the cannonball to slow and fall to the ground.

Solving Problems with Newton's Laws – Free-Body Diagrams

1. Draw a sketch.

2. For every object, draw a free-body diagram showing all the forces acting *on* the object.

   Make the magnitudes and directions as accurate as possible.

   Label each force.

   If there are multiple objects, draw a separate diagram for each object.

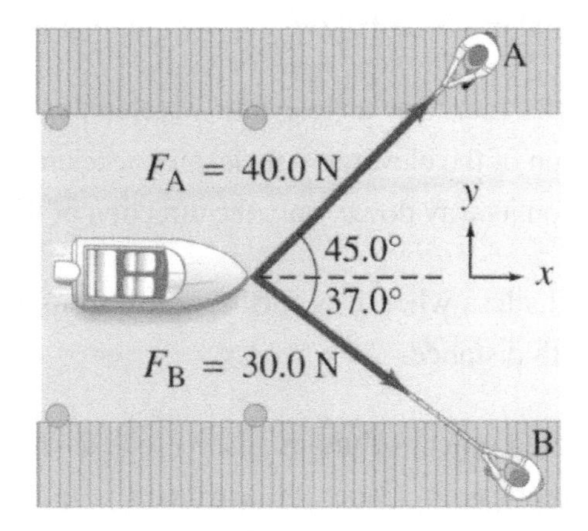

3. Resolve the vectors into components.

4. Apply Newton's Second Law of Motion ($F = ma$) to each component.

5. Solve for the net force in each component direction and combine each component to determine the total net force.

Weight

Weight is the gravitational force (i.e., the force of gravity) that acts on a mass when an object is close to the Earth's surface. The gravitational force changes with substantial increases or decreases in altitude. It is a force, therefore, a vector quantity.

The gravitational force is:

$$\vec{F_g} = m\vec{g}$$

where g is the acceleration due to gravity at 9.8 m/s^2 (sometimes rounded to 10 m/s^2).

An object weighs more on an elevator accelerating up because $F = mg + ma$, where a is the acceleration of the elevator. An elevator accelerating up has the same force as an elevator decelerating on its way down, only the direction of the acceleration changes.

An object weighs less when it is further away from the Earth because the force of gravity decreases with distance.

However, when an object orbits in space (such as an astronaut in a spacecraft), it is not genuinely weightless in space. An astronaut is falling toward the Earth at the same rate as the spacecraft she resides in. If the astronaut's spacecraft falls from space to the surface of the Earth, their weight will increase during the fall due to the increase of gravitational acceleration as they get closer to the Earth's surface.

For a given mass, its weight on Earth is different than it is on the Moon. If a person stands on the Moon, which has a gravitational acceleration of about 1/6 g, that person is 1/6 of their typical weight on Earth. Their mass, however, is the same in both locations.

From Newton's Second Law ($F = ma$), an object at rest must have no net force on it. It is also true, that an object always has a gravitational force acting upon it.

For both Newton's Second and Third Laws, there must be an equal but opposite force. If the object is resting on a flat surface, then the force of the surface pushing up on the object is the *normal force*.

The normal force is always *perpendicular* to the plane of the object.

When an object is lying still on a horizontal surface, the normal force is equal and opposite to the weight (i.e., the force of gravity).

The normal force is precisely as large as needed to balance the downward force from the object (if the required force gets too big, something breaks).

If a statue is placed on a table, the normal force from the table is equal and opposite to the force of gravity.

As a result, the statue experiences no net force and does not move.

$$-\vec{F}_G = \vec{F}_N$$

Center of Mass

Every object has a center of mass. The *center of mass* is a single, approximate point at which all an object's mass is concentrated. It is also at this point that external forces can be considered to act on the object. If an object is balanced on its center of mass, it spins uniformly about this axis.

If an object is a uniform along its entire length, the center of mass is the center of the object. If the object is not uniform, the center of mass is the point obtained by taking an average of all the positions weighted by their respective masses:

$$x_{cm} = \frac{\sum x_i m_i}{\sum m_i} \qquad y_{cm} = \frac{\sum y_i m_i}{\sum m_i} \qquad z_{cm} = \frac{\sum z_i m_i}{\sum m_i}$$

It is not necessary to have absolute coordinates when calculating the center of mass. Set the point of reference for a convenient location and use relative coordinates. It is important to understand that the center of mass need not be within the object. For example, a doughnut's center of mass is in the center of its hole. For two objects, the center of mass lies closer to the one with the most mass.

Imagine two balls of an unequal mass attached to a massless stick. The equation and image below describe the situation and the center of mass calculation:

$$m_A > m_B$$

$$x_{CM} = \frac{m_A x_A + m_B x_B}{m_A + m_B} = \frac{m_A x_A + m_B x_B}{\sum M}$$

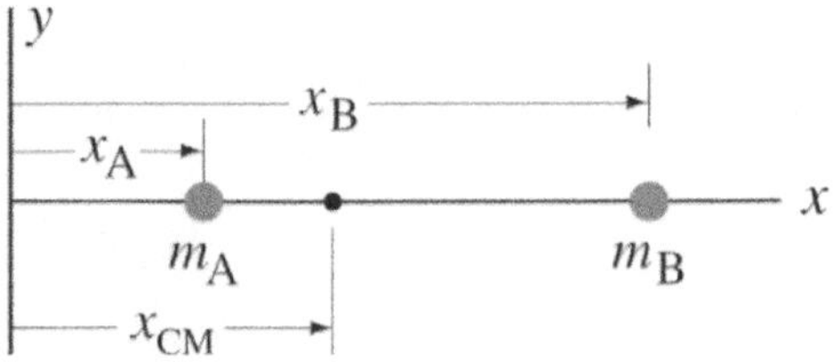

If the difference in mass between two objects is substantial, then the center of mass can be assumed to be that of the more massive object.

For example, the center of mass of the Earth and a man in space is going to be almost at the Earth's center because the man is comparatively tiny, and therefore his coordinate, weighted with respect to his mass, is almost negligible.

Translational Motion using Center of Mass

When problems involve directional forces on an object and the subsequent path that the object takes, these directions and paths are based on the object's center of mass.

The object is effectively treated as a point or particle that has the same mass as the object but is located at the object's center of mass.

The sum of all forces acting on a system is equal to the total mass of the system multiplied by the acceleration of the center of mass:

$$ma_{\text{CM}} = F_{\text{net}}$$

This simplifies calculations because calculating the motion of an irregularly shaped object involves complex mathematics.

Center of Gravity

The center of gravity (CG) is the average location of the weight of an object, and it is the point at which the gravitational force is considered to act. The center of gravity is the same as the center of mass only if the gravitational force does not vary between different parts of the object.

The center of gravity is found experimentally (see image below) by suspending an object from different points. A piece of string about the length of the object is securely attached to a point near the object's edge. The object is then held by the string near the point of connection, and a line is traced along the hanging string. The string hangs straight down, perpendicular to the ground, because gravitational force acts purely in the y-direction.

The object is then rotated, and the steps are repeated.

The point at which the drawn lines cross is the center of gravity.

If hundreds of lines are drawn onto the object, all lines intersect at this point.

Like the center of mass, an object rotates perfectly around its center of gravity.

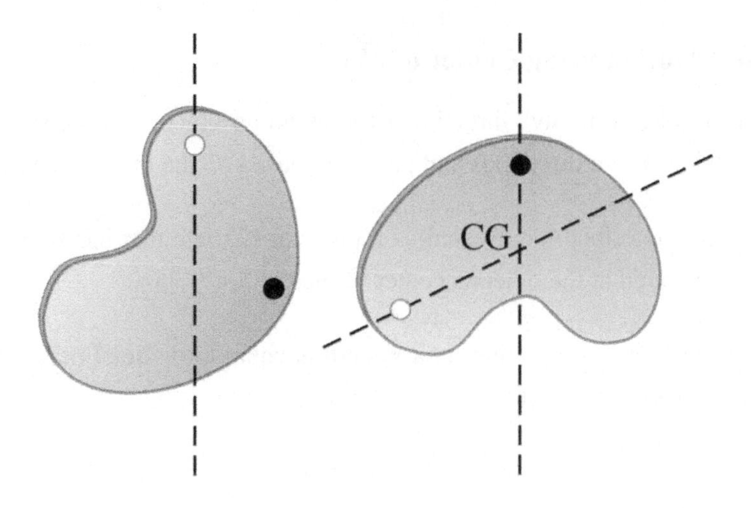

Drawing imaginary lines to determine the center of gravity (CG)

Friction: Static and Kinetic

Friction is a force that always opposes the direction of motion. Like other forces, friction is a vector (i.e., has both magnitude and direction). A dog can walk, and cars can drive because of friction. Lubricants (e.g., oil, grease) reduce friction because they change surface properties and decrease the coefficient of friction. Heat is produced as a by-product of the force of friction.

What is friction? Friction has to do with multiple factors, and it changes depending on the surface of contact. On a microscopic scale, most surfaces are rough. As one surface moves over the other, the crevices of the two surfaces catch and release. This is where the force resisting motion (i.e., friction) comes from.

In the figure below, the block is moving to the right, and as its rough surface moves over the rough surface of the table, the force of friction acts to the left. Friction always acts in the *direction opposing the motion*.

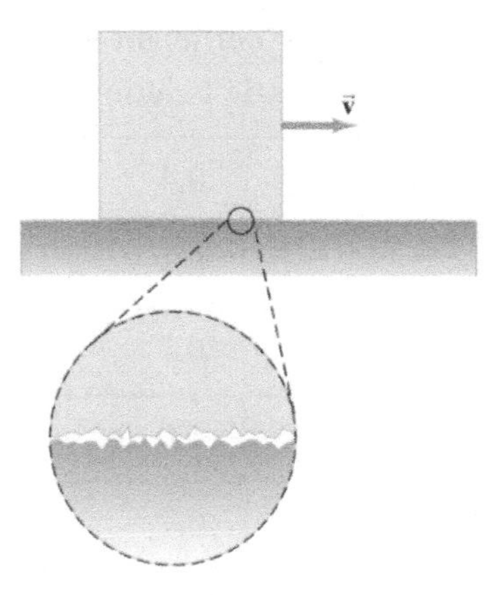

The frictional force is calculated as the product of the normal force of an object and a dimensionless value for the *coefficient of friction*. The coefficient of friction is determined empirically and is intrinsic to the material properties of the surface and the object. Friction occurs in two types: static and kinetic.

Static friction pertains to objects sitting still (an object can remain stationary on an inclined plane because of static friction).

Kinetic friction pertains to objects in motion (a vase sliding across a table has its motion resisted by a force of kinetic friction).

Both the force and the coefficient of static friction are always greater than the force and coefficient of kinetic friction. It takes a greater force to start an object moving on a rough surface than it does to keep the object moving on the same surface.

Equations to determine friction:

$$\text{Static friction:} \quad F_s = \mu_s F_N$$

$$\text{Kinetic friction:} \quad F_k = \mu_k F_N$$

where μ is the coefficient of friction, and F_N is the normal force.

$$F_s > F_k$$

$$\mu_s > \mu_k$$

This table lists the typical values of coefficients of friction. The coefficients depend on the characteristics of both surfaces.

Coefficients of Static and Kinetic Friction		
Surface	**Coefficient of Static Friction μ_s**	**Coefficient of Kinetic Friction μ_k**
Wood on wood	0.4	0.2
Ice on ice	0.1	0.03
Steel on steel (unlubricated)	0.7	0.6
Rubber on dry concrete	1.0	0.8
Rubber on wet concrete	0.7	0.5
Rubber on other solid surfaces	1 – 4	1
Teflon® on Teflon® in air	0.04	0.04
Teflon® on steel in air	0.04	0.04
Lubricated ball bearings	< 0.01	< 0.01
Synovial joints (in human limbs)	0.01	0.01
Values are approximate and intended only as an illustration of the concepts.		

An object at rest on a frictional surface has the force of static friction acting on it.

The force of this friction is $F_s \leq \mu_s F_N$, and it is only equal when the object is just about to move, meaning that the static friction is on the verge of changing over to kinetic friction.

For an object that starts at rest and then has a force applied to it, the force of static friction increases as the applied force increases, until the applied force is enough to overcome the maximum value of static friction.

Then the object starts to move, and the force of kinetic friction takes over.

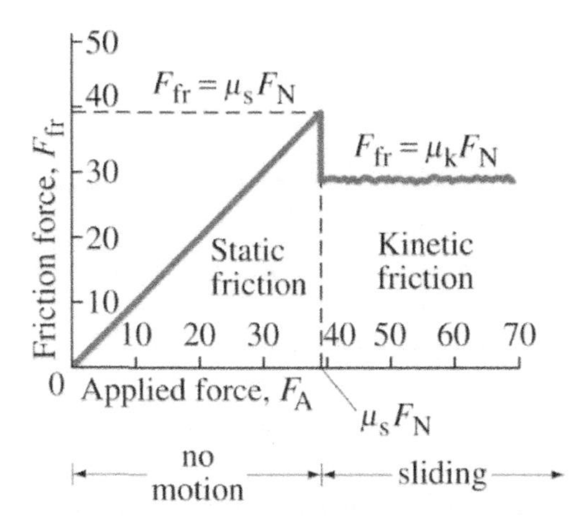

Note: the normal force of an object on a horizontal surface is equal to the weight ($w = mg$) of the object.

However, on an inclined plane, the normal force is perpendicular to the inclined plane, and its magnitude is equal to the weight of the object times the cosine of the incline angle ($F_N = w \cos \theta$; see the chapter on inclined planes).

Motion on an Inclined Plane

When an object is on an inclined plane, the force of gravity is divided into two components: one component is normal (i.e., perpendicular) to the plane surface, and the other component is parallel to the plane surface. The image and equations describe this:

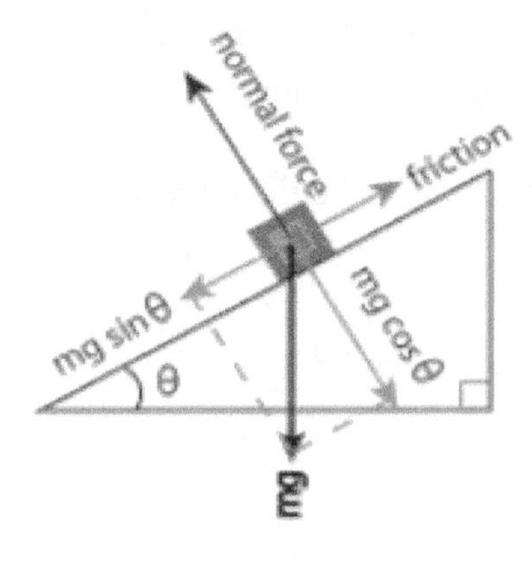

Parallel Force: $F_{\parallel} = mg \sin \theta$

Perpendicular Force: $F_{\perp} = F_N$

$F_N = mg \cos \theta$

where θ is the angle between the horizontal axis and the surface on which the object rests.

How does the force of gravity affect the force of friction? An object on an incline has three forces acting on it: the normal force, the gravitational force, and the frictional force:

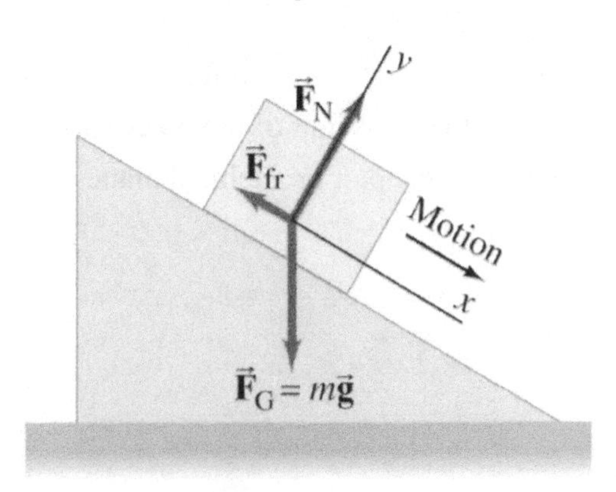

However, the gravitational force is split into parallel and perpendicular components concerning the incline. For an object moving down the inclined plane at a constant velocity, the parallel component of gravity is equal and opposite to the force of kinetic friction:

$$F_{\parallel} = F_k$$

$$F_{\parallel} = \mu_k F_N$$

$$mg \sin \theta = \mu_k mg \cos \theta$$

If the object is not moving, the parallel component of gravity is equal to and opposite to the force of static friction:

$$F_{\parallel} = Fs$$

$$F_{\parallel} = \mu_s F_N$$

$$mg \sin \theta = \mu_s mg \cos \theta$$

Note: frictional forces are always less than or equal to the forces causing an object to move; they are never greater than the forces causing movement. If a frictional force exceeds the other forces involved, it causes the object to move in the opposite direction. Friction does not produce movement; it only inhibits it.

When an object is pushed or pulled up an inclined plane, the parallel component of gravity and the force of friction must be overcome; these forces need to be broken up into their *x* and *y* components to perform calculations.

Only the force component parallel to the plane contributes to the motion.

For example, if a block of mass *m* is accelerating down an inclined plane of angle θ with friction present, what is its acceleration along the *x*-axis?

$$F_{\parallel} = mg \sin \theta$$

$$F_N = mg \cos \theta$$

$$F_{fr} = \mu_k F_N$$

$$F_x = F_{\parallel} - F_{fr}$$

$$ma_x = mg \sin \theta - \mu_k mg \cos \theta$$

$$a_x = g \sin \theta - \mu_k g \cos \theta$$

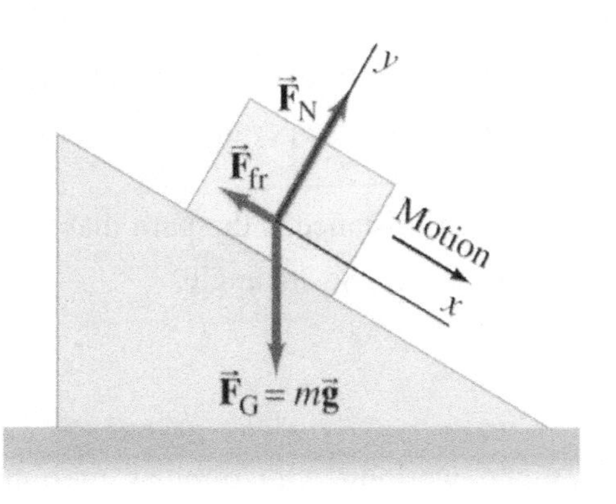

Uniform Circular Motion and Centripetal Force

An object moving in a circle at constant speed is in a *uniform circular motion*; this is an example of simple harmonic motion.

The *instantaneous velocity* is always tangent to the circle, and the *centripetal* (or radial) acceleration points inward along the radius.

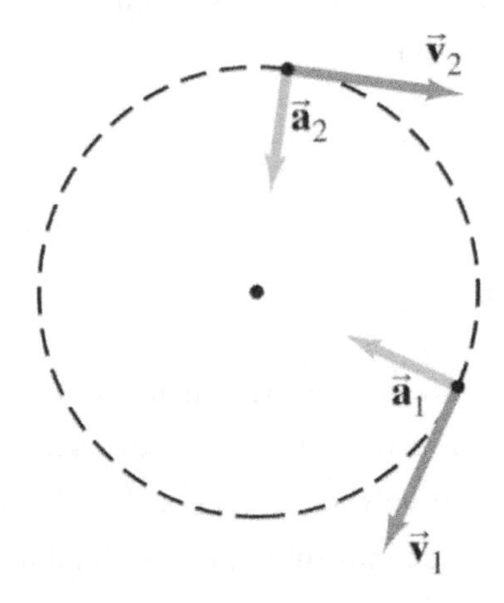

The equation for centripetal (radial) acceleration for uniform circular motion is:

$$a_c = \frac{v^2}{r}$$

Distinguishing between velocity and speed is essential.

Velocity is displacement over time, while speed is the distance over time.

For uniform circular motion, displacement is the shortest straight-line distance between two points on the perimeter of a circle (i.e., a *chord*).

However, distance is the path that the object travels and can be calculated as the product of the radius and arc angle:

$$s = \theta r$$

When working with circular motion problems, the instantaneous velocity is almost always given and equal to the speed.

Some typical issues for circular motion:

- For displacements and distances that approach zero, the instantaneous velocity equals the speed.
- For a quarter of a circle (π/2 radians or 90°), displacement is the hypotenuse of a right-angled triangle with the radius as the other two sides.

  Using the Pythagorean Theorem, the displacement is $\sqrt{2r^2}$. The distance is the length of the arc or ¼ of the circumference.
- For halfway around the circle, the displacement is the diameter, and the distance is half the circumference.
- For three-quarters around the circle, the displacement is obtained by the Pythagorean Theorem. The magnitude of the displacement is the same as that for a quarter of a circle ($\sqrt{2r^2}$), but the direction is different. The distance is ¾ of the circumference.
- Going around the circle entirely has a displacement of zero, which means that the average velocity is also zero. The distance is equal to the circumference.
- The velocity is always less than or equal to the speed. The displacement is always less than or equal to the distance.

  Displacement and velocity are vectors.

  Distance and speed are scalars.

Frequency (f) is the number of times an object makes a revolution in one second, measured in Hertz (Hz = s^{-1}).

The *period* (T) of an object in a circular motion is the time it takes the object to make one revolution, measured in seconds (s).

The frequency and period of an object are inversely related:

$$f = 1 / T$$

Formulas to know:

Centripetal Acceleration: $a_c = \frac{v^2}{r}$

Circumference: $C = 2\pi r$

Arc: $arc = \frac{\theta}{2\pi} \times C = r\theta$

Area of a Circle: $A = \pi r^2$

Note that theta (θ) is in radians and that 2π radians = 360 degrees.

Centripetal Force ($F = \frac{mv^2}{r}$)

In some situations, more forces are acting on an object than in others.

For example, a puck gliding across an air table experiences almost no frictional force and no pushing force.

A block being pulled up a rough incline experiences both a frictional force and a force of tension. Both objects experience a normal force and a gravitational force.

For circular motion, the number of forces (and whether they are changing or constant) depends on the situation.

Consider circular motion: centripetal force is due to centripetal acceleration.

Centripetal acceleration is due to changes in velocity for an object revolving around a circle. Changes in velocity (vector) are due to a constant change in direction.

The diagrams demonstrate a change in velocity due to change in direction, and the resulting centripetal acceleration.

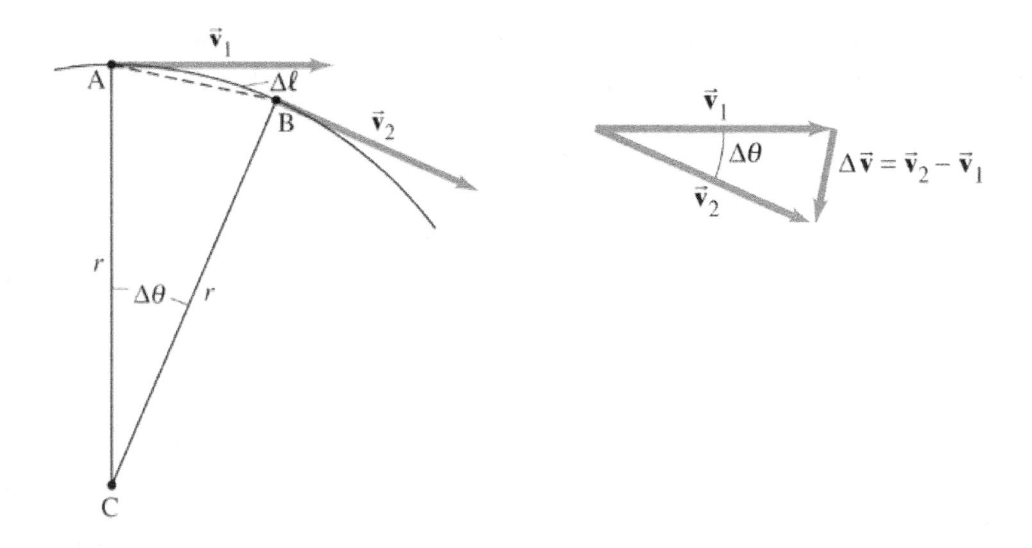

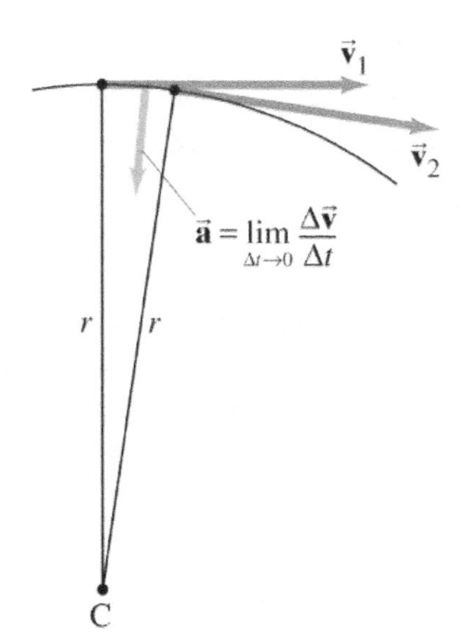

The centripetal force is found when the mass is multiplied by the centripetal acceleration. The centripetal force may be provided by friction, gravity, tension, the normal force, or other forces.

Centripetal force: $F = ma_c = \frac{mv^2}{r}$ Centripetal acceleration: $a_c = \frac{v^2}{r}$

A negative sign is used for the centripetal force to indicate that the direction of the force is toward the center of the circle.

In centripetal motion, the acceleration is always inward. Since the force is always in the same direction as the acceleration, the direction of both the acceleration and the force is toward the center of the circle.

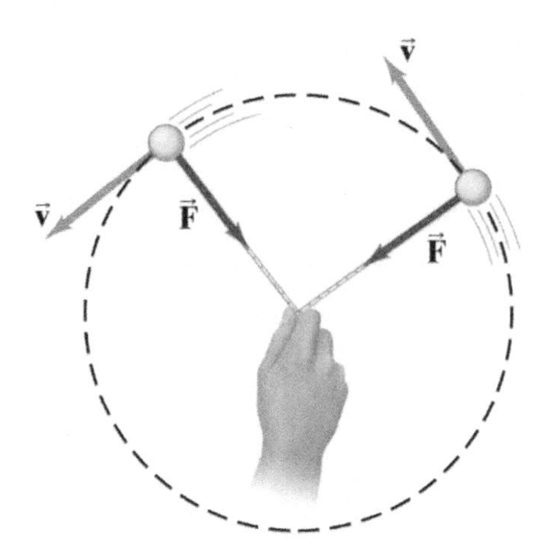

Suppose a ball attached to a string is spun around, and suddenly the string breaks. The ball flies off, following a line of motion tangent to its previous circular path. Why is this true? For an object to be in a uniform circular motion, there must be a net force (string) acting on it in the same direction as the acceleration (inward).

If gravity is ignored, the centripetal force is the only force acting on the ball and keeping it the same distance from the center. If the centripetal force vanishes (the string breaks), the object retains its instantaneous velocity at the point in time at which the string broke (following Newton's laws) and flew off, tangent to the circle.

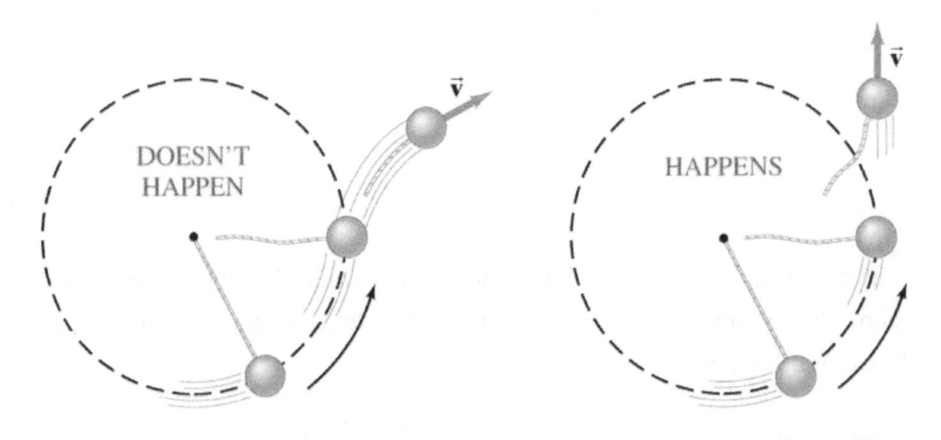

Banked curves (sloped inwards) provide another example of centripetal force. When a car goes around a curve, there must be a net force pointing toward the center of the circle, of which the curve is an arc. If the road is flat, that force is supplied by friction. If the frictional force is insufficient, the car tends to move more closely in a straight line.

If the tires do not slip, the friction is static. If the tires do start to slip, the friction is kinetic. This loss of static friction is bad in two ways:

1. The kinetic frictional force is smaller than the static frictional force, and the tires have less grip on the road.

2. The static frictional force can point toward the center of the circle. However, the kinetic frictional force opposes the direction of motion, making it difficult to regain control of the car and continue around the curve.

Banking the curve on the road can help keep cars from skidding. In fact, for every banked curve, there is one speed where the entire centripetal force is supplied by the horizontal component of the normal force, and no friction is required.

This occurs when:

$$F_N \sin\theta = \frac{mv^2}{r} = ma_c$$

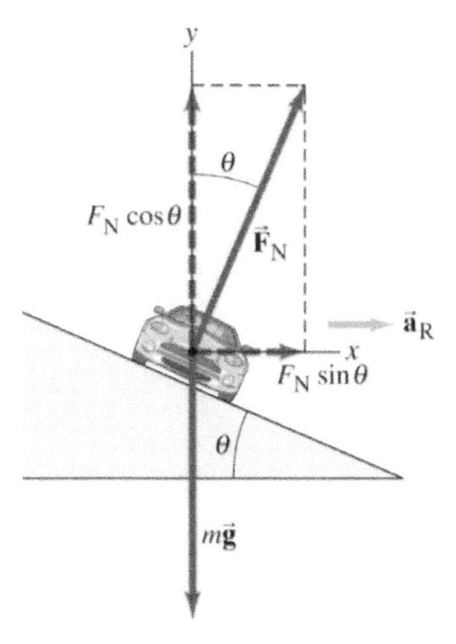

Non-uniform Circular Motion

It is important to note that uniform circular motion is usually an idealized scenario, and most systems exhibit some degree of non-uniform circular motion.

An object experiencing non-uniform circular motion has varying speeds. It has an instantaneous velocity tangent to its circular motion, but the acceleration will not point inward along the radius. Because of changing speed, it has a tangential component to its acceleration (and to its force) as well as a radial one.

The direction of net acceleration, and thus net force, depends on the combination of these two components.

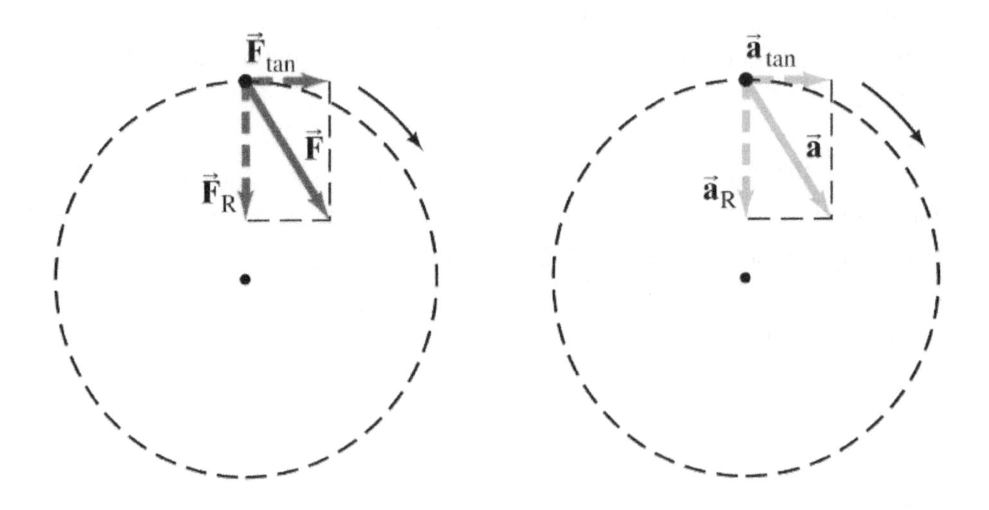

Law of Gravitation

Newton's Third Law of Motion states that every force has an equal and opposite reaction force, indicating that every object pulls on every other object. If the force of gravity is being exerted on objects on Earth, what is the origin of that force? Newton realized that the force must come from the Earth itself and that this force must be what keeps the Moon in its orbit.

Newton developed the *Law of Gravitation* to explains this relationship.

The Law of Gravitation states that any two bodies in the universe attract with force related to the square of the distance between them, the product of their two masses, and the gravitational constant *G*.

Force, for the Law of Gravitation, is defined as:

$$F_G = G \times \frac{m_1 m_2}{r^2}$$

The force of gravity on an object acts through its center of gravity.

How did Newton derive this equation?

The gravitational force is one half of an action-reaction pair: Earth exerts a downward force on a person, and they exert an upward force on the Earth. The reaction force that they exert on the Earth is undetectable because the mass of a person is negligible compared to the mass of the Earth.

However, for bodies with less disparity in mass (such as the Earth and the Moon), the reaction force can be significant.

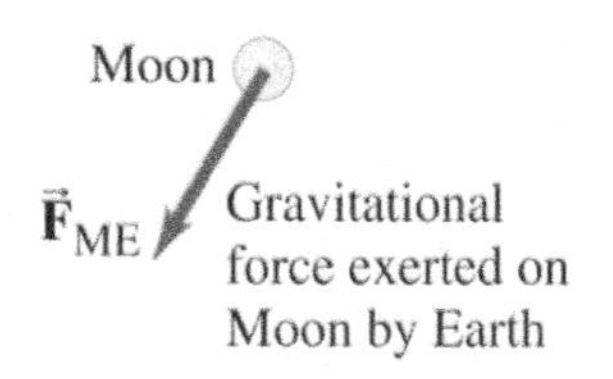

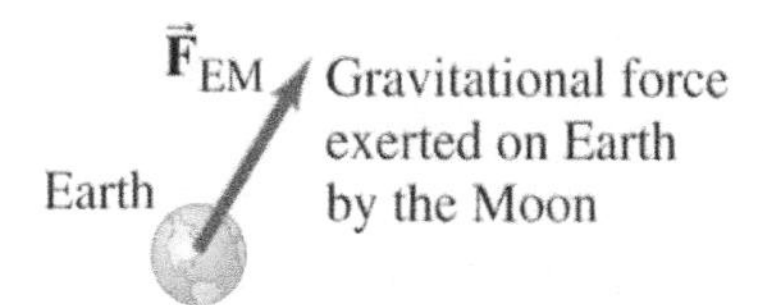

Therefore, the gravitational force must be proportional to both masses. By observing planetary orbits, Newton also concluded that the gravitational force must decrease as the inverse of the square of the distance between the masses.

How was the constant G derived?

Earlier in this chapter, the acceleration of an object toward the Earth's surface due to the force of Earth's gravity equaled 9.8 m/s^2 and was represented by the variable g.

The Law of Gravitation uses the variable G, which is different from lowercase g.

G is a constant of proportionality, as the *universal gravitational constant*. Note the term "universal" because all objects in the universe are pulling on all other objects. Examine the value of G and how it affects the equation.

$$F_G = G \times \frac{m_1 m_2}{r^2}$$

Since Newton could measure force, he was able to assign a value for G by rearranging the equation and using known values. For two 1 kg masses, 1 m apart of known force:

$$G = 0.000000000067 \text{ N·m}^2/\text{kg}^2$$

$$G = 6.67 \times 10^{-11} \text{ N·m}^2/\text{kg}^2$$

The magnitude of the gravitational constant G can also be measured in the laboratory through the Cavendish experiment (1797-1798), pictured below:

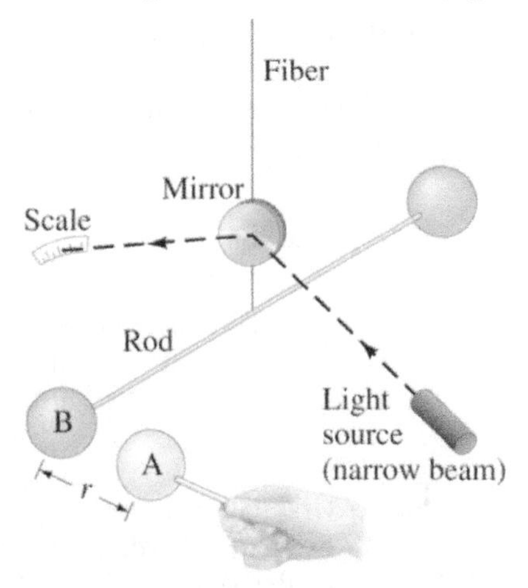

G is an exceedingly small value; it minimizes the value for the term: $\frac{m_1 m_2}{r^2}$

For example, a person pulls on the stars and vice versa. The sheer fact that the stars are a considerable distance away makes m_1m_2 / r^2 a small value (mainly since the distance is squared in this equation).

Multiplying this value by G makes it infinitely small, to the point of approaching zero. Even when considering a person's pull on their neighbor's house, where the value of "r" is relatively small, G diminishes the value.

If that person weighs 0.7 kg, the house weighs 700 kg, and the distance between them is 70 m, then:

$$\frac{m_1 m_2}{r^2} = \frac{(0.7\text{ kg})\cdot(700\text{ kg})}{(70\text{ m})^2} = 0.1\text{ kg}^2/\text{m}^2$$

Multiplied by 6.67×10^{-11} N·m^2/kg^2 (G), the resulting force equals 6.67×10^{-12} N, an undetectable value.

The only real gravitational force felt on Earth is that which is caused by the Earth. The concept of G was used to determine the mass of the Earth mathematically. The force of gravity at the Earth's surface was known (9.8 N), and the radius of the Earth was known (which approximates the distance between the center of the Earth and the center of a person standing on the surface). An object can be substituted for m and solved for the Earth's mass as m_E:

On the surface of the Earth, F = ma, so:

$$F_G = mg = G\frac{mm_E}{r_E^2}$$

Solving for g gives:

$$g = G\frac{m_E}{r_E^2}$$

Knowing g and the radius of the Earth ($r_E = 6.38 \times 10^6$ m), the mass of the Earth is:

$$m_E = \frac{gr_E^2}{G} = \frac{(9.8\text{ m/s}^2)\cdot(6.38 \times 10^6\text{ m})^2}{6.67 \times 10^{-11}\text{ N}\cdot\text{m}^2/\text{kg}^2}$$

$$m_E = 5.98 \times 10^{24}\text{kg}$$

The Earth has a mass of about 6×10^{24} kg. Since this is a large number, multiplying it by G does not decrease it to a negligible amount. The resulting value is the gravitational force felt by objects on Earth.

Like the value of G, the value of r has a significant impact on the calculations.

There are two essential aspects to note about the distance component, r, of this equation:

$$F_G = G\frac{m_1 m_1}{r^2}$$

First, it refers to the distance between the centers of mass of the two objects (this becomes critical in the discussion of the Earth's pull on objects). The two objects are treated as *point masses*; that is, as if they were the size of one-dimensional points at the locations of their centers of mass, but with the same amount of mass.

Second, the distance is squared, which means that small changes in distance result in substantial changes in F, as will be demonstrated with the *inverse square law.*

According to the Law of Gravitation equation, the force is inversely proportional to the square of the distance. Therefore, by increasing the distance between two objects, the force between those two objects is decreased by the inverse square.

If distance increases by a factor of two (i.e., doubles), the force of gravity decreases by a factor of $2^2 = 4$.

This is the primary reason why the gravitational forces of other planets are not felt on Earth, even though they have large masses that are equal to or greater than Earth's mass.

The distance between Earth and another planet is considerable. The distance is squared in the calculating gravitational force. Therefore, the force felt on Earth due to another planet is negligible and essentially eliminated.

For example, Mars is 2.25×10^{11} m away, and it has a mass of 6.4×10^{23} kg.

For a 1 kg mass on Earth, how much force would it exert on Mars?

$$F_G = G\frac{m_1 m_1}{r^2}$$

$$F = (6.67 \times 10^{-11}\ \text{Nm}^2/\text{kg}^2) \times \frac{(1\ \text{kg})\cdot(6.4 \times 10^{23}\ \text{kg})}{(2.25 \times 10^{11}\text{m})^2}$$

$$F = 6.83 \times 10^{-10}\ \text{N}$$

In most calculations thus far, a gravitational acceleration value of 9.8 m/s^2 has been used. For most calculations, this is value of 9.8 m/s^2 is acceptable.

The actual acceleration due to gravity varies over the Earth's surface because of altitude, local geology, and the nonspherical shape of the Earth.

Acceleration Due to Gravity at Specific Locations		
Location	**Elevation (m)**	**g (m/s^2)**
New York	0	9.803
San Francisco	0	9.800
Denver	1,650	9.796
Mount Everest	8,800	9.770
Sydney	0	9.798
Equator	0	9.780
North Pole (calculated)	0	9.832

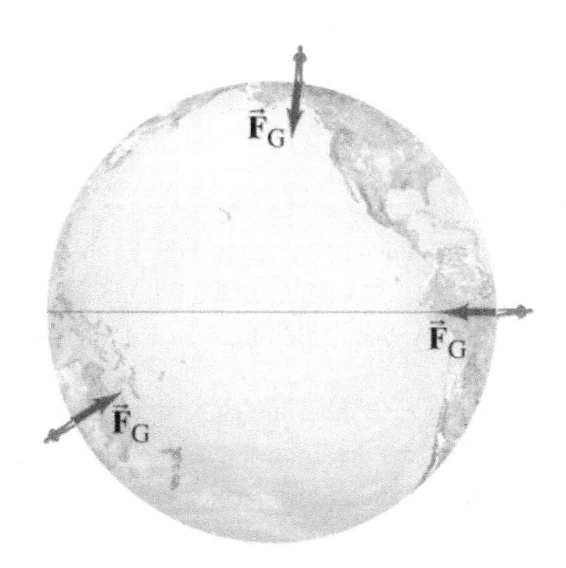

The force of gravitational acceleration points to the Earth's core

Concept of a Field

Forces often act on a body act in a single direction, like a car that accelerates in a straight-line path. However, the forces can act in a multitude of directions, either from a singular point outward or all directions inward towards a point (e.g., gravitational force).

On Earth, people experience gravity in a single direction (downward) because people are so small relative to the size of the Earth that the Earth is a flat surface below them. Since the Earth is an approximate sphere, its gravitational pull comes from all directions towards its center of mass.

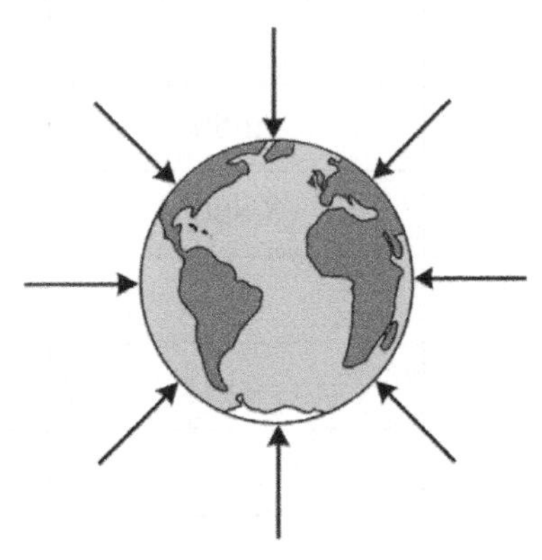

Gravitational field lines radiate towards the Earth's core

These are field lines, more specifically, gravitational field lines. Field lines are often used to depict situations in which multiple vectors are acting in a plane or 3D space.

Notice above that only eight arrows is used to depict the gravitational field around the Earth, even though gravity acts continuously all around the Earth. It would be impossible to draw enough field lines to represent the continuity of Earth's gravitational pull accurately, so it suffices to draw enough lines to show that Earth's gravity acts from all directions.

The arrows are closer to each other the closer they are to the Earth and farther apart, the farther away they are from the Earth. Field lines that are drawn close together represent a stronger field, and field lines drawn farther apart represent a weaker field.

From the figure above, the field lines are closer nearer to Earth, signifying the stronger gravitational force closer to the surface of the Earth. This supports the Law of Gravitation, which was discussed in-depth in the previous section.

In general, the gravitational field lines represent a field of vectors, typically these are force vectors.

The force vectors used in calculations, such as projectile motion, a mass-pulley system, or a mass on a pendulum, are usually single arrows that indicate the direction of a single force. This is because the masses involved in these problems are treated as point masses or masses that have an infinitesimally small size, which reduces the complexity of the calculations.

A boulder-sized projectile or car-sized object on a pendulum would have a force vector field acting on its entire volume, with all the vectors pointing in the same direction.

For example, when a car accelerates in a straight-line path, typically, this action is represented by a single force vector originating from the center of the car's mass in the direction of motion.

More accurately, a force vector would be drawn from every point in the car in the direction of motion, creating a force vector field. Since this creates a far more complex calculation to determine a negligibly more accurate value of the force involved, the vector field is simplified to a single vector originating from the object's center of mass.

Chapter Summary

- Newton's First Law: *An object at rest will remain at rest, and an object in motion (with constant velocity) will remain in motion (with that same constant velocity) unless acted upon by an external force.*

- Newton's Second Law: *Force is the vector product of mass and acceleration*:

$$F = ma$$

- Newton's Third Law: *For every action, there is an equal and opposite reaction.*

  When a person interacts with an object by exerting a force on it, that object exerts the same amount, or magnitude, of force back on them.

- Solving Problems with Newton's Laws – Free-Body Diagrams

  1. Sketch the scenario described in the problem.

  2. For every object, draw a free-body diagram showing all the forces acting *on* the object.

     Make magnitudes and directions representative to the scenario.

     Label each force.

     If there are multiple objects, draw separate diagrams.

  3. Resolve vectors into directional components.

  4. Apply Newton's Second Law to each component.

  5. Solve for the net force in each component direction and combine each component to determine the total net force.

- Gravitational force: $F_g = mg$, where g is the acceleration due to gravity at 9.8 m/s^2.

- Equations to find friction, the force that always opposes the direction of motion:

  Static friction: $F_s = \mu_s F_N$

  Kinetic friction: $F_k = \mu_k F_N$

where μ is the coefficient of friction, and F_N is the normal force.

- Motion on an Inclined Plane:

  Parallel Force: $F_{\parallel} = mg \sin \theta$

  Perpendicular Force: $F_{\perp} = F_N$

  $F_N = mg \cos \theta$

where θ is the angle between the horizontal axis and the surface the object rests.

- An object moving in a circle at constant speed is in *uniform circular motion*, an example of simple harmonic motion.

  The instantaneous velocity is always tangent to the circle, and the *centripetal* acceleration (or radial acceleration) points inward along the radius.

- Centripetal acceleration for the uniform circular motion:

  $a_c = v^2 / r$

- Centripetal force:

  $$F = ma_c = \frac{mv^2}{r}$$

- Law of Gravitation:

  $$F_G = G \times \frac{m_1 m_2}{r^2}$$

Practice Questions

1. Two bodies of different masses are subjected to identical forces. Compared to the body with a smaller mass, the body with greater mass experiences:

A. less acceleration, because the ratio of force to mass is smaller
B. greater acceleration, because the ratio of force to mass is greater
C. less acceleration, because the product of mass and acceleration is smaller
D. greater acceleration, because the product of mass and acceleration is greater
E. equal acceleration, because the forces are identical

2. An object is propelled along a straight-line path by force. If the net force were doubled, the object's acceleration would:

A. halve
B. stay the same
C. double
D. quadruple
E. none of the above

3. A 500 kg rocket ship is firing two jets at once. The two jets are at right angles with one firing with a force of 500 N and the other firing with a force of 1,200 N. What is the magnitude of the acceleration of the rocket ship?

A. 1.4 m/s^2
B. 2.6 m/s^2
C. 3.4 m/s^2
D. 5.6 m/s^2
E. 4.2 m/s^2

4. Assume the strings and pulleys in the diagram below have negligible masses, and the coefficient of kinetic friction between the 2 kg block and the table is 0.25. What is the acceleration of the 2 kg block? (Use the acceleration due to gravity $g = 9.8$ m/s^2)

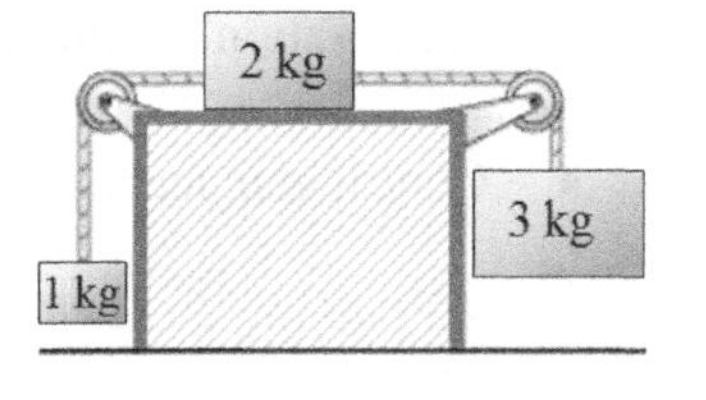

A. 3.2 m/s^2
B. 4 m/s^2
C. 0.3 m/s^2
D. 1.7 m/s^2
E. 2.5 m/s^2

5. Which of the statements must be valid for an object moving with constant velocity in a straight line?

A. The net force on the object is zero
B. No forces are acting on the object
C. A constant force is being applied in the direction opposite of motion
D. A constant force is being applied in the direction of motion
E. There is no frictional force acting on the object

6. A person gives a shopping cart an initial push along a horizontal floor to get it moving and then releases the cart. The cart travels forward along the floor, gradually slowing as it moves. Consider the horizontal force on the cart while it is moving forward and slowing. Which of the statements is correct?

A. Only a forward force is acting, which diminishes with time
B. Only a backward force is acting; no forward force is acting
C. Both a forward and a backward force are acting on the cart, but the forward force is larger
D. Both a forward and a backward force are acting on the cart, but the backward force is larger
E. No forces are acting because it has been released

7. A truck is using a hook to tow a car whose mass is one quarter that of the truck. If the force exerted by the truck on the car is 6,000 N, then the force exerted by the car on the truck is:

A. 1,500 N
B. 24,000 N
C. 6,000 N
D. 12,000 N
E. Need to know if the car is accelerating

8. How large is the force of friction impeding the motion of a bureau when a 120 N bureau is pulled across the sidewalk at a constant speed by a force of 30 N?

A. 0 N **B.** 30 N **C.** 120 N **D.** 3 N **E.** 15 N

9. An object maintains its state of motion because it has:

A. mass **B.** acceleration **C.** speed **D.** weight **E.** all of the above

10. The Earth and the Moon attract with the force of gravity. The Earth's radius is 3.7 times that of the Moon, and the Earth's mass is 80 times greater than the Moon's. The acceleration due to gravity on the surface of the Moon is 1/6 the acceleration due to gravity on the Earth's surface. If the distance between the Earth and the Moon decreases by a factor of 4, how would the force of gravity between the Earth and the Moon change?

A. Remain the same
B. Increase by a factor of 16
C. Decrease by a factor of 16
D. Decrease by a factor of 4
E. Increase by a factor of $1/\sqrt{4}$

11. Two forces acting on an object have magnitudes $F_1 = -6.6$ N and $F_2 = 2.2$ N. Which third force causes the object to be in equilibrium?

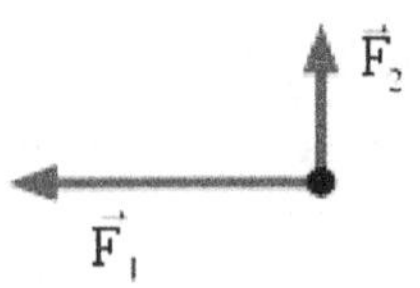

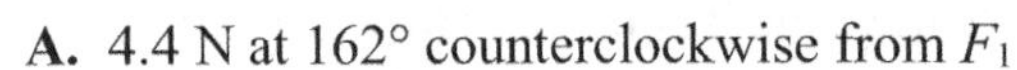

A. 4.4 N at 162° counterclockwise from F_1
B. 4.4 N at 108° counterclockwise from F_1
C. 7 N at 162° counterclockwise from F_1
D. 7 N at 108° counterclockwise from F_1
E. 3.3 N at 162° counterclockwise from F_1

12. What are the readings on the spring scales when a 17 kg fish is weighed with two spring scales if each scale has negligible weight?

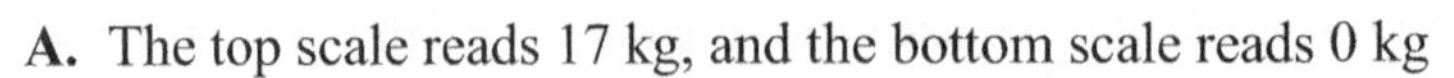

A. The top scale reads 17 kg, and the bottom scale reads 0 kg
B. Each scale reads greater than 0 kg and less than 17 kg, but the sum of the scales is 17 kg
C. The bottom scale reads 17 kg, and the top scale reads 0 kg
D. The sum of the two scales is 34 kg
E. Each scale reads 8.5 kg

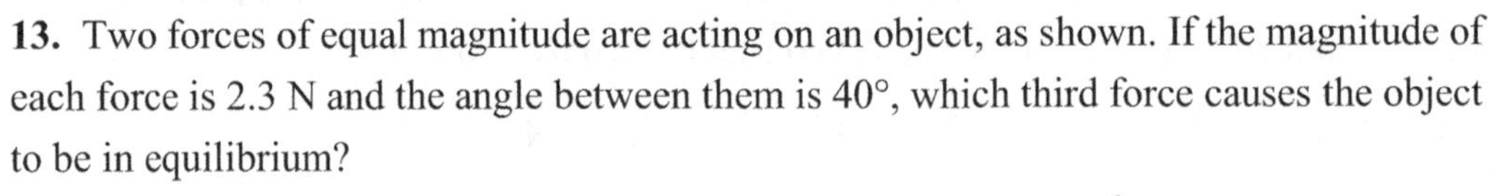

13. Two forces of equal magnitude are acting on an object, as shown. If the magnitude of each force is 2.3 N and the angle between them is 40°, which third force causes the object to be in equilibrium?

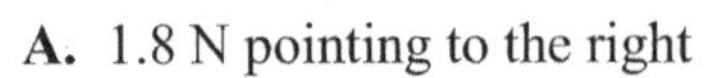

A. 1.8 N pointing to the right
B. 2.2 N pointing to the right
C. 3.5 N pointing to the right
D. 6.6 N pointing to the right
E. 4.3 N pointing to the right

14. An object at rest on an inclined plane starts to slide when the incline is increased to 17°. What is the coefficient of static friction between the object and the plane? (Use acceleration due to gravity $g = 9.8$ m/s^2)

A. 0.37 **B.** 0.43 **C.** 0.24 **D.** 0.31 **E.** 0.17

15. What is the force exerted by the table on a 2 kg book resting on it? (Use acceleration due to gravity $g = 10$ m/s^2)

A. 100 N **B.** 20 N **C.** 10 N **D.** 0 N **E.** 40 N

Solutions

1. A is correct. Newton's Second Law ($F = ma$) is rearranged:

$$a = F / m$$

If F is constant, then a is inversely proportional to m.

A larger mass implies a smaller F / m ratio; the ratio is the acceleration.

2. C is correct.

$F = ma$, so doubling the force doubles the acceleration.

3. B is correct.

Pythagorean Theorem ($a^2 + b^2 = c^2$) to calculate the net force:

$$F_1^2 + F_2^2 = F_{net}^2$$

$$(500 \text{ N})^2 + (1{,}200 \text{ N})^2 = F_{net}^2$$

$$250{,}000 \text{ N}^2 + 1{,}440{,}000 \text{ N}^2 = F_{net}^2$$

$$F_{net}^2 = 1{,}690{,}000 \text{ N}^2$$

$$F_{net} = 1{,}300 \text{ N}$$

Newton's Second Law:

$$F = ma$$

$$a = F_{net} / m$$

$$a = 1{,}300 \text{ N} / 500 \text{ kg}$$

$$a = 2.6 \text{ m/s}^2$$

4. E is correct. The acceleration of the 2 kg block is the acceleration of the system because the blocks are linked together. Balance forces and solve for acceleration:

$$F_{net} = m_3g - m_2g\mu_k - m_1g$$

$$(m_3 + m_2 + m_1)a = m_3g - m_2g\mu_k - m_1g$$

$$a = (m_3 - m_2\mu_k - m_1)g / (m_3 + m_2 + m_1)$$

$$a = [3 \text{ kg} - (2 \text{ kg})\cdot(0.25) - 1 \text{ kg}]\cdot(9.8 \text{ m/s}^2) / (3 \text{ kg} + 2 \text{ kg} + 1 \text{ kg})$$

$$a = 2.5 \text{ m/s}^2$$

5. A is correct. Objects moving at constant velocity experience no acceleration and, therefore, no net force.

6. B is correct. The cart decelerates, which is an acceleration in the opposite direction caused by the force of friction in the opposite direction.

7. C is correct. Newton's Third Law states that when two objects interact by a mutual force, the force of the first on the second is equal in magnitude to the force of the second on the first.

8. B is correct. If the bureau moves in a straight line at a constant speed, its velocity is constant. Therefore, the bureau is experiencing zero acceleration and zero net force.

The force of kinetic friction equals the 30 N force that pulls the bureau.

9. A is correct. Newton's First Law of Motion (Law of Inertia) states that an object at rest tends to stay at rest, and an object in motion tends to maintain that motion unless acted upon b an unbalanced force. This law depends on a property of an object's inertia, which is inherently linked to the object's mass. More massive objects are more difficult to move and manipulate than less massive objects.

10. B is correct.

$$F_g = Gm_{Earth}m_{moon} / d^2$$

d is the distance between the Earth and the Moon.

If d decreases by a factor of 4, F_g increases by a factor of $4^2 = 16$

11. C is correct. Find equal and opposite forces:

$$F_{Rx} = -F_1$$

$$F_{Rx} = -(-6.6 \text{ N})$$

$$F_{Rx} = 6.6 \text{ N}$$

$$F_{Ry} = -F_2$$

$$F_{Ry} = -2.2 \text{ N}$$

Pythagorean Theorem ($a^2 + b^2 = c^2$) to calculate the magnitude of the resultant force:

The magnitude of F_R:

$$F_R^2 = F_{Rx}^2 + F_{Ry}^2$$

$$F_R^2 = (6.6\ \text{N})^2 + (–2.2\ \text{N})^2$$

$$F_R^2 = 43.6\ \text{N}^2 + 4.8\ \text{N}^2$$

$$F_R^2 = 48.4\ \text{N}^2$$

$$F_R = 7\ \text{N}$$

The direction of F_R:

$$\theta = tan^{-1}\ (–2.2\ \text{N} / 6.6\ \text{N})$$

$$\theta = tan^{-1}\ (–1 / 3)$$

$$\theta = 342°$$

The direction of F_R with respect to F_1:

$$\theta = 342° – 180°$$

$$\theta = 162°\ \text{counterclockwise of}\ F_1$$

12. D is correct. Each scale weighs the fish at 17 kg, so the sum of the two scales is:

$$17\ \text{kg} + 17\ \text{kg} = 34\ \text{kg}$$

13. E is correct. If θ is the angle with respect to a horizontal line, then:

$$\theta = ½(40°)$$
$$\theta = 20°$$

Therefore, for the third force to cause equilibrium, the sum of all three forces' components must equal zero. Since F_1 and F_2 mirror are equal in the y-direction:

$$F_{1y} + F_{2y} = 0$$

Therefore, for F_3 to balance the forces in the y direction, its y component must also equal zero:

$$F_{1y} + F_{2y} + F_{3y} = 0$$

$$0 + F_{3y} = 0$$

$F_{3y} = 0$

Since the y component of F_3 is zero, the angle that F_3 makes with the horizontal is zero:

$\theta_3 = 0°$

The x component of F_3:

$F_{1x} + F_{2x} + F_{3x} = 0$

$F_1 \cos\theta + F_2 \cos\theta + F_3 \cos\theta = 0$

$F_3 = -(F_2 \cos\theta_2 + F_3 \cos\theta_3)$

$F_3 = -[(2.3 \text{ N}) \cos 20° + (2.3 \text{ N}) \cos 20°]$

$F_3 = -4.3 \text{ N}$

$F_3 = 4.3$ N to the right

14. D is correct.

At $\theta = 17°$, the force of static friction is equal to the force due to gravity:

$F_f = F_g$
$\mu_s mg \cos\theta = mg \sin\theta$
$\mu_s = \sin\theta / \cos\theta$
$\mu_s = \tan\theta$
$\mu_s = \tan 17°$
$\mu_s = 0.31$

15. B is correct.

The force of the table on the book, the normal force (F_N), is a result of Newton's Third Law of Motion, which states for every action there is an equal and opposite reaction.

A book sitting on the table experiences a force from the table equal to the book's weight:

$\text{W} = mg$
$F_N = \text{W}$
$F_N = mg$
$F_N = (2 \text{ kg})\cdot(10 \text{ m/s}^2)$
$F_N = 20 \text{ N}$

Chapter 5

Waves and Periodic Motion

PERIODIC MOTION

- **Amplitude, Period, Frequency**
- **Simple Harmonic Motion, Displacement as a Sinusoidal Function of Time**
- **Hooke's Law ($F = -kx$)**
- **Energy of a Mass-Spring System**
- **Motion of a Pendulum**

WAVE CHARACTERISTICS

- **Transverse and Longitudinal Waves**
- **Phase, Interference and Wave Addition**
- **Reflection and Transmission**
- **Refraction and Diffraction**
- **Amplitude and Intensity**

PERIODIC MOTION

Amplitude, Period and Frequency

Periodic motion (i.e., sinusoidal motion) is a motion that repeats. If an object moves back and forth over the same path at a constant speed, each cycle takes the same amount of time; the resulting motion is periodic. In the periodic motion, this back-and-forth activity is *oscillation* or *vibration*.

Many systems can display oscillatory or vibratory motion; there are several types of periodic motion as well. Periodic motion can be damped (e.g., friction is present) or undamped, driven (an outside force is acting upon the system), or free.

Regardless of the motion, any oscillatory system has several inherent characteristics that define its periodic motion. These characteristics are the amplitude, frequency, and period.

In general, most questions refer to several types of oscillatory systems, but only one motion: undamped and free periodic motion (simple harmonic motion).

Amplitude (A): In periodic motion, the *amplitude* expresses the displacement of the system about its equilibrium (center or resting) position. Amplitude may have different units depending on the system.

Frequency (f): *Frequency* is the rate of oscillation, or the number of cycles per second, where a cycle is one complete vibration (when the displacement is zero). The value for frequency is usually measured in Hertz (Hz) and can be calculated by:

$$f = \frac{\#\, of\ cycles}{time}$$

In some situations, the frequency is expressed as angular frequency (ω). This value is related to frequency; however, angular frequency is measured in radians per second (rad/s).

The equation for angular frequency ω is:

$$\omega = 2\pi f$$

Period (T): The *period* is the time required to complete one cycle or one oscillation. The period is often measured in seconds and can be calculated by:

$$T = \frac{time}{\#\ of\ cycles}$$

The period is sometimes measured in degrees or radians.

One rotation around a circle requires 360° or 2π radians.

The value of 360° or 2π radians represents one complete cycle of an oscillating system.

From the above equation, the period (T) and frequency (f) are inversely related.

The relationship between the period and frequency is expressed as:

$$T = \frac{1}{f}$$

Simple Harmonic Motion, Displacement as a Sinusoidal Function of Time

An object or system exhibiting *simple harmonic motion* is a particular case of a periodic motion that is undamped and free. There is no force of friction present and no external, time-dependent force acting upon it. The system requires a restoring force (not a driving force), which must be proportional to the displacement (amplitude). This proportionality allows the system to oscillate about its equilibrium position while retaining the same amplitude.

Simple harmonic motion can be depicted as a sine or cosine graph, as below:

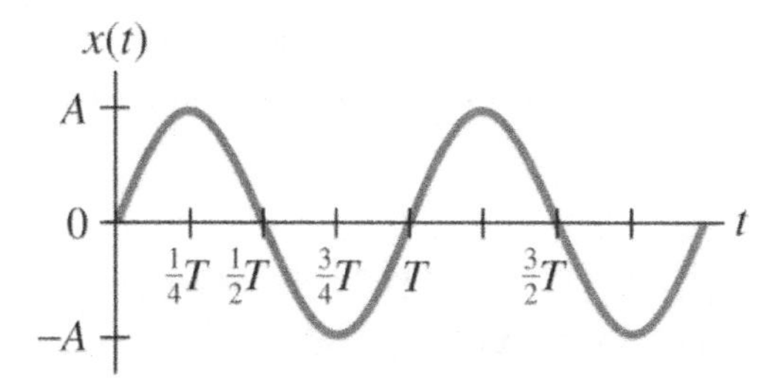

The system oscillates about its equilibrium position with the same amplitude and a fixed period (and thus constant frequency). Another important distinction is that, regardless of the magnitude of the amplitude, the period and frequency do not change. In simple harmonic motion, frequency and period are independent of amplitude. Any object's simple harmonic motion may be determined using the equation:

$$x = A \sin(\omega t)$$

$$x = A \sin(2\pi f t)$$

where x is displacement, A is amplitude, t is time, ω is the angular frequency, and f is frequency.

Sometimes the equation is written as:

$$x = A \cos(\omega t)$$

$$x = A \cos(2\pi f t)$$

Sine or cosine may be used interchangeably if the translation of these graphs is accounted for. A cosine graph has a y-intercept of A (maximum amplitude), whereas a sine graph has a y-intercept of 0 (equilibrium). These are translations, but the translations must be accounted for in the equation. Otherwise, the solution will be incorrect.

Hooke's Law

One classic example of simple harmonic motion is the oscillatory motion of a mass-spring system. When a mass is attached to a spring, the only forces acting upon it are tension and weight of the mass. In such systems, there is no driving force, and it is assumed to be completely undamped.

An important characteristic of a mass-spring system is the *spring constant*. Robert Hooke (1635-1703) discovered that, given a certain spring, there is a constant that can be found when different masses are attached, and the equilibrium position is measured. This constant is an inherent constant of all springs and relates displacement of the spring (change in length) to the restoring force the spring exerts. This relationship is Hooke's Law and is expressed as:

$$F = -k\Delta x$$

where k is the spring constant measured in (N/m), Δx is the displacement, and F is the restoring force (N).

The minus sign in Hooke's Law is by convention, and the restoring force is in the opposite direction to the displacement (because force and displacement are vectors).

In the diagram shown below:

A spring is attached to a wall, and the other side is attached to a mass (m).

For an ideal system (i.e., no friction present), when the mass is displaced, and the spring is stretched, the direction of the restoring force is opposite to the displacement and acts to the left (middle image).

Conversely, if the spring were compressed the same distance, then the restoring force would be of equal magnitude but in the opposite direction (bottom image).

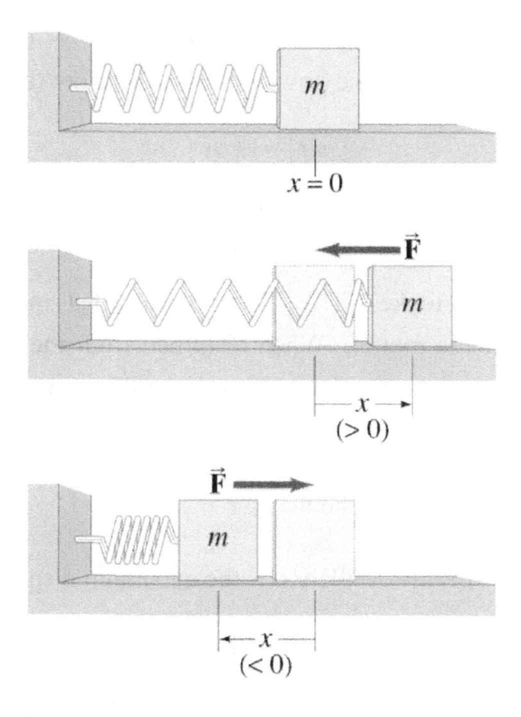

If the spring is hung vertically (shown below), the only change is in the equilibrium position, which is now at a point where the restoring spring force is equal and opposite the weight of the mass.

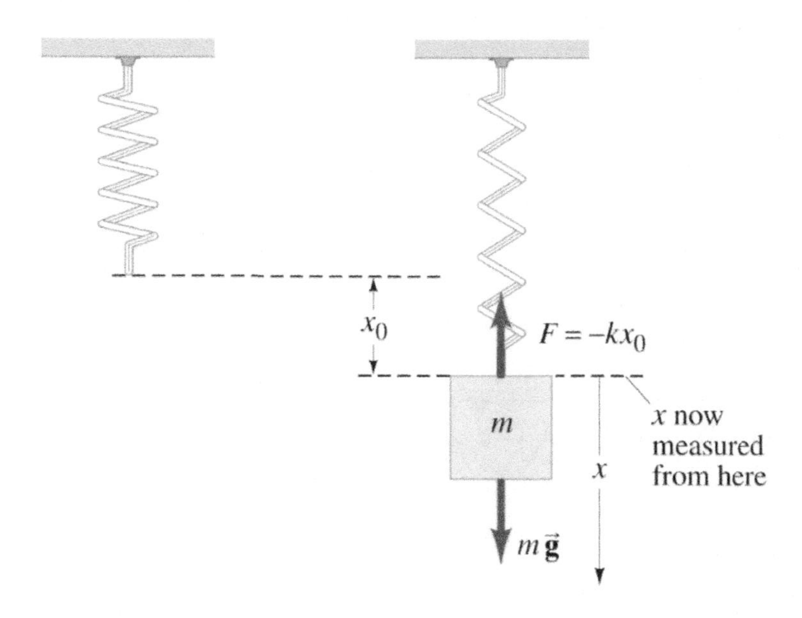

Hooke's Law ($F = -k\Delta x$) can be used to find the acceleration of the mass attached to the spring.

Regardless if the mass-spring system is horizontally- or vertically-positioned, forces always follow Newton's Second Law ($F = ma$), therefore (assuming no friction):

$$| ma | = | k\Delta x |$$

$$| a | = \left| \frac{k\Delta x}{m} \right|$$

More importantly, Hooke's Law demonstrates that mass-spring systems are examples of simple harmonic motion. When disturbed from the equilibrium position, a restoring force acts to bring the spring back towards equilibrium.

The restoring force (F) is proportional to the stretched or compressed distance, which makes the system a simple harmonic.

The amplitude (A) of this oscillation is the maximum displacement (Δx):

$$A = \Delta x$$

For example, if a mass-spring system is oriented vertically, and the mass is displaced a certain distance, the resulting motion follows a simple harmonic oscillator.

A simple harmonic oscillator is seen if a pen is attached to a mass and a roll of paper is moved across the surface of the pen during the motion of the mass:

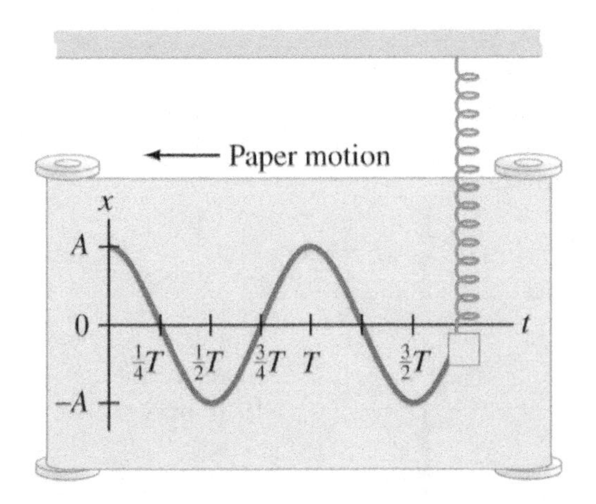

$$x = A \cos(\omega t)$$

$$F = -Ka$$

In the simple harmonic motion of a mass-spring system, the angular frequency (ω) and frequency (f) of the system can be calculated by:

$$\omega = \sqrt{\frac{k}{m}}$$

$$f = \frac{\sqrt{k/m}}{2\pi}$$

or

$$f = 2\pi\sqrt{\frac{m}{k}}$$

The period (T) can be calculated as:

$$T = \frac{2\pi}{\sqrt{k/m}}$$

or

$$T = 2\pi\sqrt{\frac{m}{k}}$$

The frequency (f) and the period (T) of a mass-spring system are only based on the mass and spring constant.

It is important to note that a vertically-angled mass-spring system has the same frequency and period, regardless of the value for gravitational acceleration.

The frequency (f) and period (T) do not change, even if the system were on the moon, in space or on Earth.

Energy of a Mass-Spring System

An oscillating mass accelerates, decelerates, changes direction, and repeats. An oscillating mass contains motion energy and stationary energy (its velocity is zero at the peaks and troughs of its movement).

The equations for the *PE* and *KE* of a mass on a spring:

$$\text{Potential energy} = PE = \frac{1}{2}kx^2$$

$$\text{Kinetic energy} = KE = \frac{1}{2}mv^2$$

When the mass is in motion and passes through the equilibrium position, the displacement is zero, as is the potential energy. However, the kinetic energy will be at its maximum because the mass is at its maximum velocity. This occurs at the point where the net force is zero.

At the maximum displacement, the value of x equals the amplitude (A). The potential energy is at its maximum, and the kinetic energy is zero (because the velocity is zero as the object changes its direction of motion). This occurs at the moment of maximum force, which is also the point of maximum acceleration of an oscillating particle.

At any point along its motion, the potential energy plus the kinetic energy of the particle will always be the same value. This value of total energy (potential energy plus kinetic energy) is constant. The sum is the same as the total energy present in the system:

$$\text{Total Energy} = E$$

$$E = PE + KE$$

$$E = \frac{1}{2}kx^2 + \frac{1}{2}mv^2$$

When potential energy is at its maximum (at the amplitude), kinetic energy is zero. All the energy in the system must be equal to only the potential energy.

The same relationship applies when kinetic energy is at its maximum:

$$PE_{max} = KE_{max} = E = \text{Total Energy}$$

The equations for maximum potential and kinetic energy are summed by:

$$PE_{max} = \frac{1}{2}kA^2$$

$$KE_{max} = \frac{1}{2}mv^2 \text{ at } x = 0$$

In the following diagram, a spring has been compressed, but it is not currently in motion. All the energy is stored in potential energy (PE):

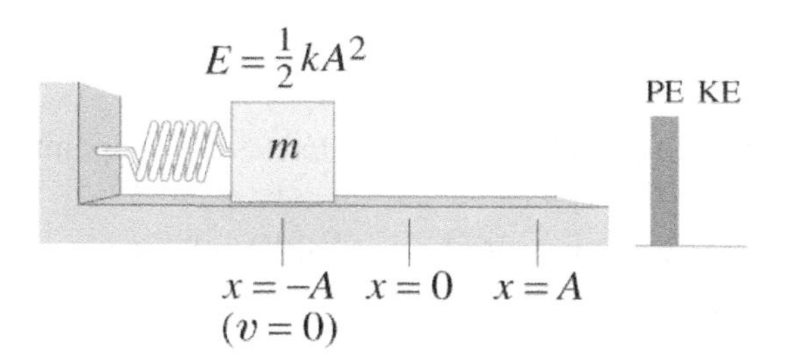

In the following diagram, if the mass (m) is allowed to move, the spring expands towards its equilibrium position. Once it has reached its equilibrium position, all its potential energy (PE) is transformed into kinetic energy (KE):

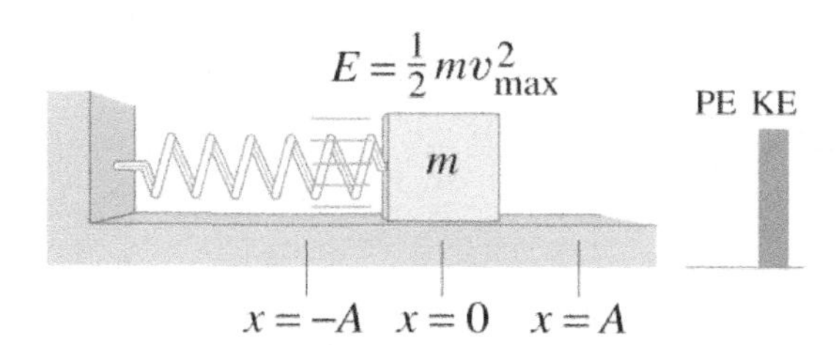

In the following diagram, the mass (m) reaches the maximum stretched amplitude. Kinetic energy (KE) is converted back to potential energy (PE):

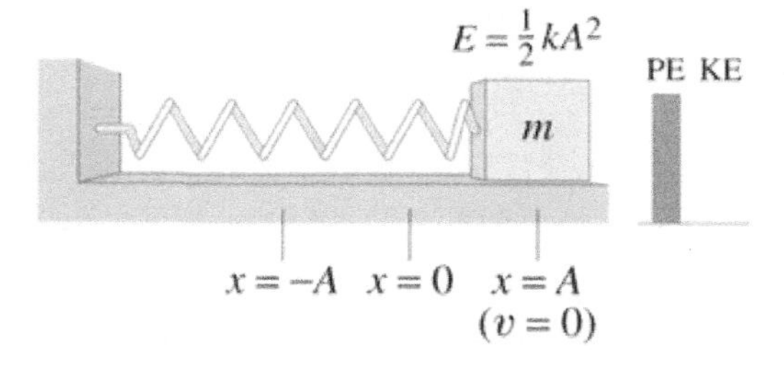

The mass (m) then moves back toward equilibrium but does not yet reach the equilibrium position. Therefore, most of the energy is kinetic (KE), but some energy is stored as potential energy (PE).

In the following diagram, as the mass (m) gets closer to the equilibrium position, the amount of potential energy decreases, and the amount of kinetic energy increases:

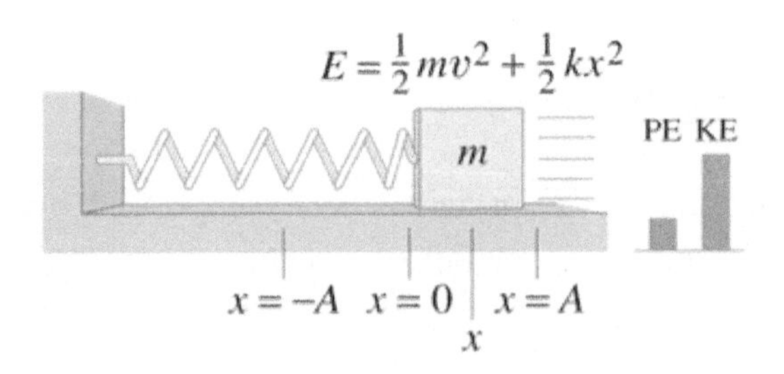

The velocity of the mass (m) can also be calculated at various points during its motion.

By setting the maximum potential energy (PE) and kinetic energy (KE) equal, the maximum velocity of mass (m) is calculated as:

$$v_{max} = \sqrt{\frac{k}{m}} \times A$$

Velocity can be solved as a function of position:

$$v = \pm v_{max}\sqrt{1 - \frac{x^2}{A^2}}$$

Motion of a Pendulum

Thus far, the examples have been oscillating masses attached to springs. In this chapter, the simple harmonic motion of a pendulum is discussed. A pendulum undergoing simple harmonic motion (i.e., simple pendulum) consists of a mass at the end of an assumed massless cord or rod.

To be in simple harmonic motion, the restoring force on the pendulum must be proportional to the negative of the displacement. For example, see the diagram below:

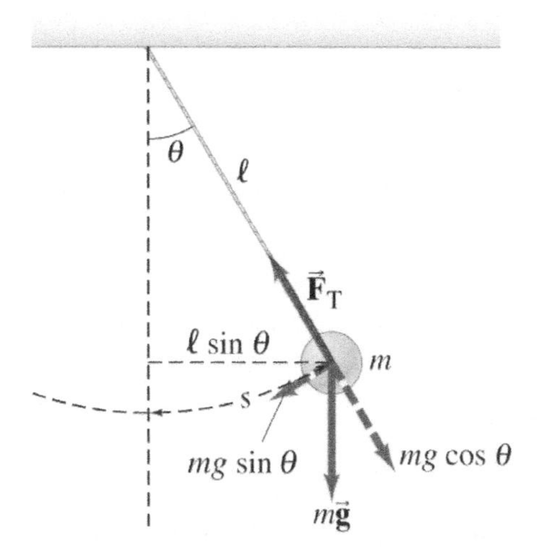

The diagram above shows that the angle θ measures the displacement of the pendulum.

The restoring force is:

$$F = -mg \sin \theta$$

$$|a| = |g \sin \theta|$$

For small displacement angles, $\sin \theta \approx \theta$.

Thus, the restoring force becomes:

$$F = -mg\theta$$

$$|a| = |g\theta|$$

Unlike mass-spring systems, the frequency (f) of a simple pendulum is dependent upon the acceleration due to gravity (g).

The frequency of a simple pendulum is also dependent upon the length of the cord (or rod) attached to the mass:

$$\omega = \sqrt{\frac{g}{l}}$$

$$f = 2\pi\sqrt{\frac{g}{l}}$$

where g is the acceleration of gravity (9.8 m/s^2), and l is the length of the cord (or rod).

The period (T) of a simple pendulum's oscillation is:

$$T = 2\pi\sqrt{\frac{l}{g}}$$

Like mass-spring systems, simple pendulums conserve energy. The total energy of the pendulum oscillation is the sum of potential energy and kinetic energy:

$$E = PE + KE$$

Unlike the mass-spring system, PE is dependent upon gravity:

$$PE = mgh$$

where h is the height above the equilibrium position.

PE is at its maximum when the height (or x) is equal to the amplitude.

$$PE_{max} = mgl(1 - \cos\theta)$$

Maximum KE is the same for a simple pendulum as it is for mass-spring systems:

$$KE_{max} = \frac{1}{2}mv^2$$

Wave Characteristics

Waves are periodic disturbances that transport energy through a medium but do not transport matter. Many examples of waves exist, including the light one sees, the sound one hears, and ocean waves breaking on the beach. While these examples may seem different, they share characteristics intrinsic to all waves.

Like periodic motion, the motion of a wave is defined by:

- Amplitude, A
- Frequency f and period T
- Wavelength, λ
- Wave velocity

Amplitude, frequency, and period were discussed earlier; their definition has not changed, only the context of what they describe. Wavelength (λ) and wave velocity are new terms but are understood.

Observe the diagram of a simple sine wave shown below:

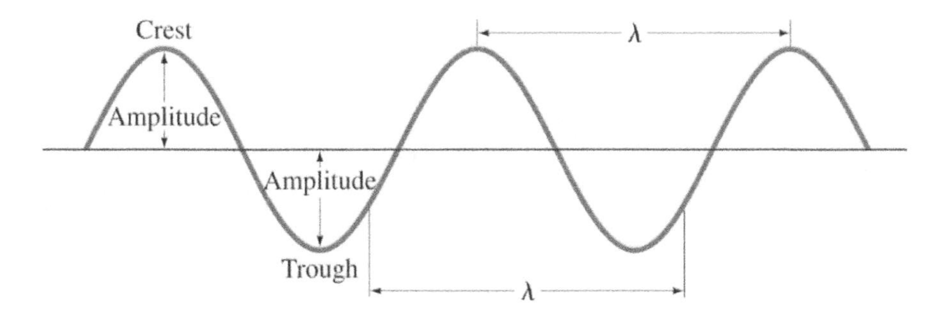

Like before, the amplitude is a representation of energy. The magnitude of the energy is not dependent upon frequency or period. The *wavelength* (λ) is the distance of one cycle of the wave and is measured in meters (m). The *wave velocity* (v) is the speed and direction of the wave. These two characteristics are related to the frequency and period by the equations:

$$f = \frac{v}{\lambda}$$

$$T = \frac{1}{f} = \frac{\lambda}{v}$$

where v is the wave velocity (units of m/s), and λ is the wavelength (units of m).

Transverse and Longitudinal Waves

Three main types of wave motion are transverse, longitudinal, and surface waves. These types can be described by the characteristics above but differ in the motion of particles within the wave. The wave is not transporting the particles. Instead, as the wave passes through a particle in a medium, the particle undergoes a specific periodic motion.

Transverse waves and longitudinal waves are shown below. For example, imagine a hand moving a slinky (figures below).

A *transverse wave* (top diagram below) oscillates particles with motion perpendicular to the wave direction.

A *longitudinal wave* (bottom diagram below) oscillates particles with motion parallel to the wave direction.

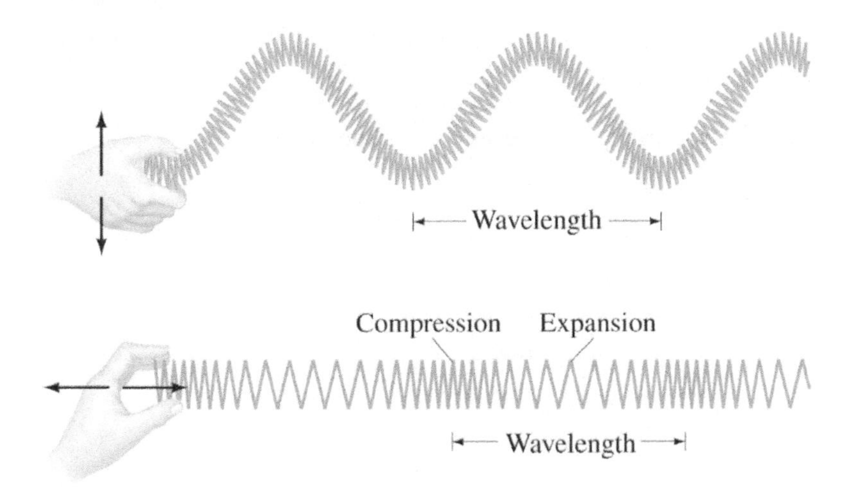

Another way to think about these types of waves is to ask, "What is oscillating, and in what direction?" Imagine there is a mark on one of the crests in the first diagram above.

For the top image above, the imaginary mark moves up and down, while energy travels through the medium from left to right.

The wave displacement is *perpendicular* to the direction of the wave motion; the imaginary mark is experiencing a transverse wave.

Imagine the mark on top of a crest in the bottom diagram above. As the hand pushes the slinky through the medium from left to right and back, the imaginary mark travels left-to-right and vice versa along with the propagated wave.

There is no vertical motion along the wave. The wave displacement is *parallel* to the direction of motion; the imaginary mark experiences a longitudinal wave. The particles in this medium move closer and then farther apart. This *compression and expansion* only occur in longitudinal waves.

Transverse waves have *crests and troughs*.

Longitudinal waves have *compressions and expansions*.

Surface waves have particles that oscillate in a circular motion, rather than vertically or horizontally. Surface waves only occur at the interface of two different mediums, where the wave is traveling in only one of the mediums.

An example of surface waves is the waves seen on the ocean. The diagram below shows a wave propagating through the ocean with a velocity *v*.

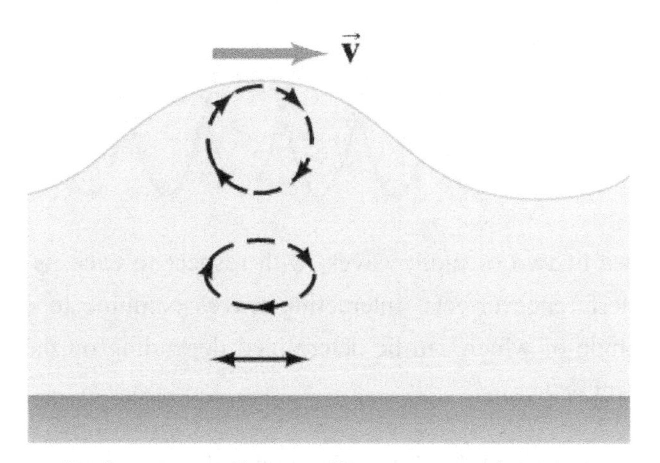

Surface waves that oscillate in a circular motion

For surface waves, the interface of the two mediums (e.g., air and water) causes the particles to move in a circular pattern (diagram above). Deeper in the ocean, the particles are further from the interface, and the motion resembles that of an ellipse.

At greater depths, the particle motion is only horizontal, and the wave behaves as a longitudinal wave. For surface waves, the further (i.e., deeper) from the interface of the two mediums, the more longitudinal the wave becomes.

Phase, Interference and Wave Addition

Phases occur when two waves interact. A phase difference between two waves means that the crests (or troughs) do not occur at the same points in space.

The diagrams below have three waves: one in-phase and two out-of-phase:

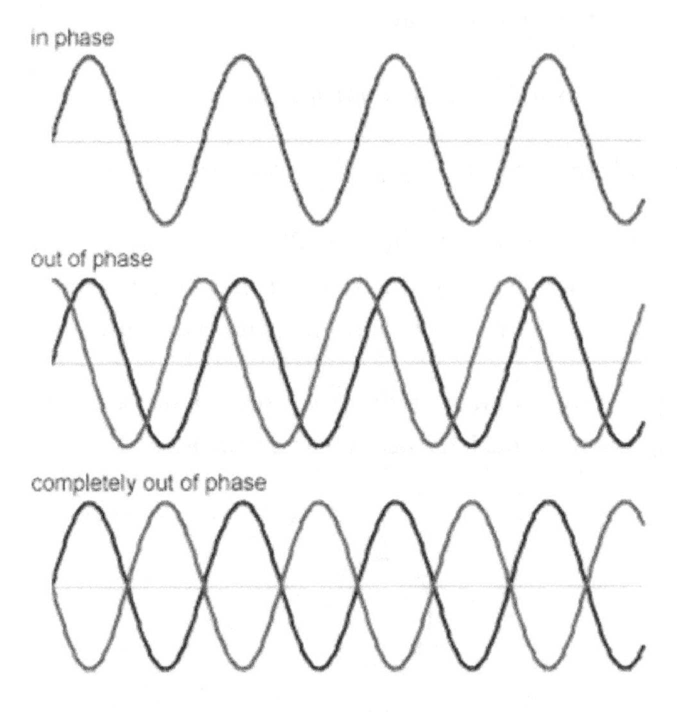

The phase of two or more waves, with respect to each, is important because it gives rise to interference effects. Interacting waves combine to create a combination wave, the amplitude of which can be determined depending on the interaction between the two-component waves.

The resulting amplitude of the interference depends upon the phase of the waves: *constructive interference* (waves are in-phase) and *destructive interference* (waves are out-of-phase).

In the diagram on the left below, two transverse waves of equal magnitude travel across a rope until they meet. At the moment that the two waves overlap, they are entirely out-of-phase and destructively interfere to produce no wave.

In the diagram on the right below, two waves of equal magnitude overlap and are completely in-phase. This produces constructive interference and results in a wave with double the magnitude of the two incoming waves.

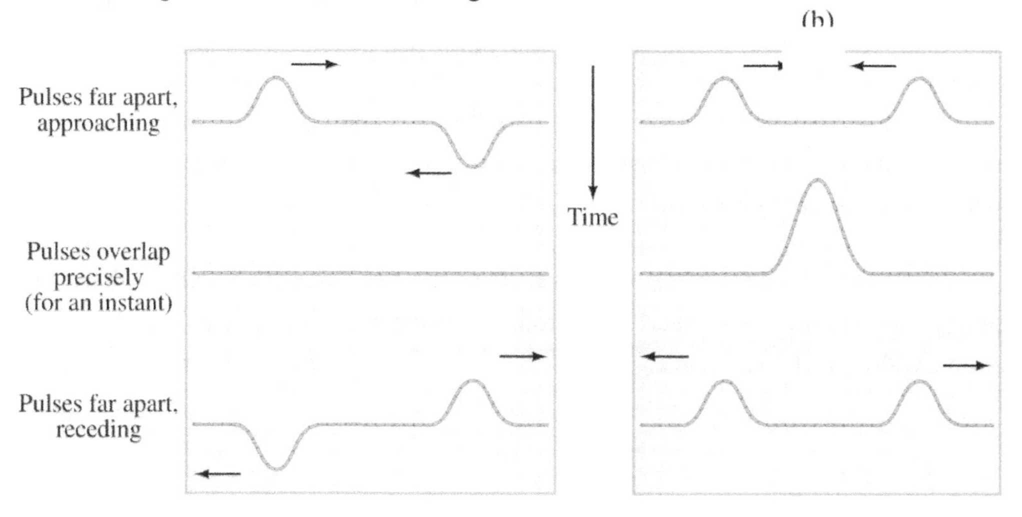

In-phase occurs when the waves are 0 or 2π radians (0 or 360°) apart in their oscillations. The peaks and troughs of one wave are perfectly aligned with the peaks and troughs of another. When two waves are in-phase and interfere, they combine. The amplitude of the resulting wave is the sum of the amplitudes of the interfering waves.

For constructive interference.

$$A_1 + A_2 = A_{net}$$

Out-of-phase occurs when the waves are not are 0 or 2π radians (0 or 360°) apart in their oscillations. All out-of-phase waves produce destructive interference. The resulting amplitude of the combined waves is less than the sum of the interfering waves.

$$0 < A_{net} < A_1 + A_2$$

Completely out-of-phase occurs when the waves are π radians (180°) apart in their oscillations. The crests of one wave are aligned with the troughs of the other wave. When two waves are completely out-of-phase, the amplitude of the resulting wave is the absolute value of the difference between the two separate amplitudes.

If the waves have the same amplitude, then the interfering waves completely destructively interfere, and no wave exists (amplitude = 0).

$$A_{net} = | A_1 - A_2 |$$

The figures below show all three examples of wave interference discussed above.

Diagram (a) below demonstrates two in-phase waves and exhibits constructive interference. The resulting amplitude is the sum of the two individual amplitudes.

Diagram (b) demonstrates two waves that are entirely out-of-phase and exhibit destructive interference. In this example, the two individual amplitudes are equal, and the result is that no wave is produced.

Diagram (c) illustrates two waves that are not completely out-of-phase. The resulting waves destructively interfere to produce a wave with an amplitude less than that of the wave with a greater amplitude.

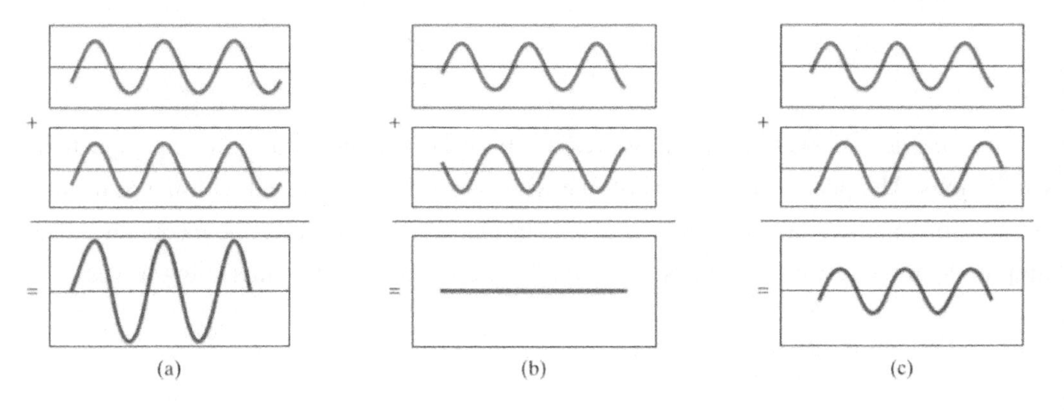

Figure a illustrates constructive interference, figure b illustrates entirely out-of-phase waves, and figure c illustrates waves that are not completely out-of-phase.

Reflection and Transmission

Reflection

Reflection occurs when waves bounce off the surface of an object or medium. All waves exhibit reflection. Examples of reflection are the reflection of light off a mirror to see an image, echoes produced by the reflection of sound waves, and the glare from sunlight reflecting off the water.

These reflections occur when a wave encounters an obstacle or an interface between two different mediums; the wave is forced to change direction.

Reflections of waves depend upon the object or medium the wave reflects off.

Imagine a wave traveling down a string, as in the figure below.

In the diagram on the left, if the wave hits an immobile object, it is reflected, but its reflection is inverted.

In the diagram on the right, the end of the string is connected to a point, so it is free to move. The wave on the string is reflected in the opposite direction, but its reflection is upright instead of inverted.

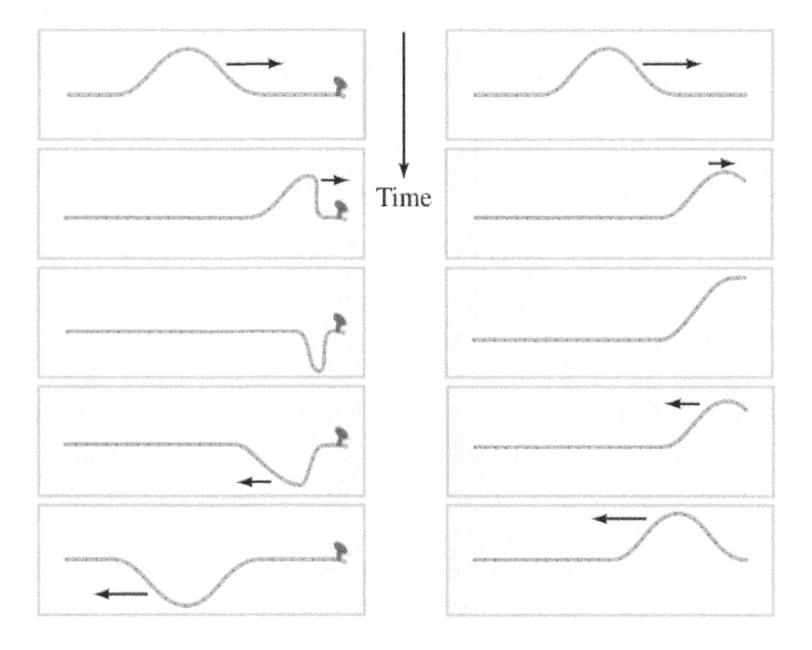

Law of Reflection

The Law of Reflection governs the reflection of light off surfaces and interfaces of two different mediums. The Law of Reflection states that *the angle of incidence of the incoming light ray is equal to the angle of reflection of the outgoing light ray, concerning the normal line of the reflecting surface.*

$$\theta_{incident} = \theta_{reflected}$$

Billiards is an example, in which the balls are waves and the edges of the table surface from which they reflect. A billiards player understands the reflection angle of the ball off an edge equals the incoming angle of the ball, concerning the normal of the surface.

The diagram below represents an incident light ray on a reflecting surface:

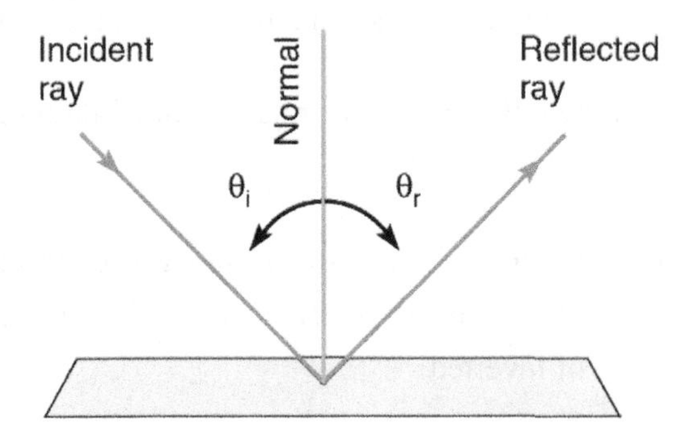

From the diagram, the incident angle is equal to the reflected angle.

The Law of Reflection only applies to smooth surfaces, which are assumed to be ideal reflectors. Plane mirrors are modeled as ideal reflectors.

If the surface is rough and not smooth, the angle of reflection of one wave is not equal to the angle of reflection of the other wave, even if they left the same point with the same angle. Rough surfaces scatter incoming waves.

The figure below is an ideal reflector and a rough surface. The ideal reflector obeys the Law of Reflection, and thus the angle of reflection is equal to the angle of incidence, allowing an observer to see the source of the light (e.g., a mirror).

The rough surface scatters the incoming light rays, and the angle of reflection is not equal to the angle of incidence. The observer would not see the source of light but instead see only the surface of the rough-surfaced reflector.

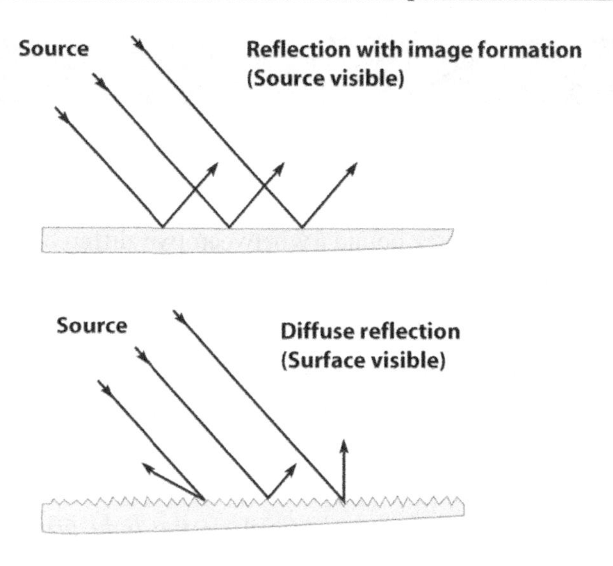

Transmission

Transmission occurs when a wave encounters a different medium. Unlike reflection, the transmission is the phenomenon where some or all the wave energy is transmitted through a different medium. Transmission is usually not total. An incoming wave is usually partially reflected from an interface of two different mediums, while the remaining portion is transmitted through the new medium.

Transmission and reflection of water is an example. When light travels through the atmosphere and encounters the surface of a lake, some of the light is reflected, but some of the light is transmitted through the water. Glare on lakes indicates reflection. However, when a person jumps into the water, it becomes see-through, indicating transmission of light.

From the figure below, consider two strings of different densities that are attached. For a wave traveling down the lower-density section, it both reflects off and transmit some of its energy to the higher-density section.

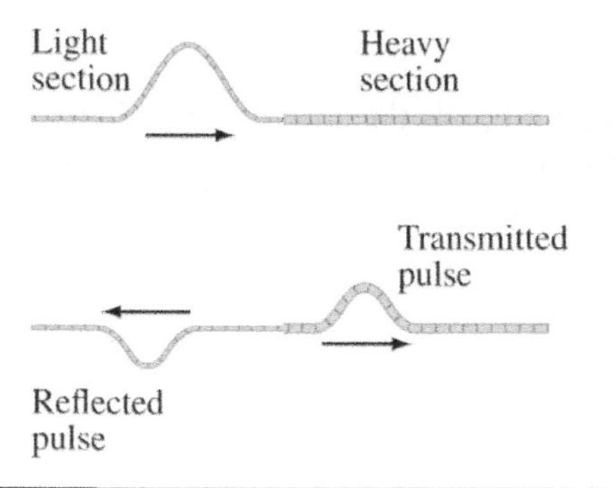

Refraction and Diffraction

Refraction

When a wave encounters a boundary between two different mediums or enters a medium where the wave speed is different, the wave is refracted. Specifically, when waves are transmitted from one medium to another, the different physical properties cause the light to bend and therefore change direction within the new medium. This is *refraction* and is commonly observed in light waves. Refraction is possible with any wave that encounters such a boundary.

For light, the speed that the wave travels is the *light propagation speed.* This speed depends upon the medium through which the light propagates. In a vacuum, light travels at a constant 3×10^8 m/s. In different mediums, such as glass or water, the light propagating speed is less than the speed in a vacuum. Which direction the light bends during refraction depends on the light speed for the different mediums.

Snell's Law expresses the direction the light bends when traveling through mediums with different wave propagation speeds. Snell's Law relates the speed and angle of the incoming wave to that of the refracted wave:

$$n_1 \sin(\theta_1) = n_2 \sin(\theta_2)$$

where n is the index of refraction, and θ is the angle of the wave concerning the normal of the medium interface.

The index of refraction is the relation of the light propagation speed in a medium to that of the speed of light propagation in a vacuum.

n is always greater than 1 because light cannot travel faster than in a vacuum.

$$n = \frac{c}{v}$$

where n is the index of refraction, c is the speed of light in a vacuum, v is the wave propagation speed for that medium.

Indices of refraction (*n*) for some common materials. *c* is the speed of light in a vacuum.

Substance	Index of refraction	Light speed
Air	Approx. 1	~c
Water	1.333	0.75c
Glass	1.5	0.67c
Diamond	2.4	0.42c

The figures below depict the refraction of light from water to air and vice versa.

When a wave moves to a denser medium (i.e., higher refractive index), it bends *toward the normal*.

When a wave moves to a less dense medium (i.e., smaller refractive index), it bends *away from the normal*.

When light refracts from water to air (left diagram), it bends *away* from the normal.

When light refracts from air to water (right diagram), it bends *toward* the normal.

Given the angle of incidence, the angle of refraction is computed by Snell's Law.

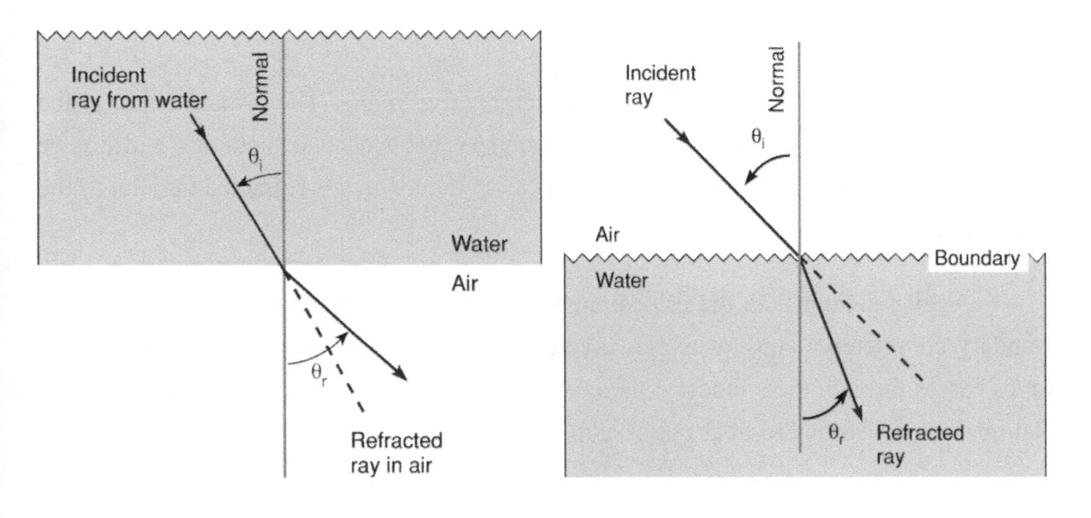

Consider the refraction of white light through a prism. The prism is an exciting example of refraction because it separates the mixture of colors in sunlight (white light) into a rainbow. The observation of a rainbow through a prism is only possible because the index of refraction (*n*) varies with speed (and therefore with wavelength).

Different wavelengths refract at different angles. This difference in refraction produces an assortment of colored light (e.g., a rainbow), where the refraction occurs in suspended water droplets instead of through a prism.

Violet light has the shortest wavelength for visible light and is refracted the most. This refraction gives the sky a blue color.

Red light has the longest wavelength for visible light and is refracted the least. When the angle of refraction is greater, red sunsets are observed in the sky.

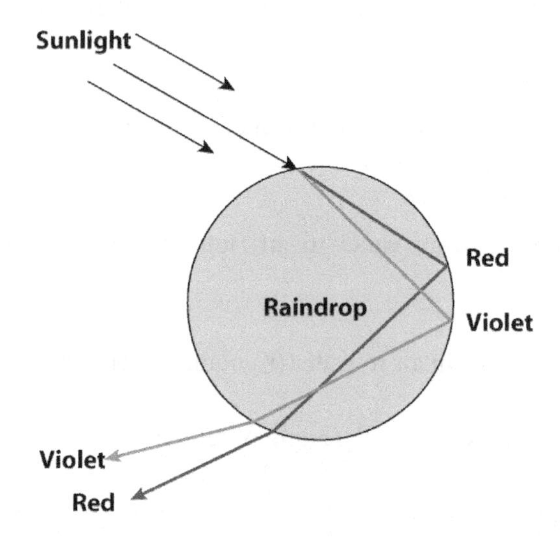

An essential aspect of refraction is the *critical angle*. The critical angle is the greatest angle a wave can make with the boundary between two mediums and not be refracted into the second medium. At the critical angle, a wave is refracted parallel to the surface of the interface boundary.

Angles greater than the critical angle causes the incident wave to reflect off the boundary completely. Thus, at angles greater or equal to the critical angle, none of the waves passes through the boundary; there is "*total internal reflection.*" Some examples of total internal reflection are fiber optics and gemstone brilliance.

$$n_1 \sin(\theta_{critical}) = n_2 \sin(90°)$$

Diffraction

Diffraction is refraction that occurs when a wave encounters an obstacle in the medium within which it is traveling. A wave then spreads out (dissipates), to move around the obstacle, or through an opening (an *aperture*) in the obstacle, producing a "*shadow region.*"

The figures demonstrate diffraction around an obstacle and through an aperture:

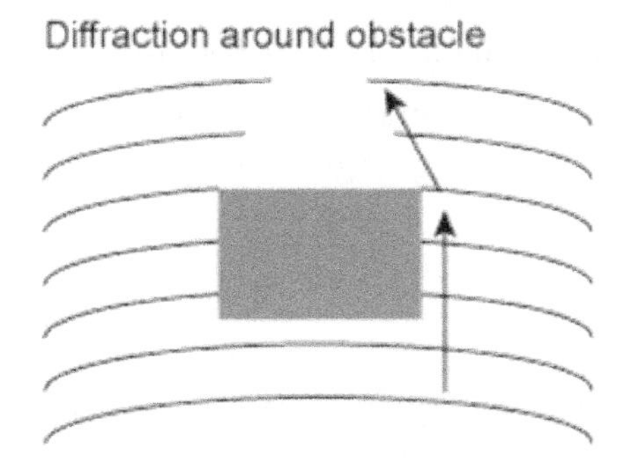

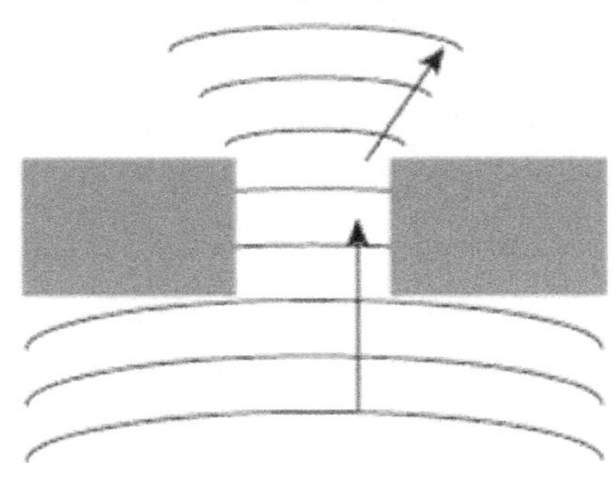

Some types of waves diffract more efficiently than others. For example, consider a sound wave compared to a light wave.

Music (sound waves) can be heard from around the corner of a building, but the speaker (light waves as the visual image) that it is emitting cannot be seen around the corner. Conversely, shining light through a hole does not produce a dot of light, but a diffused circle of light. The amount of diffraction depends on how the wavelength and the size of the obstacle (or aperture) compare.

If the opening is larger than the wavelength, diffraction is minimal, and for the most part, the wave continues as before.

If the opening is much smaller than the wavelength, the diffraction is noticeable, and the wave bends into a circular orientation.

If the wave encounters an obstacle, its reaction is the opposite. For example, the figures below depict different waves diffracting off different objects.

If the obstacle is much smaller than the wavelength, the wave barely diffracts, as in diagram for water waves passing blades of grass (top left).

If the object is comparable to or larger than the wavelength, diffraction is more significant. This is seen in the diagrams for a stick in the water (top right), short-wavelength waves passing a large log (bottom left), or long-wavelength waves passing the same log (bottom right).

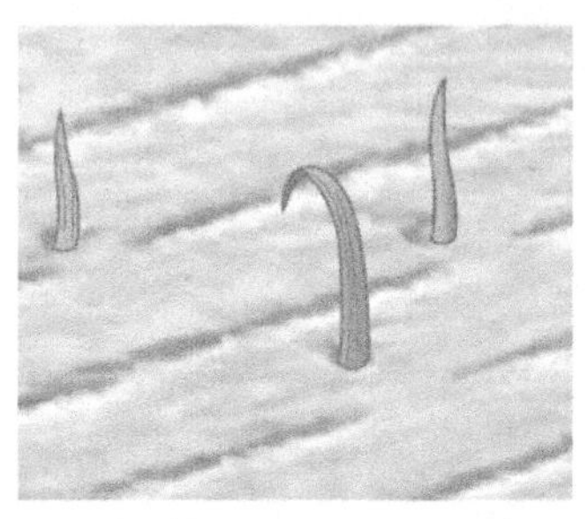

Water waves passing blades of grass

Stick in water

Short-wavelength waves passing log

Long-wavelength waves passing log

Amplitude and Intensity

Like the oscillation that starts the wave, the energy transported by a wave is proportional to the square of the amplitude.

Recall that the amplitude is the maximum height of a crest or the maximum displacement from the equilibrium position.

Amplitude correlates with the energy of the wave.

A greater amplitude means the wave has higher energy.

A lesser amplitude means the wave has lower energy.

Intensity is the energy per area per time, which translates to power (Watts) per area (m^2).

$$I = \frac{P}{A}$$

Thus, amplitude and intensity are correlated similarly to amplitude and energy.

A higher amplitude (and therefore greater energy) leads to a higher intensity.

Chapter Summary

Periodic Motion Summary

- Periodic motion is a motion that repeats
- Amplitude relates the displacement and potential energy
- Frequency gives cycles per second
- The period is the time to complete one cycle
- At the equilibrium position, $PE = 0$, $KE =$ maximum
- At the maximum displacement (amplitude) $x = A$, $PE =$ maximum, $KE = 0$
- At any point, $PE + KE =$ maximum $PE =$ maximum $KE =$ constant
- From the velocity at the equilibrium position, the amplitude is calculated by setting the maximum $KE =$ maximum PE
- From the amplitude, the velocity at the equilibrium position is calculated by setting the maximum $PE =$ maximum KE

Wave Characteristics Summary

- The speed of a wave depends on the medium through which it travels.

  Wave speed is determined by $v = f\lambda$, where f is the frequency ($f = \#$ cycles/time), and λ is the wavelength.

  $f = 1/T$, the speed $v = \lambda/T$.
- Changing the period, frequency, or wavelength of a wave affects the other two quantities. This change does not affect the speed of the wave if it remains in the same medium.
- Superimposition is when waves interact constructively or destructively (e.g., interference creates larger or smaller amplitudes).

- Waves reflect and transmit when traveling through different mediums.

- Waves transmitted through two different mediums with different indices of refraction (n) bend. Snell's Law: $n_1 \sin(\theta_1) = n_2 \sin(\theta_2)$.

- Angles equal to or greater than the critical angle exhibit total internal reflection: $n_1 \sin(\theta_{critical}) = n_2 \sin(90°)$.

- Diffraction occurs for all waves around objects or through apertures.

- Standing waves on a string fixed at both ends form with wavelengths of

$$\lambda_n = 2L/n$$

Standing waves on a string fixed at both ends form with frequencies of

$$f_n = nv/2L$$

where L is the length of the string, v is the speed of the wave, and n is a whole positive number.

Practice Questions

1. Consider a vibrating mass on a spring, what effect on the system's mechanical energy is caused by doubling the amplitude only?

A. Increases by a factor of two
B. Increases by a factor of four
C. Increases by a factor of three
D. Produces no change
E. Increases by a factor of $\sqrt{2}$

2. Which of the following is an accurate statement?

A. Tensile stress is measured in N·m
B. Stress is a measure of external forces on a body
C. Stress is inversely proportional to strain
D. Tensile strain is measured in meters
E. The ratio stress/strain is the elastic modulus

3. The efficient transfer of energy taking place at a natural frequency occurs in a phenomenon is:

A. reverberation
B. the Doppler effect
C. beats
D. resonance
E. the standing wave phenomenon

4. A simple pendulum and a mass oscillating on an ideal spring both have period T in an elevator at rest. If the elevator now accelerates downward uniformly at 2 m/s^2, what is true about the periods of these two systems?

A. The period of the pendulum increases, but the period of the spring remains the same
B. The period of the pendulum increases and the period of the spring decreases
C. The period of the pendulum decreases, but the period of the spring remains the same
D. The periods of the pendulum and the spring both increase
E. The periods of the pendulum and the spring both decrease

5. All of the following are true of a pendulum that has swung to the top of its arc and has not yet reversed its direction, EXCEPT:

A. The *PE* of the pendulum is at a maximum
B. The displacement of the pendulum from its equilibrium position is at a maximum
C. The *KE* of the pendulum equals zero
D. The velocity of the pendulum equals zero
E. The acceleration of the pendulum equals zero

6. The explanation for refraction must involve a change in:

I. frequency II. speed III. wavelength

A. I only **B.** II only **C.** III only **D.** I and II only **E.** I and III only

7. Consider the wave shown in the figure. The amplitude is:

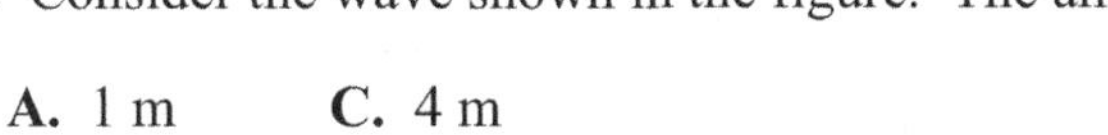

A. 1 m **C.** 4 m
B. 2 m **D.** 8 m
E. $\sqrt{2}$ m

8. Increasing the mass *m* of a mass-and-spring system causes what kind of change on the resonant frequency *f* of the system?

A. The *f* decreases
B. There is no change in the *f*
C. The *f* decreases only if the ratio k / m is < 1
D. The *f* increases
E. The *f* decreases only if the ratio k / m is > 1

9. A simple pendulum that has a bob of mass *M* has a period T. What is the effect on the period if *M* is doubled while all other factors remain unchanged?

A. T/2 **B.** $T/\sqrt{2}$ **C.** 2T **D.** $\sqrt{2}T$ **E.** T

10. A skipper on a boat notices wave crests passing the anchor chain every 5 s. The skipper estimates that the distance between crests is 15 m. What is the speed of the water waves?

A. 3 m/s **C.** 12 m/s
B. 5 m/s **D.** 9 m/s **E.** Requires more information

11. For an object undergoing simple harmonic motion, the:

A. maximum potential energy is larger than the maximum kinetic energy
B. acceleration is greatest when the displacement is greatest
C. displacement is greatest when the speed is greatest
D. acceleration is greatest when the speed is greatest
E. total object energy oscillates at frequency $f = \frac{1}{2\pi}(\sqrt{k} / m)$

12. As the frequency of a wave increases, which of the following must decrease?

A. The speed of the wave
B. The velocity of the wave
C. The amplitude of the wave
D. The cycles per second
E. The period of the wave

13. What is the period for a weight on the end of a spring that bobs up and down one complete cycle every 2 s?

A. 0.5 s **B.** 1 s **C.** 2 s **D.** 3 s **E.** 2.5 s

14. After rain, one sometimes sees brightly colored oil slicks on the road. These are due to:

A. selective absorption of different λ by oil
B. diffraction effects
D. interference effects
C. polarization effects
E. birefringence

15. If two traveling waves with amplitudes of 3 cm and 8 cm interfere, which of the following best describes the possible amplitudes of the resultant wave?

A. Between 5 and 11 cm
B. Between 3 and 8 cm
C. Between 3 and 5 cm
D. Between 8 and 11 cm
E. Less than 8 cm

Solutions

1. B is correct.

$PE = ½kx^2$

Doubling the amplitude x increases the PE by a factor of 4.

2. E is correct.

The elastic modulus is given by:

E = tensional strength / extensional strain

$E = \sigma / \varepsilon$

3. D is correct.

Resonance is the phenomenon where one system transfers its energy to another at that system's resonant frequency (natural frequency). It is a forced vibration that produces the highest amplitude response for a given force amplitude.

4. A is correct.

Period of a pendulum:

$T_P = 2\pi\sqrt{(L / g)}$

Period of a spring:

$T_S = 2\pi\sqrt{(m / k)}$

The period of a spring does not depend on gravity and is unaffected.

5. E is correct.

At the top of its arc, the pendulum comes to rest momentarily; the KE and the velocity equal zero.

Since its height above the bottom of its arc is at a maximum at this point, its (angular) displacement from the vertical equilibrium position is at a maximum also.

The pendulum continually experiences the forces of gravity and tension and is, therefore, continuously accelerating.

6. B is correct.

Refraction is the bending of a wave when it enters a medium where its speed is different. Refraction occurs in sound waves and light waves.

7. C is correct.

The amplitude of a wave is the magnitude of its oscillation from its equilibrium point.

8. A is correct.

$$f = (1/2\pi)\sqrt{(k/m)}$$

An increase in m causes a decrease in f.

9. E is correct.

$$T = 2\pi[\sqrt{(L / g)}]$$

No effect on the period because T is independent of mass.

10. A is correct.

speed = wavelength × frequency

period = 1 / frequency

$$v = \lambda f$$

$$v = \lambda / T$$

$$v = 15 \text{ m} / 5 \text{ s}$$

$$v = 3 \text{ m/s}$$

11. B is correct.

When the displacement is greatest, the force on the object is greatest.

When force is maximized, then acceleration is maximum.

12. E is correct.

$\lambda = vf$

$f = v / \lambda$

$f = 1 / T$

The period is the reciprocal of the frequency.

If f increases, then T decreases.

13. C is correct.

The definition of the period is the time required to complete one cycle.

Period = time / # cycles

$T = 2$ s / 1 cycle

$T = 2$ s

14. D is correct.

When it rains, the brightly colored oil slicks on the road are due to thin-film interference effects. This is when light reflects from the upper and lower boundaries of the oil layer and form a new wave due to interference effects. These new waves are perceived as different colors.

15. A is correct.

Constructive interference: the maximal magnitude of the amplitude of the resultant wave is the sum of the individual amplitudes:

3 cm + 8 cm = 11 cm

Destructive interference: the minimal magnitude of the amplitude of the resultant wave is the difference between the individual amplitudes:

8 cm – 3 cm = 5 cm

The magnitude of the amplitude of the resultant wave is between 5 and 11 cm.

Chapter 6

Sound

- **Production of Sound**
- **Relative Speed of Sound in Solids, Liquids, and Gases**
- **Intensity of Sound**
- **Attenuation: Damping**
- **Doppler Effect**
- **Pitch**
- **Ultrasound**
- **Resonance in Pipes and Strings**
- **Beats**
- **Shock Waves**

Production of Sound

Sound is a wave that is experienced in everyday life. Everything that is being heard is the result of the sound wave. This includes the noise of traffic, the wind through trees, and a jet flying by overhead. However, many sounds are not audible.

Like other waves, the sound is produced by vibrations in a medium.

Specifically, the sound is a mechanical longitudinal wave that travels through a medium as vibrations.

Vibrations produce pressure waves, which oscillate parallel to the direction of propagation.

If the vibrations have a frequency too low to hear, they are *infrasound.*

Comparatively, if the frequency is too high to hear, they are *ultrasound.*

Musical instruments produce sounds through vibrations in various ways—strings, membranes, metal or wood shapes, or air columns. Vibrations may be started by plucking, striking, bowing, or blowing. They are transmitted to the air and then to the perceivers' ears.

For example, a drum produces sound by vibrating a thin membrane when struck. The vibrating membrane creates a longitudinal wave through the air to a person's ears, which results in the perception of sound.

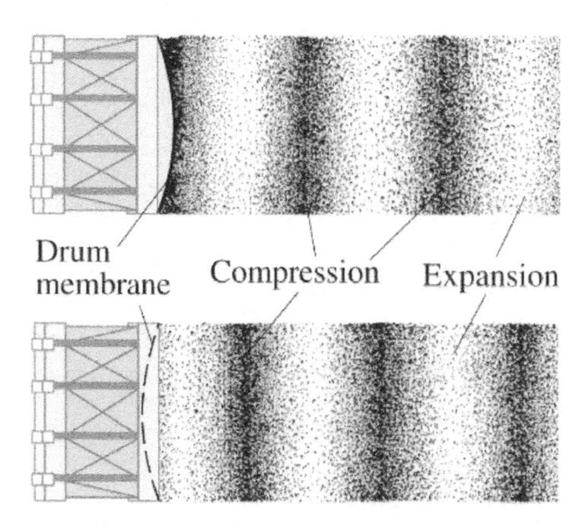

Longitudinal waves produced by the beating of a drum

The strings on a guitar can be effectively shortened by fingering, which raises the fundamental pitch.

The pitch of a string at a given length can be altered by using a different density string.

The strings on stringed instruments produce a fundamental tone whose wavelength is twice the length of the string. There are also various harmonics present.

A piano uses both methods (string length and string density) to cover its more than seven-octave range—the lower notes are produced by strings that are much longer and much thicker than the strings that produce the higher notes.

Wind instruments create sound through standing waves in a tube, and they have a vibrating column of air when played.

During a storm, lightning is seen before thunder is heard, even though they occur at the same time because the speed of light (300,000,000 m/s or 3.0×10^8 m/s) is much faster than the speed of sound (300 m/s or 3.0×10^2 m/s).

Relative Speed of Sound in Solids, Liquids, and Gases

Like all waves, sound waves do not transport matter. Even though a sound wave travels, the medium through which the sound wave propagates does not travel. The particles within the medium vibrate around their initial position (horizontally because the wave is longitudinal) and return to their initial position after the wave has propagated. It is important to remember that waves carry energy, not matter.

Two values affect the speed of sound: the *elasticity* and *density* of the medium.

Sound travels through a medium by transferring energy from one particle to the other. A rigid medium, or one with little elasticity, will have strong attractions between its molecules. Therefore, the molecules can vibrate at a much higher speed and transfer energy faster than a more elastic material.

The speed of sound is faster through solids than through liquids, and faster through liquids than through gases. Sound cannot travel in a vacuum; sound requires a medium, and in a vacuum, there are no particles to vibrate and transmit the wave.

As previously stated, the speed of sound depends upon the density and elasticity of the medium. The equation gives the speed of sound is:

$$v = \sqrt{\frac{\beta}{\rho}}$$

Where β is the bulk modulus, which indicates how the medium responds to compression (how elastic or inelastic it is), and ρ is the density of the medium.

Generally, a substance with a large density is made of larger molecules.

Larger molecules mean more mass, thus requiring more kinetic energy to vibrate them. Since waves are made of kinetic energy, they travel slower through a medium with larger molecules and a higher mass and faster through a medium with smaller molecules and a lower mass.

If the medium is a gas, the temperature can also change the speed of sound. Say two equally sized jars contain the same gas but with different densities. Since there is so much space between particles in a gas, the ideal gas laws play a much larger role than individual particle characteristics, such as mass.

A higher density at the same volume means an increase in temperature.

An increase in temperature means the addition of heat energy, or kinetic energy, to the system.

More kinetic energy means the gas molecules bounce around faster than at cooler temperatures, and transfer energy at a quicker rate.

Sound waves travel faster through the warmer medium.

$$v = \sqrt{\frac{\gamma RT}{M}}$$

where γ is the adiabatic constant of the gas, R is the ideal gas constant (8.314 J/mol K), T is the temperature in Kelvin (K), and M is the molecular mass of the gas (kg/mol).

Some examples of speed comparison through different media:

- Sound will travel faster through wood than it will through the air. Wood is inelastic, whereas air is exceptionally compressible.

- Sound travels faster through American Redwood, which has a density of 28 lb/ft^3, than through African Teak, which has a density of about 45 lb/ft^3.

  Their elasticities are similar, which is why density matters.

- Even though gases are less dense than solids, sound travels slower in a gas (e.g., helium) because it is too compressible.

- Sound travels much faster over a hot desert than it does over a cold tundra; the temperature is higher. Therefore, the speed of sound is higher is the hot desert.

At room temperature and normal atmospheric pressure, the speed of sound is approximately 343 m/s.

Below is a table of sound speed through various mediums, at constant temperature and pressure, unless otherwise noted:

Speed of Sound in Various Materials (20 °C and 1 atm)

Material	Speed (m/s)
Air	343
Air (0 °C)	331
Helium	1,005
Hydrogen	1,300
Water	1,440
Seawater	1,560
Concrete	≈3,000
Hardwood	≈4,000
Glass	≈4,500
Iron and steel	≈5,000
Aluminum	≈5,100

Intensity of Sound

The *intensity* of a wave is equal to the energy transported per unit time across a unit area. The human ear detects sounds with intensity between 10^{-12} W/m^2 and 1 W/m^2.

However, perceived loudness is not proportional to the intensity; the two are related, though, as in the table below.

The Intensity of Sample Sounds

Source of the Sound	Sound Level (dB)	Intensity (W/m^2)
Jet plane at 30m	140	100
Threshold of pain	120	1
Siren at 30m	100	1×10^{-2}
Heavy street traffic	80	1×10^{-4}
Noisy restaurant	70	1×10^{-5}
Talk, at 50cm	65	3×10^{-6}
Quiet radio	40	1×10^{-8}
Whisper	30	1×10^{-9}
Rustle of leaves	10	1×10^{-11}
Threshold of hearing	0	1×10^{-12}

The loudness of a sound is much more closely related to the logarithm of the intensity. Sound level is measured in decibels (dB) and is defined as:

$$\beta = 10log\left(\frac{I}{I_0}\right)$$

where β is the sound level in decibels (dB), and I is the intensity in power per area, or energy per unit time per unit area, given in Watts per meters squared (W/m^2).

I_0 is the threshold of hearing at 10^{-12} W/m^2.

The intensity follows an inverse square law of $I \alpha 1 / r^2$, where r is the distance between the source and detector (observer).

In conjunction with the above equation, as the distance increases, the decibels decrease.

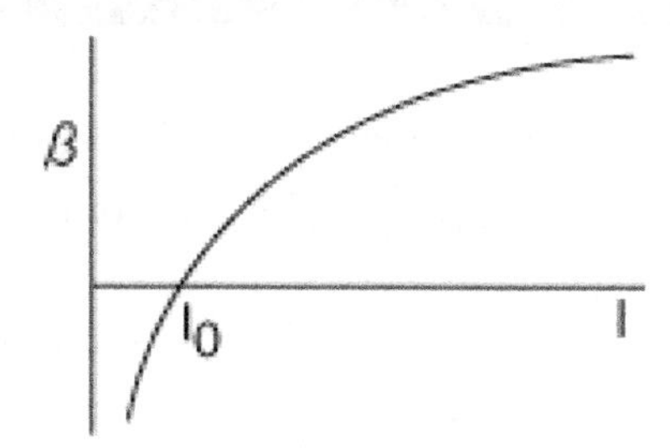

Sound vs. intensity graph where horizontal line signifies audible sounds

Intensity	Decibels
I_0	0
10 I_0	10
100 I_0	20
1,000 I_0	30

An increase in sound level of 3 dB, which is a doubling in intensity, is a small change in loudness. In a large area (substantial value of *r*), this change in intensity is insignificant.

However, in a smaller space (smaller value of *r*), any change in intensity will be extremely noticeable.

The decibel system is based on human perception.

The decibel value for sound with an intensity of I_0 is zero. Below this intensity, the sound is not audible.

As intensity increases, the perception of loudness increases, but to a much lesser degree.

Attenuation: Damping

Sound attenuation is the gradual loss of intensity as sound travels through a medium, and is the greatest for soft, elastic, viscous, or less dense material.

Vibration causes the sound, and the characteristics of this vibration determine how the sound wave acts as it travels through the medium. A wave can oscillate in free harmonic motion, in which the amplitude remains unchanged throughout the vibration.

A wave can also exhibit damped harmonic motion, in which a frictional or dragging force inhibits the vibrations oscillation.

If the damping is small, it can be treated as an "envelope" that modifies the undamped oscillation:

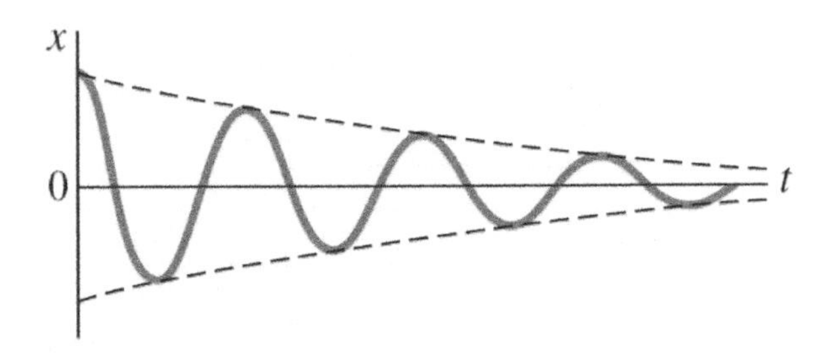

However, if the damping is significant, it no longer resembles a general oscillation (Simple Harmonic Motion, or SHM) wave:

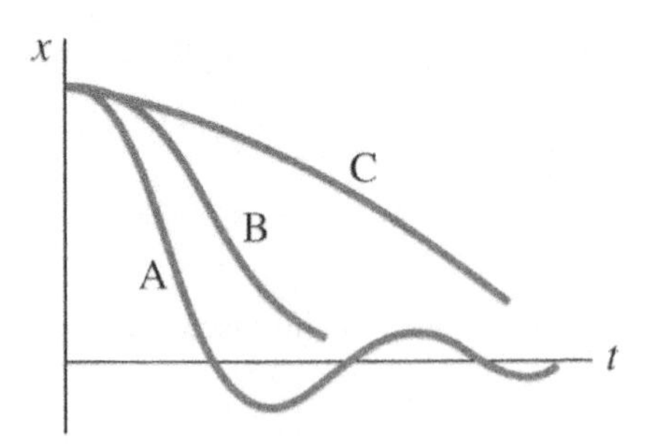

There are a few representative shapes on a graph distance vs. time graph when there is a large damping force acting on the system.

Underdamping (labeled as line A in the graph above) is when there are a few small oscillations before the oscillator comes to rest.

Critical damping (labeled B in the graph above) is the fastest way to get to equilibrium.

Over-damping (labeled C in the graph above) occurs when the system is slowed so much that it takes a long time to get to equilibrium.

There are systems where damping is unwanted, such as clocks and watches.

Moreover, there are systems in which dampening is desirable, and often need to be as close to critical damping as possible.

A system for which dampening is desirable includes earthquake protection for buildings and automobile shock absorbers (pictured below).

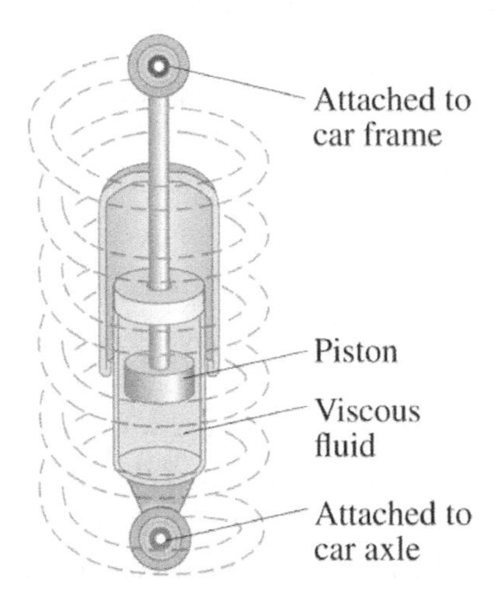

Automobile shocks dampen the vibrations from the bumps on the road surface

Doppler Effect

The Doppler shift describes the relationship between frequency detected and frequency emitted by a source when a source and detector are in relative motion.

$$f_d = f_e[(v_s \pm v_d) / (v_s \mp v_e)]$$

where f_d is the frequency detected by the observer, f_e is the frequency emitted by the source v_d is the velocity of the detector, v_e is the velocity of the source, and v_s is the speed of the wave.

When the source and a detector are moving with respect to each other, the f of the detected wave is shifted from the f of the emitted wave (Doppler shift).

The f_d increases when the source and detector are approaching.

The f_d decreases when the source and detector are receding.

When calculating the Doppler shift between a source and detector, the individual velocities are with respect to each other. If the source and detector are moving away at equal speeds, this is equivalent to the source moving at double its speed away from a stationary detector. This allows one of the velocities to be set to zero in the calculation.

The Doppler Effect states that *the frequency of the sounds varies if there is relative motion between the source of the sound and the observer of the sound.*

If the relative motion between the sound source and observer is towards, the frequency is perceived as higher.

If the relative motion between the sound source and observer is away, the frequency is perceived as lower. Therefore, oncoming sirens seem to sound louder than when they have already passed.

The equation expresses this:

$$f' = \frac{v \pm v_o}{v \mp v_s} \times f$$

where f' is the perceived frequency, f is the wave frequency, v_o is the velocity of the observer or detector, v_s is the velocity of the source, and v is a reference value equal to the velocity of sound through the given medium.

First, decide whether to add or subtract the velocities from the speed of sound through the medium. Think about the problem logically.

If the observer is walking away from the source, the wavelength becomes longer, and the perceived pitch decreases. The velocity of the observer must be subtracted in the numerator.

If the source is getting closer to the observer, the wavelengths are getting shorter, and the perceived frequency increases. The denominator must get smaller, and the velocity of the source must be subtracted in the denominator.

Use care when analyzing the movement of the source and remember how fractions react as their denominator is either increased or decreased.

Situations where the observed frequency is higher than the actual:

- Source moving toward stationary observer: $f' = \frac{v}{v - v_S} \times f$
- Observer moving toward stationary source: $f' = \frac{v + v_O}{v} \times f$
- Source and observer both moving toward each other: $f' = \frac{v + v_O}{v - v_S} \times f$

Situations where the observed frequency is lower than the actual:

- Source moving away from the stationary observer: $f' = \frac{v}{v + v_S} \times f$
- Observer moving away from the stationary source: $f' = \frac{v - v_O}{v} \times f$
- Source and observer both moving away from each other: $f' = \frac{v - v_O}{v + v_S} \times f$

Situations where the observed frequency could be higher or lower than the actual:

- Source is moving toward the observer; the observer is moving away from the source:

$$f' = \frac{v - v_o}{v - v_s} \times f$$

- Source is moving away from the observer; the observer is moving toward the source:

$$f' = \frac{v + v_o}{v + v_s} \times f$$

When the source and observer are moving at the same speed, and in the same direction, there is no frequency shift. Their relative motion is zero. This instance shows that reference frames are important.

When the source of waves and a detector are moving with respect to each other, the frequency of the detected wave shifts from the frequency of the emitted wave (*Doppler shift*).

The detected frequency increases when the source and detector are approaching.

The detected frequency decreases when the source and detector are receding.

The f_{det} is the detected frequency, v_s is the speed of the wave in the medium, v_{det} is the speed of the detector, v_{em} is the speed of the emitter, and f_{em} is the emitted frequency.

Choose the sign in the numerator to reflect the direction the detector is moving (positive if approaching, negative if receding).

Choose the sign in the denominator to reflect the direction the source is moving (negative if approaching, positive if receding).

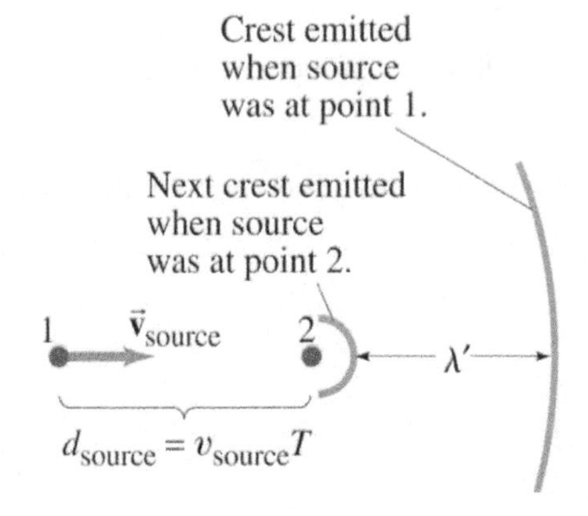

The change in the wavelength is given by:

$$\lambda' = d - d_{source}$$

$$\lambda' = \lambda - v_{source}$$

$$\lambda' = \lambda - v_{source} \times \frac{\lambda}{v_{snd}}$$

$$\lambda' = \lambda(1 - \frac{v_{source}}{v_{snd}})$$

If the observer is moving concerning the source, the wavelength remains the same, but the wave speed is different for the observer:

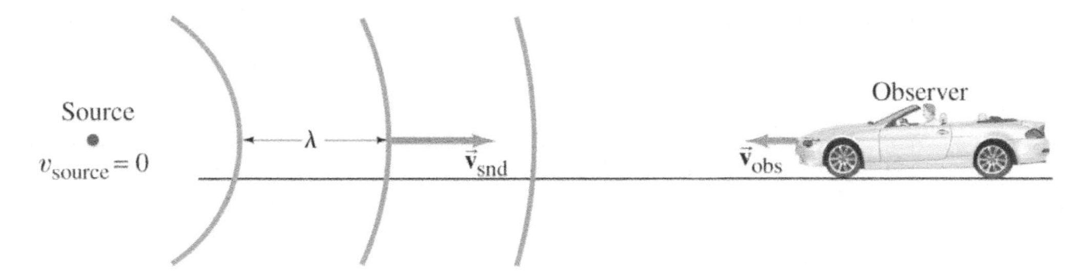

Pitch

Pitch is the human perception of the frequency of sound.

High pitch results from high vibrations and shorter wavelengths; a higher frequency.

Low pitch results from low vibrations and longer wavelength; a lower frequency.

Humans hear sound waves with frequencies between 20-20,000 Hz. The upper limit of this range decreases with age.

Waves with frequencies above 20,000 Hz are *ultrasonic*.

Waves with frequencies below 20 Hz are *infrasonic*.

Ultrasound

The sound has three fundamental properties: reflection, refraction, and diffraction.

Ultrasound imaging uses the reflective property of sound. A source emits a sound wave with an ultrasonic frequency, which reflects off a surface (e.g., internal organ) and travels back into the detector to form an image.

Ultrasound is sound with frequencies of 20,000 Hz or higher.

Ultrasound is used for medical imaging.

Repeated traces are made as the transducer is moved, and a complete picture is built.

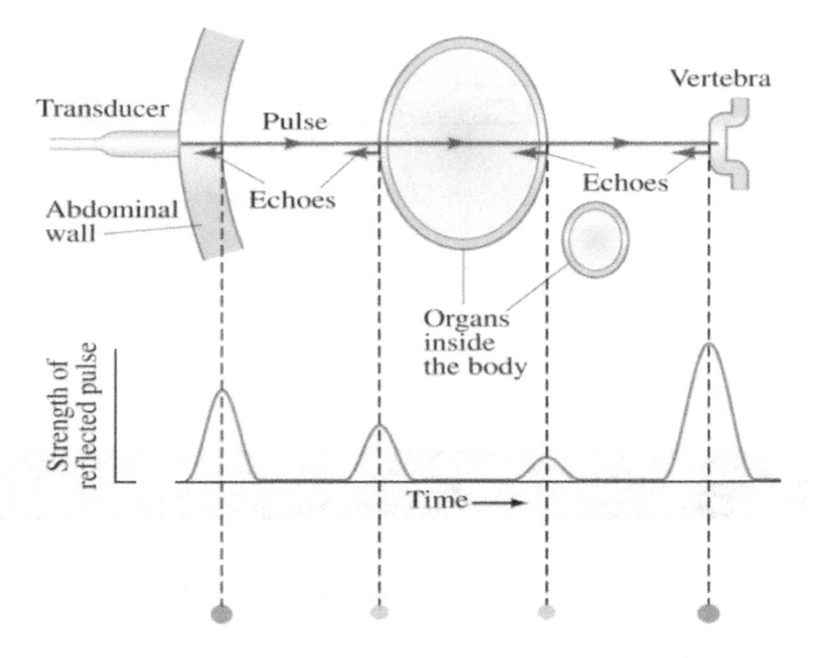

Sonar is used to locate objects underwater by measuring the time it takes a sound pulse to reflect to the receiver.

Sonar techniques are used to learn about the internal structure of the Earth.

Sonar usually uses ultrasound waves, as the shorter wavelengths are less likely to be diffracted by obstacles.

Resonance in Pipes and Strings

In standing wave patterns, there are points along the medium that appear to be still.

These still points are *nodes* and are described as points of no displacement.

There are other points along the medium that undergo the maximum displacement during each vibrational cycle. These maximum displacement points are *antinodes,* and they are opposite of nodes.

A standing wave pattern consists of an alternating pattern of nodes and antinodes.

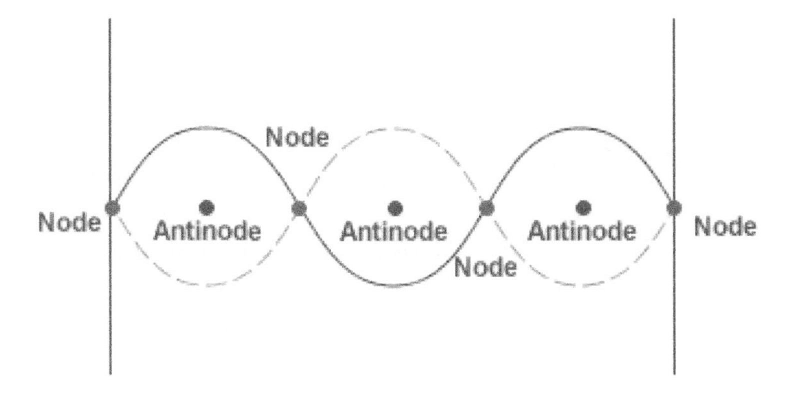

If a tube is open at both ends, like most wind instruments, and a wave of sound passes through it, the characteristics of the wave can be analyzed if the amount of nodes or antinodes the wave has can be determined.

In a tube, an open-end always has a displacement antinode, and a closed-end in the tube always has a displacement node. The pressure in the air varies in the same way air displacement does but shifted.

Everywhere there is a *displacement antinode*; there is a *pressure node.*

Everywhere there is a *displacement node*; there is a *pressure antinode.*

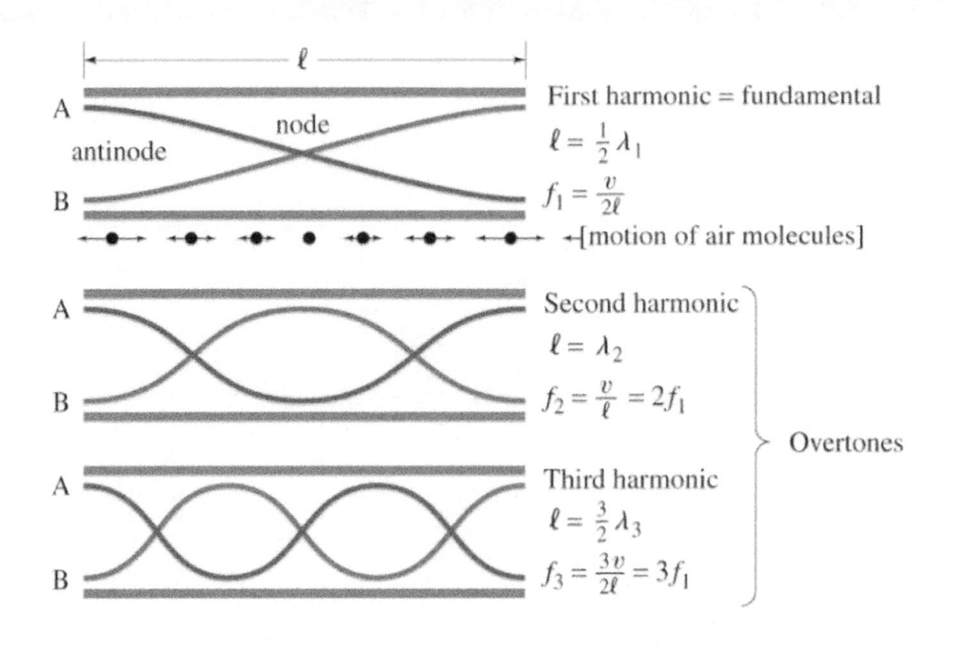

Tube open at both ends illustrating the displacement of air

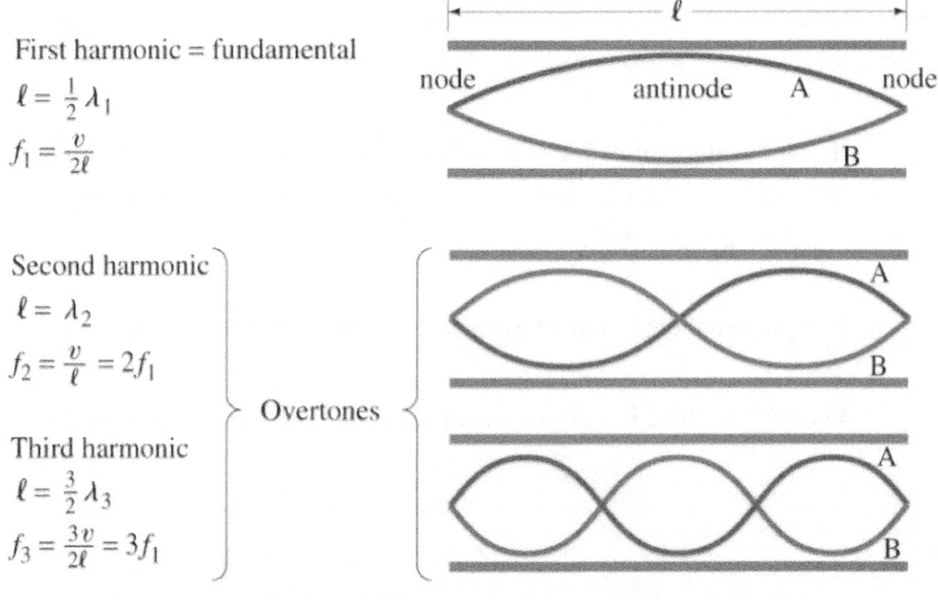

Tube open at both ends illustrating the pressure variation of air

A tube closed at one end, like some organ pipes, has a displacement node and a pressure antinode at the closed end. The number of each node and antinode contained within the pipe depends on the harmonic level (e.g., first harmonic, second harmonic, etc.).

The *fundamental frequency*, also the fundamental, is the first harmonic ($n = 1$).

The next frequency is the second harmonic ($n = 2$).

This continues to the nth harmonic.

Higher harmonics in pipes, either open or closed, have shorter wavelengths and higher frequencies, but the same wave speed.

For the nth harmonic on a string, an n number of half wavelengths fit precisely along the length of the string.

The frequency of a wave can be obtained by:

$$f = \frac{v}{\lambda}$$

If a pipe is open at both ends, or if a string is unfixed at both ends, its length is:

$$L = \frac{n}{2} \times \lambda$$

If a pipe has one closed-end, or if a string has one fixed end, its length is:

$$L = \frac{2n - 1}{4\lambda}$$

For an open-ended tube, resonance occurs when:

$$f_n = n \times \frac{v}{2L}$$ where n can be any integer

Resonance occurs when an outside force acts at a frequency close to the standard frequency of some component of the system. This depends on whether the tube is closed on one end or open at both ends.

For a tube with one closed end, resonance occurs when:

$$f_n = n\frac{v}{4L}$$ and $$\lambda = \frac{4L}{n}$$ where n must be an *odd* integer

An instrument does not play just one sound. They are known for their overtones, which combine with other instruments' overtones to create music, as opposed to merely noise. A trumpet sounds different from a flute, and the reason is overtones—which ones are present and how strong they are.

The plots below show the frequency spectra for a clarinet, a piano, and a violin. The differences in overtone strength are apparent.

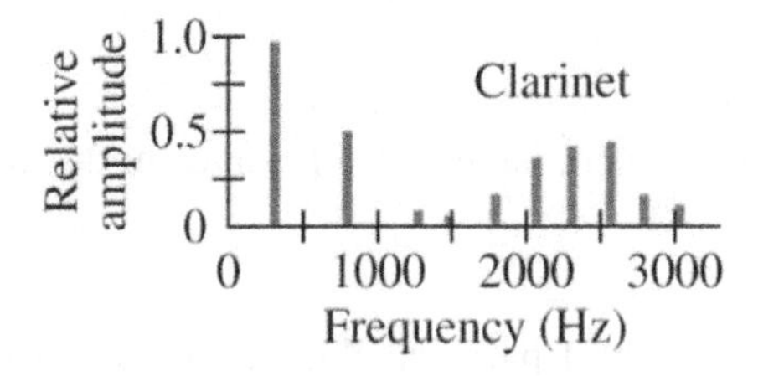

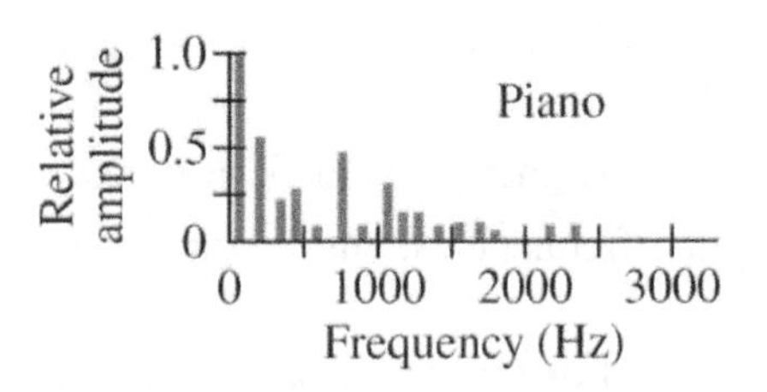

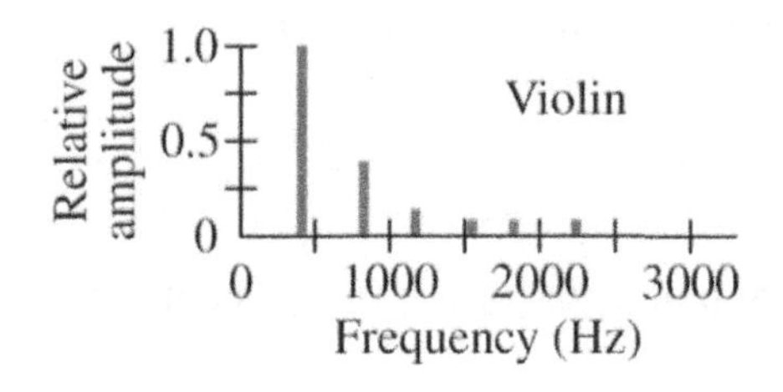

Beats

Sound waves interfere in the same way that other waves do, causing *beats*.

Beats are the slow "envelope" around two waves that are relatively close in frequency, as the two speakers exhibit below.

The region of overlapping sound waves is what produces a beat.

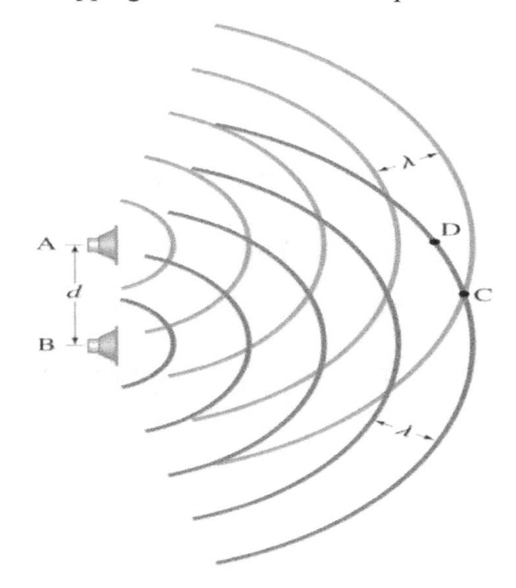

The two frequencies are graphed on a linear axis (top graph below). Although they start in the same place, one frequency is higher than the other.

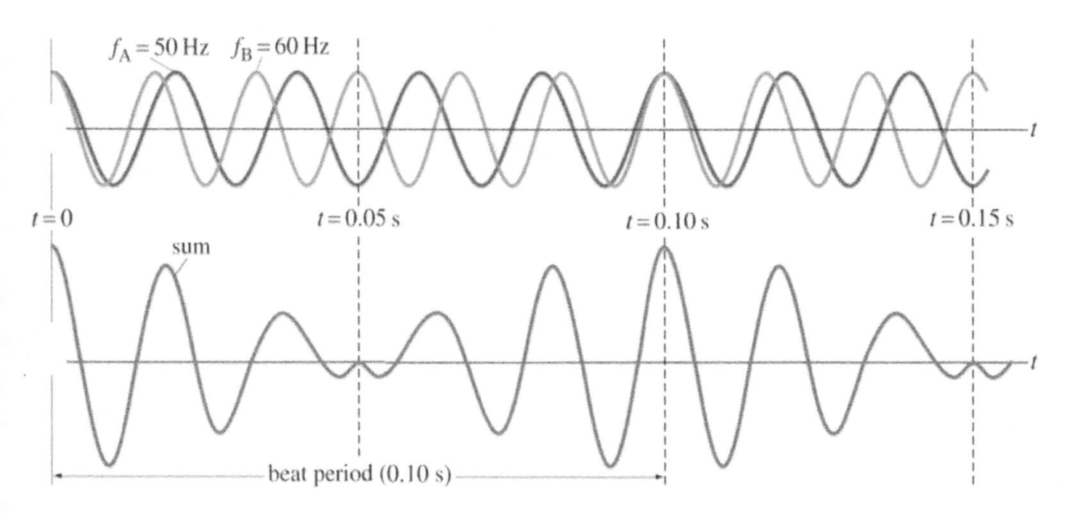

The combination, or interference, of these two results in the wave shown above in the lower graph. This is an out-of-phase interference, meaning the amplitude of this combined wave is equal to the addition of the two individual amplitudes of the two waves shown in the top graph shown above.

A *beat period* is the length of time it takes for the beat to start all over again.

The wave in the lower graph does not oscillate with a constant amplitude as the graph above; it repeats if the component waves stay constant.

The beat frequency equation is:

$$f_{beat} = |f_1 - f_2|$$

where f_1 and f_2 are the frequencies of the interfering waves.

If two notes are played at the same time that are close in pitch but not the same, a pulse or "wobble" is heard. This is the beat, and its frequency is f_{beat}.

Musicians are tuning their instruments by listening for this beat. They try to eliminate it, thereby matching the frequency of their instrument with the frequency of the reference pitch.

Shock Waves

If a source is moving faster than the speed of the waves that it is producing in a given medium, then a *shock wave* is formed. The diagram shows an object with a velocity of zero (top left), a velocity less than the speed of sound (top right), a velocity equal to the speed of sound (bottom left), and velocity higher than the speed of sound (bottom right). This is the only situation in which a shock wave occurs. The object in the bottom left creates a shock wave front, but the waves do not pile up as in the bottom right.

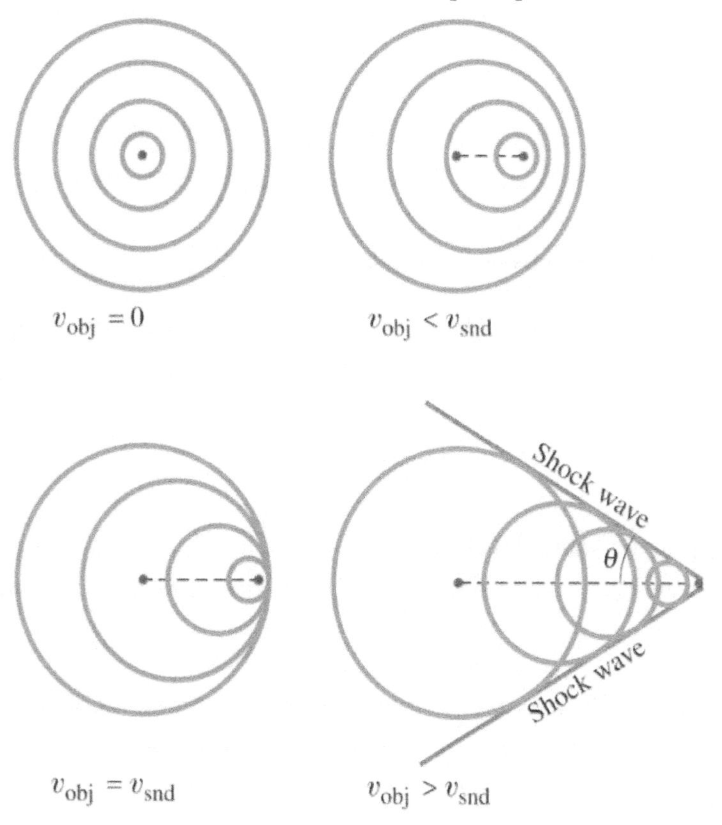

Shock waves are analogous to the bow waves produced by a boat going faster than the wave speed of the water. An airplane exceeding the speed of sound in air produces two sonic booms, one from the nose and one from the tail of the plane.

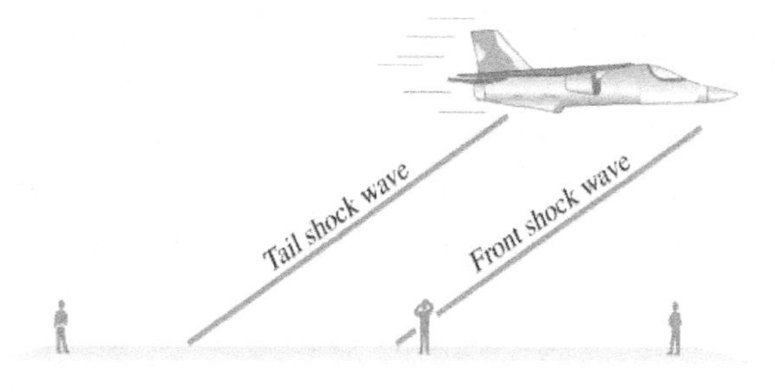

Chapter Summary

- Sound is a mechanical longitudinal wave that travels through a medium as vibrations. The vibrations produce pressure waves, which oscillate parallel to the direction of propagation. Two essential values affect the speed of sound: the elasticity and the density of the medium.

$$v = \sqrt{\frac{\beta}{\rho}}$$

where β is the bulk modulus, which indicates how the medium responds to compression (how elastic or inelastic it is), and ρ is the density of the medium.

- *Intensity* of a wave is equal to the energy transported per unit time per unit area.

- The loudness of a sound is much more closely related to the logarithm of the intensity. Sound level is measured in decibels (dB) and is defined as:

$$\beta = 10log\left(\frac{I}{I_0}\right)$$

where β is the sound level in decibels (dB), and I is the intensity in power per area, or energy per unit time per unit area, given in Watts per meters squared (W/m^2).

- The Doppler shift describes the relationship between the frequency detected and the frequency emitted by the source when the source and detector are in relative motion.

$$f_d = f_e[(v_s \pm v_d) / (v_s \mp v_e)]$$

where f_d is the frequency detected by the observer, f_e is the frequency emitted by the source v_d is the velocity of the detector, v_e is the velocity of the source, and v_s is the speed of the wave.

- In standing wave patterns, there are points along the medium that appear still. These no displacement points are *nodes*. Other points along the medium undergo a maximum displacement during each vibrational cycle. The maximum displacement points are *antinodes*.

- The frequency of a wave can be obtained by:

$$f = \frac{v}{\lambda}$$

- For a pipe that is open at both ends or a string that is unfixed at both ends, the length can be found by:

$$L = \frac{n}{2} \times \lambda$$

- If a pipe has one closed-end, or if a string has one fixed end, its length can be found by:

$$L = \frac{2n - 1}{4\lambda}$$

- For an open-ended tube, resonance occurs when:

$$f_n = n \times \frac{v}{2L} \qquad \text{where n can be any integer}$$

- Sound waves interfere in the same way that other waves do, causing a phenomenon of *beats*. The beat frequency is given by the equation:

$$f_{beat} = |f_1 - f_2|$$

where f_1 and f_2 are the frequencies of the interfering waves.

- If a source is moving faster than the speed of the waves that it is producing in a given medium, then a *shock wave* is formed.

Practice Questions

1. Two tuning forks have frequencies of 460 Hz and 524 Hz. What is the beat frequency if both are sounding simultaneous and resonating?

A. 52 Hz **B.** 64 Hz **C.** 396 Hz **D.** 524 Hz **E.** 588 Hz

2. Which of the following is true of the properties of a light wave as it moves from a medium of the lower refractive index to a medium of the higher refractive index?

A. Speed decreases
B. Speed increases
C. Frequency decreases
D. Frequency increases
E. None of the above

3. Resonance can be explained as forced vibration with the:

A. matching of constructive and destructive interference
B. matching of wave amplitudes
C. maximum amount of energy input
D. least amount of energy input
E. minimum beat frequency

4. The Doppler effect occurs when a source of sound moves:

I. toward the observer II. away from the observer III. with the observer

A. I only **B.** II only **C.** III only **D.** I and II only **E.** I and III only

5. Which of the following increases when a sound becomes louder?

A. Amplitude **B.** Period **C.** Frequency **D.** Wavelength **E.** Velocity

6. Sound intensity is defined as the:

A. sound power per unit volume
B. sound power per unit time
C. sound energy passing through a unit of area
D. loudness of a sound
E. sound energy passing an area per unit time

7. A violin with string length 36 cm and string density 3.8 g/cm resonates with the first overtone of an organ pipe with one end closed. The pipe length is 3 m. What is the tension in the string so that the sound wave resonates at its fundamental frequency? (Use the speed of sound $v = 340$ m/s)

A. 1,390 N **B.** 1,946 N **C.** 1,414 N **D.** 987 N **E.** 1,216 N

8. A speaker is producing a total of 10 W of sound, and Rahul hears the music at 20 dB. His roommate turns up the power to 100 W. What level of sound does Rahul now hear?

A. 15 dB **B.** 30 dB **C.** 40 dB **D.** 100 dB **E.** 120 dB

9. Seven seconds after a flash of lightning, thunder shakes a house. Approximately how far was the lightning strike from the house? (Use the speed of sound $v = 340$ m/s)

A. Requires more information
B. About five kilometers away
C. About one kilometer away
D. About ten kilometers away
E. About two kilometers away

10. The natural frequencies for a stretched string of length L and wave speed v are nv / (2L), where n equals:

A. 0, 1, 3, 5, ...
B. 1, 2, 3, 4, ...
C. 2, 4, 6, 8, ...
D. 0, 1, 2, 3, ...
E. 1, 3, 4, 5, ...

11. What is the source of all electromagnetic waves?

A. electric fields
B. vibrating charges
D. magnetic fields
E. none of the above
C. heat

12. When a radio is tuned to a radio station, the frequency of the internal electrical circuit is matched to the frequency of that radio station. In tuning the radio, what is being affected?

A. Beats
B. Reverberation
C. Forced vibrations
D. Resonance
E. Wave interference

13. A standing wave is oscillating at 670 Hz on a string, as shown in the figure. What is the wave speed?

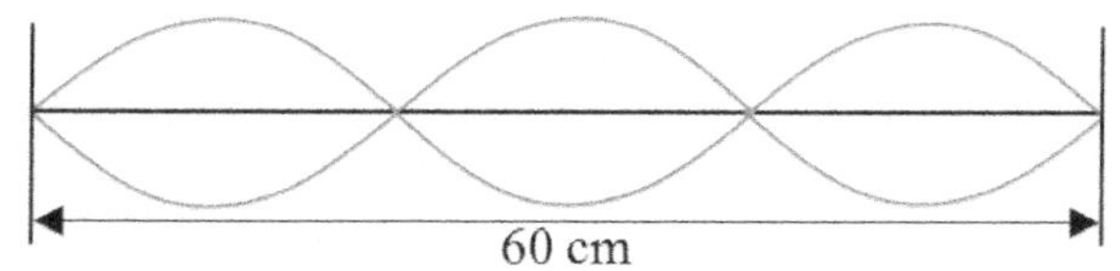

A. 212 m/s **B.** 178 m/s **C.** 268 m/s **D.** 404 m/s **E.** 360 m/s

14. A 50 m/s train is moving directly toward Carlos, who is standing near the tracks. The train is emitting a whistling sound at 415 Hz. What frequency does Carlos hear? (Use the speed of sound $v = 340$ m/s)

A. 290 Hz **B.** 476 Hz **C.** 481 Hz **D.** 487 Hz **E.** 473 Hz

15. Two tuning forks are struck simultaneously, and beats are heard every 500 ms. What is the frequency of the sound wave produced by one tuning fork, if the other produces a sound wave of 490 Hz frequency?

A. 490 Hz **B.** 498 Hz **C.** 492 Hz **D.** 506 Hz **E.** 510 Hz

Solutions

1. B is correct.

Beat frequency equation:

$f_{beat} = | f_2 - f_1 |$

$f_{beat} = | 524\text{ Hz} - 460\text{ Hz} |$

$f_{beat} = 64\text{ Hz}$

2. A is correct.

The frequency of a wave does not change when it enters a new medium.

$v = (1 / n)c$

where c = speed of light in a vacuum and n = refractive index.

The speed v decreases when light enters a medium with a higher refractive index.

3. D is correct.

Resonance is the phenomenon where one system transfers its energy to another at that system's resonant frequency (natural frequency).

Resonance is the forced vibrations with the least energy input.

4. D is correct.

The Doppler effect is the observed change in frequency when a sound source is in motion relative to an observer (away or towards). If the sound source moves with the observer, then there is no relative motion between the two, and the Doppler effect does not occur.

5. A is correct.

When a sound becomes louder, the energy of the sound wave becomes higher.

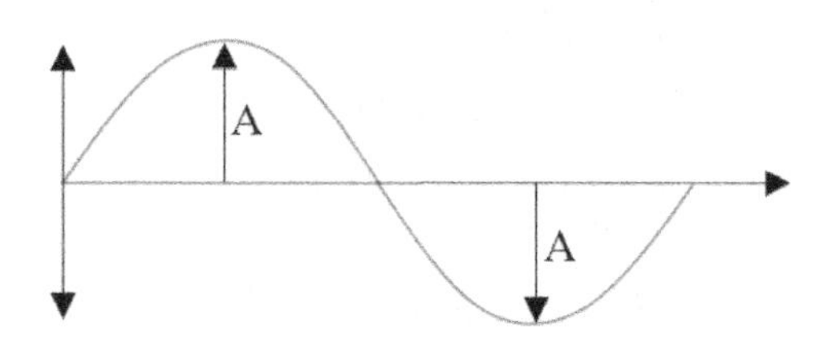

Amplitude is directly related to the energy of the sound wave:

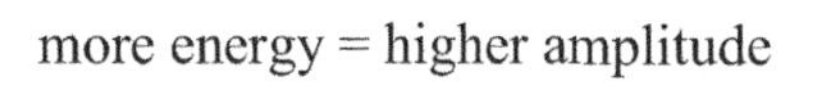

more energy = higher amplitude

less energy = lower amplitude

6. E is correct. Intensity is the power per unit area:

$I = \text{W/m}^2$

Loudness is a subjective measurement of the strength of the ear's perception to sound.

7. C is correct. An overtone is any frequency higher than the fundamental.

In a stopped pipe (i.e., open at one end and closed at the other):

Harmonic #	Tone
1	fundamental tone
3	1st overtone
5	2nd overtone
7	3rd overtone

The first overtone is the 3rd harmonic.

Find the wavelength:

$\lambda_n = (4L / \text{n})$, for stopped pipe

where n = 1, 3, 5, 7…

$\lambda_3 = [(4)\cdot(3\text{ m}) / 3)]$

$\lambda_3 = 4\text{ m}$

Find the frequency:

$f = v / \lambda$

$f = (340\text{ m/s}) / (4\text{ m})$

$f = 85\text{ Hz}$

Find the velocity of a standing wave on the violin:

$f = v / 2L$

$v = (2L)\cdot(f)$

$v = (2)\cdot(0.36\text{ m})\cdot(85\text{ Hz})$

$v = 61\text{ m/s}$

Convert the linear density to kg/m:

$\mu = (3.8\text{ g/cm})\cdot(1\text{ kg}/10^3\text{ g})\cdot(100\text{ cm}/1\text{ m})$

$\mu = 0.38\text{ kg/m}$

Find the tension:

$v = \sqrt{(T / \mu)}$

$T = v^2\mu$

$T = (61 \text{ m/s})^2 \times (0.38 \text{ kg/m})$

$T = 1{,}414 \text{ N}$

8. B is correct.

$I = \text{Power / area}$

The intensity I is proportional to the power, so an increase by a factor of 10 in power leads to an increase by a factor of 10 in intensity.

$I \text{ (dB)} = 10\log_{10} (I / I_0)$

dB is related to the logarithm of intensity.

If the original intensity was 20 dB, then:

$20 \text{ dB} = 10\log_{10} (I_1 / I_0)$

$2 = \log_{10} (I_1 / I_0)$

$100 = I_1 / I_0$

The new intensity is a factor of 10 higher than before:

$I_2 = 10\ I_1$

$1{,}000 = I_2 / I_0$

$I \text{ (dB)} = 10\log_{10} (1{,}000)$

$I \text{ (dB)} = 30 \text{ dB}$

9. E is correct.

velocity = distance / time

$v = d / t$

$d = vt$

$d = (340 \text{ m/s})\cdot(7 \text{ s})$

$d = 2{,}380 \text{ m} \approx 2 \text{ km}$

10. B is correct.

A stretched string has all harmonics of the fundamental.

11. B is correct.

All electromagnetic waves arise from accelerating charges. When a charge is vibrating it is accelerating (change in the direction of motion is acceleration).

12. D is correct.

When tuning a radio, the tuner picks up specific frequencies by resonating at those frequencies. This filters out the other radio signals, so only that specific frequency is amplified.

By changing the tuner on the radio, a frequency is chosen that the tuner resonates at that frequency, and the signal is amplified.

13. C is correct.

From the diagram, the wave is a 3rd harmonic standing wave.

Find the wavelength:

$\lambda_1 = 2L$

$\lambda_n = 2L / n$

where n = 1, 2, 3, 4…

$\lambda_n = (2 \times 0.6 \text{ m}) / 3$

$\lambda_n = 0.4 \text{ m}$

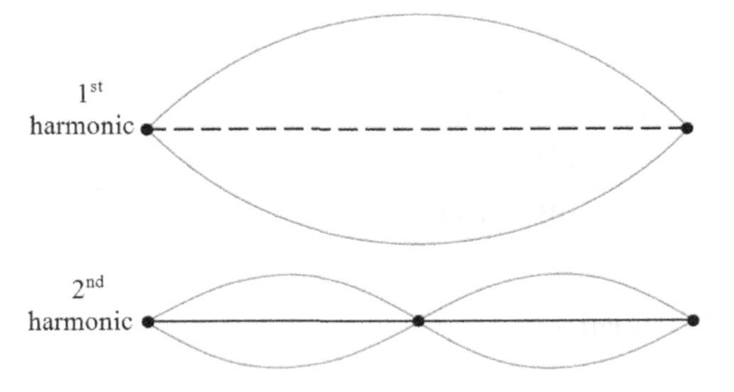

Find the speed:

$\lambda = v / f$

$v = \lambda f$

$v = (0.4 \text{ m}) \cdot (670 \text{ Hz})$

$v = 268 \text{ m/s}$

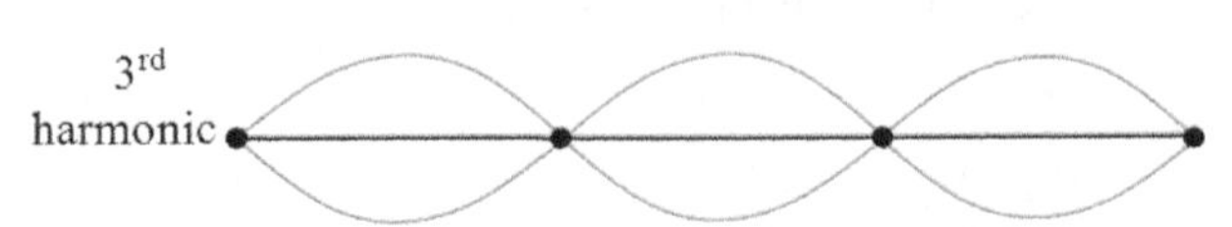

14. D is correct.

Doppler equation for an approaching sound source:

$$f_{observed} = [v_{\text{sound}} / (v_{\text{sound}} - v_{\text{source}})]f_{\text{source}}$$

$$f_{observed} = [(340 \text{ m/s}) / (340 \text{ m/s} - 50 \text{ m/s})]\cdot(415 \text{ Hz})$$

$$f_{observed} = 487 \text{ Hz}$$

The observed frequency is always higher when the source is approaching.

15. C is correct.

Beats are heard every 500 ms (i.e., ½ s), or 2 beats per second.

Since $f_{\text{beat}} = 2$ Hz, the frequencies of the two tuning forks differ by 2 Hz.

One tuning fork has an f of 490 Hz:

$$f \text{ is } (490 \text{ Hz} - 2 \text{ Hz}) = 488 \text{ Hz}$$

or

$$f \text{ is } (490 \text{ Hz} + 2 \text{ Hz}) = 492 \text{ Hz}$$

Chapter 7

Light and Optics

LIGHT

- **Visual Spectrum**
- **Double Slit Diffraction**
- **Diffraction Grating**
- **Single Slit Diffraction**
- **Thin Films**
- **Other Diffraction Phenomena, X-Ray Diffraction**
- **Polarization of Light**
- **Doppler Effect**

OPTICS

- **Reflection from Plane Surface**
- **Refraction, Refractive Index N, Snell's Law**
- **Dispersion**
- **Conditions for Total Internal Reflection**
- **Mirrors**
- **Thin Lenses**
- **Combination of Lenses**
- **Lens Aberration**
- **Optical Instruments**

Visual Spectrum

The visible light spectrum is an important topic within physics because it contributes extensively to the way humans perceive and interact with the world. For example, red stop signs, masterpiece paintings, and reading glasses are just a couple of objects that characterize the diverse range that visible light occupies in life. As stated in earlier chapters, electromagnetic radiation is a form of energy that consists of oscillating electric and magnetic fields, without any need of a medium to propagate.

Visible light is a narrow spectrum of electromagnetic radiation that humans can perceive as different colors. For perspective, the electromagnetic radiation spectrum spans from long radio waves to high energy gamma rays with wavelengths ranging from 10^4 m to 10^{-14} m, respectively. Within the electromagnetic radiation spectrum, the human eye can only perceive a narrow range of wavelengths between 400 nm to 750 nm. This range is the *visible light spectrum* and consists of all the colors that humans can see.

Light of different frequencies is perceived as different colors. The figure below depicts the visible light range with respect to frequency, wavelength, and color. Lower frequency light consists of red, orange, and yellow hues, while higher frequency light consists of greens, blues, and violets. Not only do different frequencies give different colors, but the different frequencies of light have more or less energy (i.e., $E = hf$).

A way to remember the spectrum of visible light, concerning increasing frequency (and thus energy), is by the mnemonic ROY G BIV. The lowest frequency light is Red, then Orange, Yellow, Green, Blue, Indigo up to the highest frequency (highest energy) light – Violet.

The sun, candles, and light bulbs are a few examples of objects that emit light. However, most objects absorb and reflect light, but do not emit light. When an object is perceived to be a particular color, it is usually not emitting light, but rather reflecting that specific frequency of light.

Likewise, if an object absorbs specific frequencies of light, those colors would not be visible at all. This is an important distinction because it determines the perceived colors of most objects. For example, sunlight is considered white light because it contains all the frequencies within the visible light spectrum. Therefore, it is perceived as white.

White tablecloths reflect all frequencies of the visible range and appear white.

Red tablecloths absorb most frequencies but reflect predominately red light.

Black tablecloth absorbs every color of light, and therefore, it appears black.

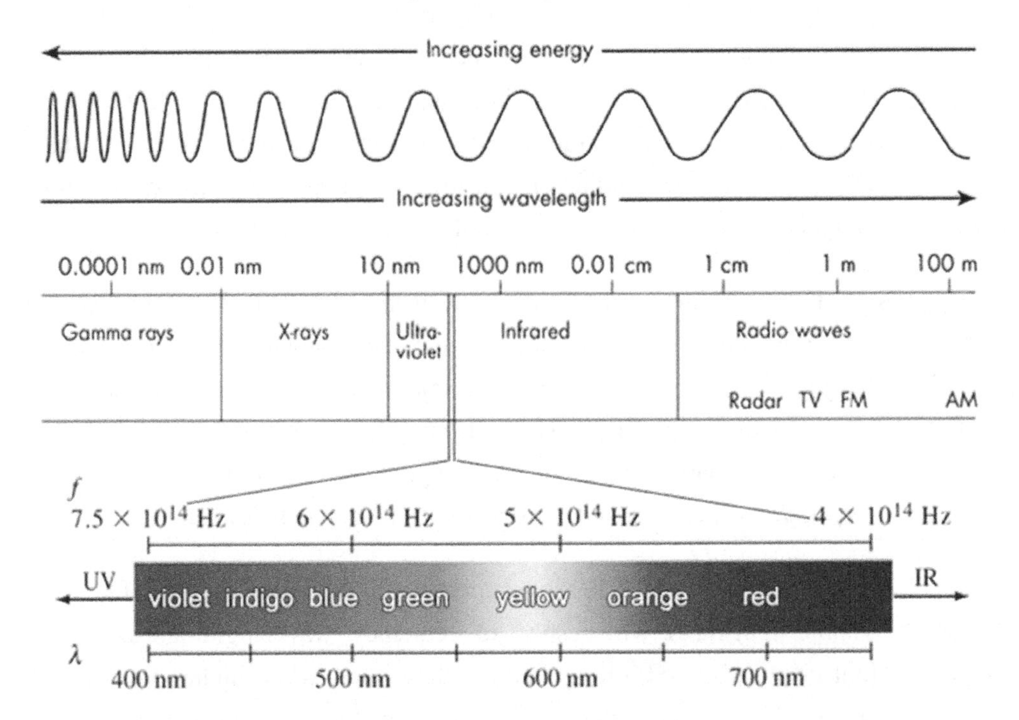

The sky is blue because of *Rayleigh scattering* (when the blue light gets scattered in the atmosphere). Sunlight hits the nitrogen and oxygen molecules and gets bounced all around; it is reemitted at a high frequency. Violet scatters the most, and then on down the spectrum to red. Because human eyes are not violet-sensitive, the next color down on the spectrum (blue) is perceived.

This scattering process is affected by the composition of the atmosphere. In the Mediterranean areas where the air is dry, a deep blue is seen.

When dust, water vapor, or other large particles are present, the light of lower frequency is scattered, and the sky appears paler or hazier. In cities, smog makes particles so large that they absorb light instead of scattering it, resulting in a brownish haze.

Sunsets appear red because red, orange, and yellow are transmitted better than colors at the other end of the visible light spectrum.

When the sun is lower on the horizon, its light must travel through the largest distance of the Earth's atmosphere.

The more atmosphere it travels through, the more that violet and blue are scattered, causing red to be perceived more intensely.

Primary Colors

Yellow, blue, and green are the "primary" colors because they are detected by the eye independent of intensity but based instead on frequencies and the ocular receptors.

If any two of these primary colors are added, all the other colors can be made.

Red + blue = magenta

Red + green = yellow

Blue + green = cyan

If two secondary colors are added, they create white if they are complementary.

Magenta + green = white (magenta = red + blue)

Double Slit Diffraction

Like all waves, visible light can constructively and destructively interfere. This was observed in an iconic experiment as Young's *double-slit diffraction experiment*. This experiment proved that light was indeed a wave, as it displayed the wave property of interference. Also, this helped establish the wave nature of light in the particle-wave duality disagreement of the time.

Observe the figure below of the double-slit diffraction experiment. Two slits, S_1 and S_2, are positioned on an opaque wall, at a known distance between them. A viewing screen is also placed at a known distance from the wall with the slits.

When a light source is shone at the wall with the slits, the expected behavior of light, according to particle theory, can be seen in the top right diagram.

The light that passes through the slits constructively and destructively interfere, producing the pattern seen on the viewing screen in the bottom diagram.

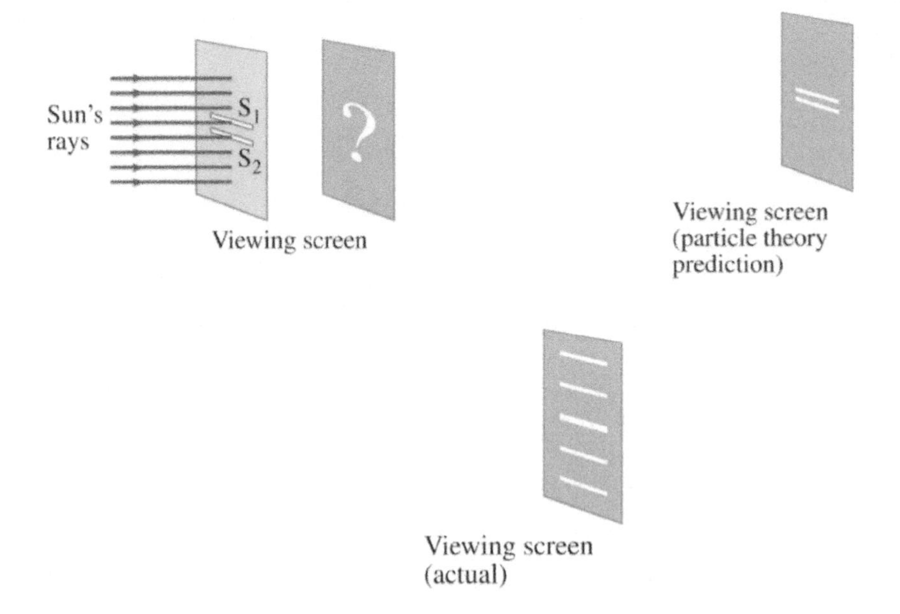

This experimental result supports the theory that light is indeed a wave because the resulting pattern is consistent with interference patterns for waves. The interference occurs because each point on the screen is not the same distance from both slits.

Depending on the path length difference, the wave can interfere constructively to produce a bright spot on the wall or destructively to produce a dark spot on the wall. This observation is seen in the figures below.

The top diagrams below illustrate constructive interference spots on the viewing screen. At these points, the light waves coming through the screen are in phase, constructively interfere, and are visible.

Notice in the top right diagram below that the wave from the lower slit traveled an extra wavelength compared to the upper wave. This is the path length difference and is important in identifying if a spot will be bright or dark (constructive or destructive).

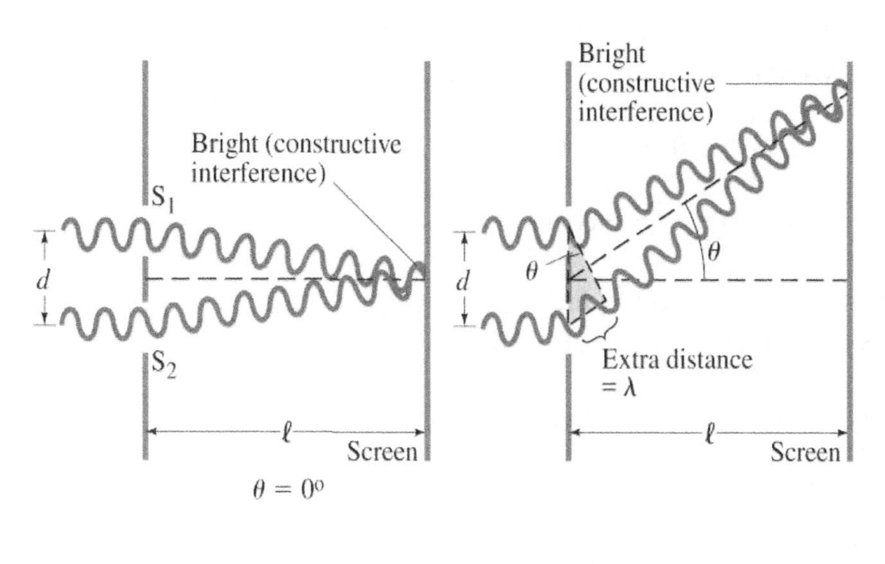

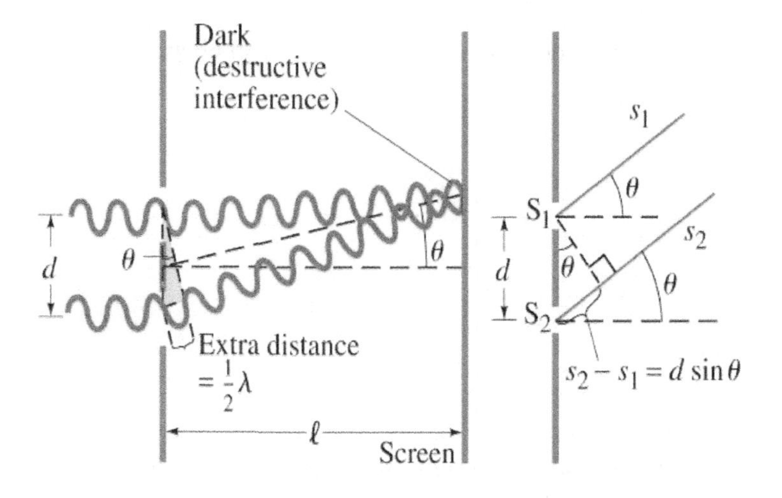

Geometry is used to find conditions for constructive and destructive interference:

The equation can predict the occurrence of bright and dark spots.

$$d \sin \theta = m\lambda$$

For constructive interference (bright):

$$d \sin \theta = m\lambda \qquad m = 0, 1, 2, 3, \ldots$$

where d is the slit separation distance (m), θ is the angle of the bright or dark spot with respect to the center between the two slits (degrees), λ is the wavelength of the light passing through the slits (m), and m is the order value and determines whether the spot appears bright (constructive interference) or dark (destructive interference).

If $m = 0, 1, 2, 3, \ldots$ then the interference is constructive, and the spot is bright.

If $m = \frac{1}{2}, \frac{3}{2}, \frac{5}{2} \ldots$ then the interference is destructive, and the spot is dark.

The figure below depicts the relation of the order number m to constructive interference and destructive interference spots.

The maxima represent constructive interference.

The minima represent destructive interference.

Only the order values m for constructive interference are shown below.

The order values m for destructive interference occurs between these.

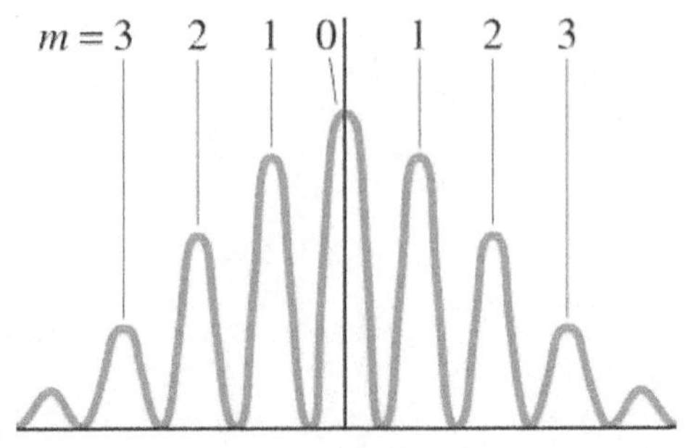

The maxima order number m represents constructive interference

Diffraction Grating

A *diffraction grating* is like the double-slit experiment except that instead of two slits, the diffraction grating has many slits positioned close together.

Diffraction gratings operate via the same principle as the double-slit experiment, and the equation for a diffraction grating interference spots is the same as the double-slit experiment, including order values for destructive and constructive interference.

However, diffraction gratings are better able to separate wavelengths of incoming light than a double slit. Additionally, the peaks of the maxima from a diffraction grate will be much sharper and more pronounced than those of a double slit.

Observe the two figures below of a double-slit maxima pattern (on the left) and a diffraction grating (below on the right) maxima pattern.

The figure on the left below is the maxima pattern produced by double slits.

The figure on the right below is the maxima pattern for the diffraction grating.

The sharper peaks of the maxima are produced by the diffraction grating (below on the right) compared to the double split (below on the left).

The order values *m* are the same for the diffraction grating as for the double slit.

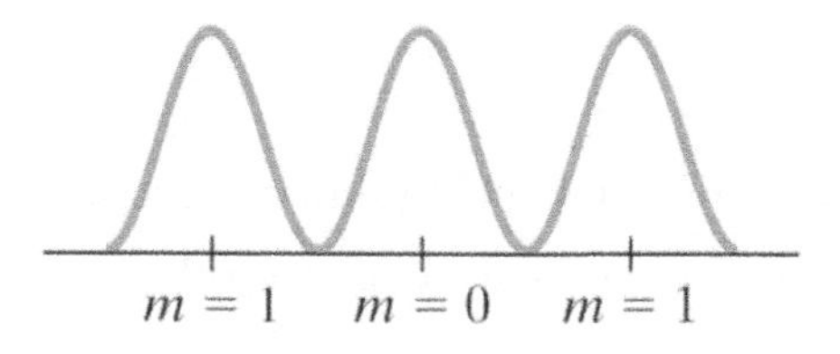

Double slit diffraction pattern

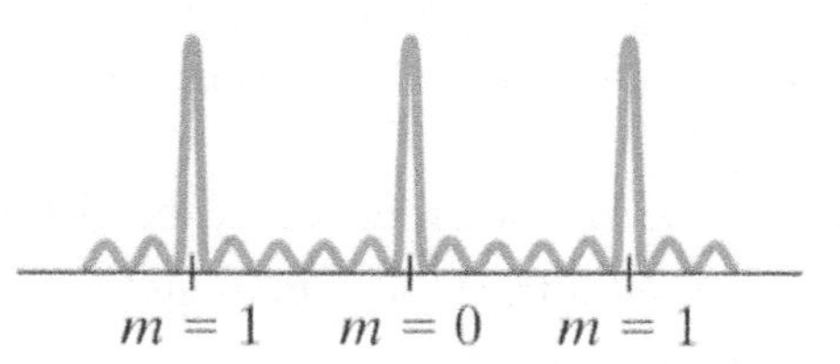

Single slit refraction pattern

Single Slit Diffraction

Single slit diffraction is another method of diffraction where light is passed through a single slit, and the resulting diffraction pattern can be observed.

Like other diffraction techniques, single slit diffraction is the result of interference of the incoming wave as it bends around the slit through which it passes.

Unlike double slit and diffraction grates, the maxima pattern produced by single slit diffraction does not follow the same order values or equation. This is because the light waves in single slit diffraction are in-phase (constructive) and out-of-phase (destructive) at different points than diffraction grates and double-slit diffraction.

The maxima for a single slit diffraction pattern is defined by:

$$a \sin \theta = m\lambda$$

where a is the width of the slit (m).

If $m = 0, \frac{3}{2}, \frac{5}{2}$ ... then the interference is constructive, and the spot is bright.

If $m = 1, 2, 3, \ldots$ then the interference is destructive, and the spot is dark.

The diffraction pattern of single-slit diffraction will also be slightly different. The figure below compares the diffraction patterns produced by the three diffraction methods.

The single slit diffraction pattern produces the widest maxima; all subsequent maxima after $m = 0$ are substantially reduced.

The pattern produced by the diffraction grating and double slit is sharper than for the single slit diffraction pattern and does not fade nearly as much after a value of $m = 0$ (see the diagram below).

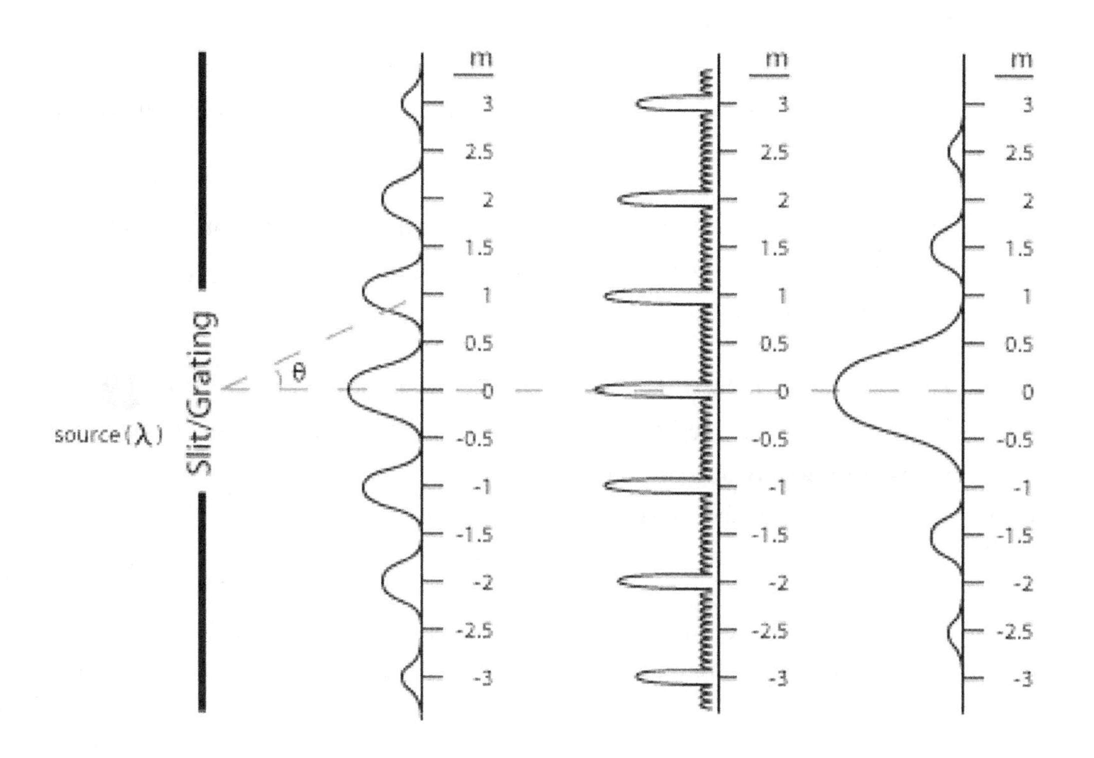

Double slit: d sin θ = mλ (shown on the left); diffraction grating: *d sin θ = mλ* (shown in the middle); single slit: *a sin θ = mλ* (shown on the right)

Thin Films

When observing an oil slick on water or the ground, various colors can be seen swirling on top of the slick. This phenomenon is *thin-film interference* because the observed colors are not the color of the oil, but rather an effect of light interference reflecting off the slick.

The figure illustrates thin-film interference. The interference of the light (both destructive and constructive) occurs because the incoming light partially reflects off the oil surface and partially transmits through the oil.

The transmitted light is then reflected off the substance below the oil (water, in this case), and the observer sees both the oil-reflected light and the water-reflected light. However, because the two reflected light rays traveled slightly different distances, according to the material they reflected off, they have a path difference and therefore interfere.

The most visible swirling colors are seen on top of the oil slick are a result of constructive interference, and the dim colors (or those not visible) are a result of destructive interference.

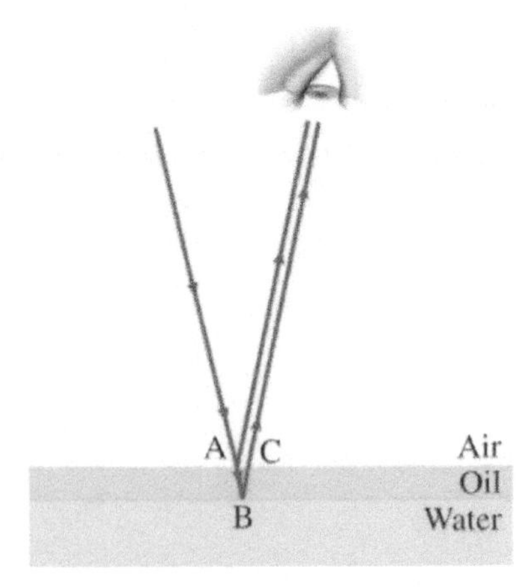

The wavelengths of light that will be displayed is found via the equation:

$$(2)\cdot(n)\cdot(d) = m\lambda$$

where n is the index of refraction for the thin film, d is the thickness of the thin film (m), λ is the wavelength of the light source (m), and m is the order value dependent upon the material underneath the thin film.

Let $n_{thin\ film}$ be the index of refraction of the thin-film material, and n_{under} be the index of refraction of the material underneath the thin-film.

If $n_{thin\ film} > n_{under}$ then:

Destructive interference: $m = 1,\ 2, 3, \ldots$

Constructive interference: $m = \frac{1}{2}, \frac{3}{2}, \frac{5}{2} \ldots$

If $n_{thin\ film} < n_{under}$ then:

Destructive interference: $m = \frac{1}{2}, \frac{3}{2}, \frac{5}{2} \ldots$

Constructive interference: $m = 1, 2, 3, \ldots$

Other Diffraction Phenomena, X-Ray Diffraction

Another exciting form of diffraction is X-ray diffraction. Although X-rays are not visible light, they are waves that can produce interference patterns caused by phase differences. This is particularly useful when investigating the structure of crystal lattices and molecular arrangements. X-rays have incredibly small wavelengths of 10 nanometers or 1.0×10^{-8} meters. Thus X-rays can diffract off these objects at the molecular level, revealing the pattern of the investigated structure.

The illustration below is an X-ray crystal diffraction experiment. When X-rays are shone upon a crystal structure, the wavelengths are small enough such that the waves reflect off individual atoms within the crystal structure. If two or more waves reflect off the crystal surface, they interfere due to their path length difference. The resulting diffraction pattern gives information into the crystal structure of the specimen being examined.

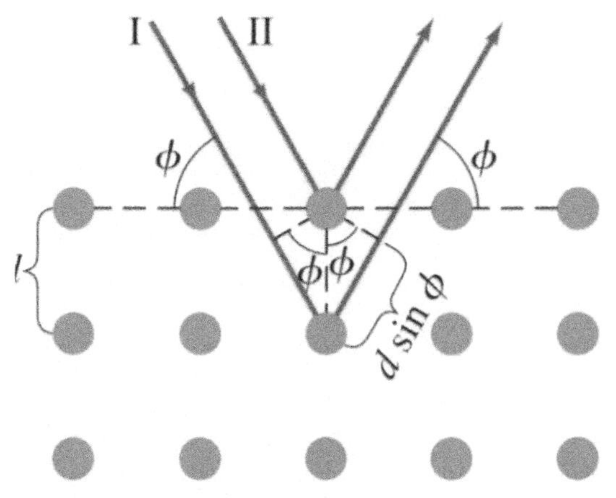

The pattern of maxima produced by X-ray diffraction is calculated via Bragg's Law (1913), which is expressed as:

$$n\lambda = 2d \sin \phi$$

where n is the order number (1, 2, 3, …), λ is the wavelength of the X-ray (m), d is the lattice spacing of the crystal (m), and ϕ is the reflection angle (degrees).

Polarization of Light

Light from the sun and most sources are un-polarized. *Polarization* refers to the planes in which the electric field and magnetic field oscillate. When light is unpolarized, as it is from the sun, all the light waves oscillate in random directions.

If the light were to become polarized, then the oscillation of the electric and magnetic field would be uniform amongst all the light waves. In the diagram below, light emitted from the sun is unpolarized, and its electric field oscillates in many directions. The arrows of the unpolarized light indicate this. When the light strikes the surface of the water, it reflects and becomes horizontally polarized. The horizontally polarized light only has its electric field oscillating in one direction, rather than all directions.

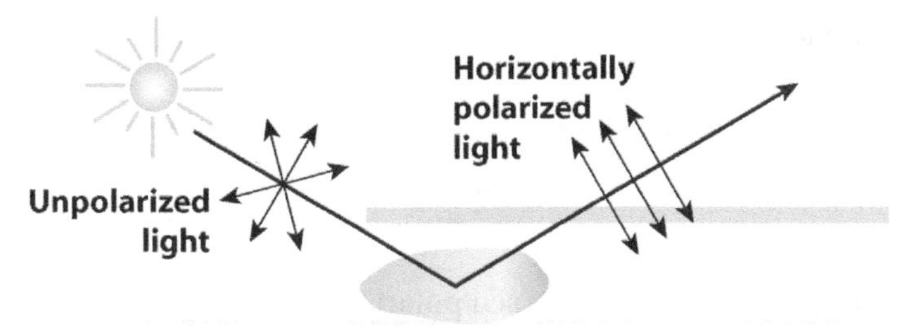

Polarized light and materials which polarize light have many applications.

In the diagram on the left below, the rope represents a vertically polarized wave, which can be transmitted through a vertically polarized filter.

In the diagram on the right below, the wave is horizontally polarized and cannot pass through the vertical polarizer. This is how polarized glasses remove glare from surfaces such as water.

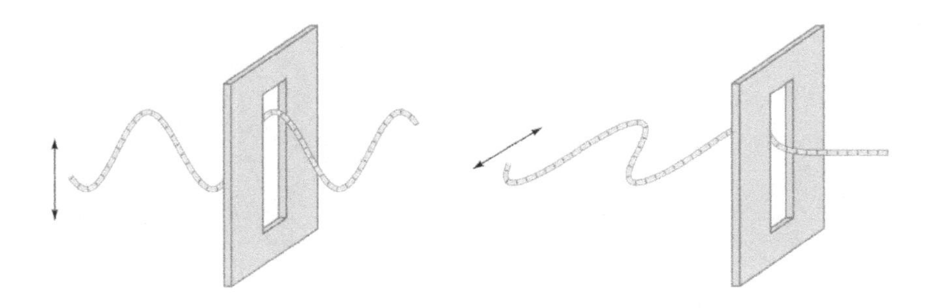

The rope represents polarized light. The rope (or light) passes through the slit on the left when aligned with the direction of oscillation. The rope (or light) does not pass when out of phase with the slit.

Light reflected off the water (i.e., glare) is usually horizontally polarized and can be eliminated by a vertical polarizer. Polarized sunglasses allow the wearer to see through water because they have a vertical polarizer, and therefore reduce or eliminate the horizontally polarized light reflecting off the water.

The intensity of light after passing through a polarizer filter is calculated by:

$$I = I_0 \cos^2(\theta)$$

where I_0 is the initial intensity of the light (W/m^2), I is the intensity of the light after passing through the polarizer, and θ is the angle of the original light wave electric field concerning the axis of the polarizer, the transmission axis (degrees).

When unpolarized light is incident upon any polarizing filter, the intensity of the light is reduced by half:

$$I = \frac{1}{2} I_0$$

If the light is incident upon crossed polarizers (90° angle between transmission axis), no light passes.

In the diagram below, the unpolarized light is initially polarized such that its electric field only oscillates vertically after it passes through a filter with a vertical transmission axis.

Then, it is passed through a polarizer with a horizontal transmission axis. As a result, no light is transmitted because none of the light's electric field oscillates in the horizontal axis.

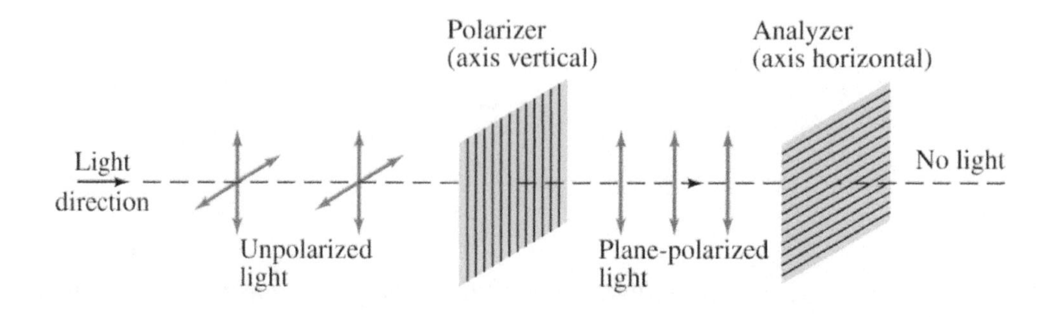

Doppler Effect

All electromagnetic radiation (light, in this case), can undergo the *Doppler effect*.

The Doppler effect for light is commonly used in astronomy. When observing stars, galaxies, and other interstellar bodies, astronomers noted that sometimes the observed colors of these bodies shifted regarding different periods of the year. This observation is because of the Doppler effect.

The Earth moves towards and away from these stellar bodies during the year. The color shifts as a result of the Earth's speed about that of the observed body.

Two types of shifts are observed: the redshift and the blueshift.

A *redshift* occurs when the source of light and the observer are *moving away* from each other. For the Doppler effect for sound, the frequency of the observed wave decreases.

Since lower frequencies of visible light are red, these sources appear to be redder than they are because the Doppler effect shifts the source frequency to a lower frequency for the observer.

Blueshift occurs when the source of light and the observer are *moving towards* each other. The observer sees a higher frequency than the source light itself.

Blues and violets are the higher frequency colors in the visible color spectrum. Thus, the sources of light appear bluer than they are.

Reflection from Plane Surface

When light strikes various surfaces, it can be absorbed, transmitted, or reflected. Mirrors are the most identifiable examples of surfaces that reflect all incoming light (ideal mirror). When a person looks in a mirror, she sees her image because the light incident upon the mirror is reflected. Many other surfaces also provide reflection, such as shiny metal or calm water.

However, these images are usually not nearly as clear as the mirror and tend to be somewhat blurry. All these effects, from the clear image presented in the mirror to the blurry image provided by other surfaces, are due to the law of reflection.

The law of reflection states that the angle of incidence of a light ray, with respect to the normal of the surface, will be equal to the angle of reflection.

When this occurs, a *specular reflection* is observed in flat mirrors.

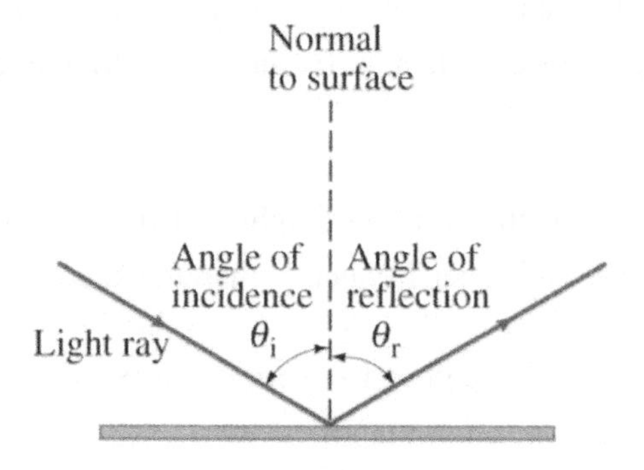

The angle of incidence θi equals the angle of reflection θr

Blurry images are also formed according to the law of reflection. When the reflecting surface is irregular rather than smooth, the light rays reflect off the irregularities and the image is not clear but appears blurry to the viewer.

However, for irregular surfaces, the reflected rays will not be parallel, but rather in every direction. This is a *diffuse reflection* and results in a blurry image being formed.

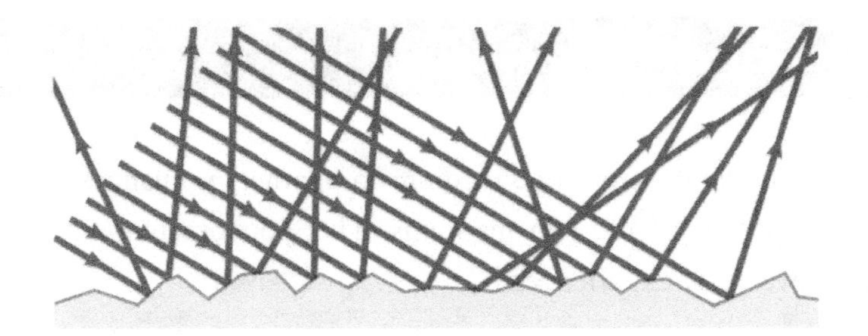

With diffuse reflection from irregular surfaces, the light is reflected at all angles.

With specular reflection (from a mirror), the eye must be in the correct position.

Visualize a person positioning themselves in the mirror to see something behind them—where they place themselves determines what they can see.

The top diagram below demonstrates *diffuse reflection*, in which the observer can see the reflected light from the source in multiple locations.

The bottom diagram below demonstrates *specular reflection*, in which the reflected light can only be seen in a specific location.

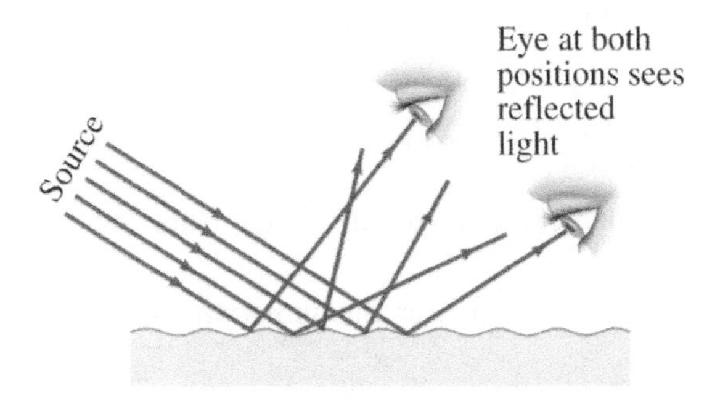

Diffuse reflection produces a clear image

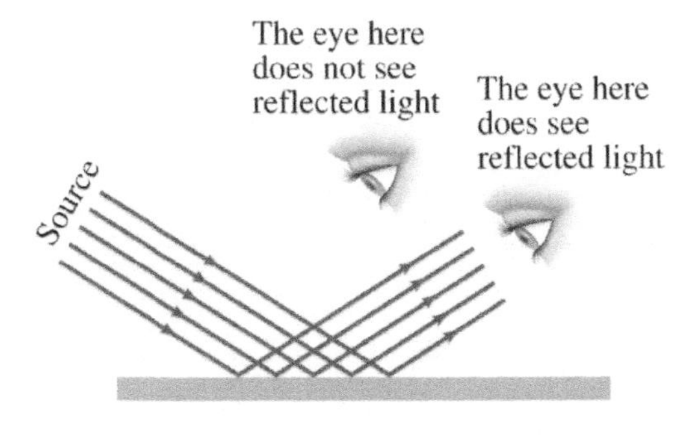

Specular reflection produces a blurry image

Refraction, Refractive Index *n*, Snell's Law

When light transmits from one medium to another, it changes direction slightly based upon the properties of the two mediums. This phenomenon is *refraction* and is observed in many everyday experiences.

The angle that the light ray bends, with respect to the normal, depends upon the indices of refraction (n) for each medium.

Snell's Law gives the angle of refraction:

$$n_1 \sin(\theta_1) = n_2 \sin(\theta_2)$$

where n is the index of their respective refraction for each medium, and θ is the angle of the wave with respect to the normal of the medium interface.

The index of refraction n is a relation of the light propagation speed v in a medium to that of the speed of light c in a vacuum. The value of n is always greater than 1 because light cannot travel faster than in a vacuum.

$$n = \frac{c}{v}$$

The figures below depict the refraction of light from water to air and vice versa.

When a wave moves to a denser medium (with a higher refractive index), it bends toward the normal.

When a wave moves to a less dense medium (with a smaller refractive index), it bends away from the normal.

In the diagram on the left below, if the light is refracted from air to water, it bends *toward* the normal ($n_2 > n_1$).

In the diagram on the right below, if the light is refracted from water into the air, it is refracted *away from* the normal ($n_1 > n_2$).

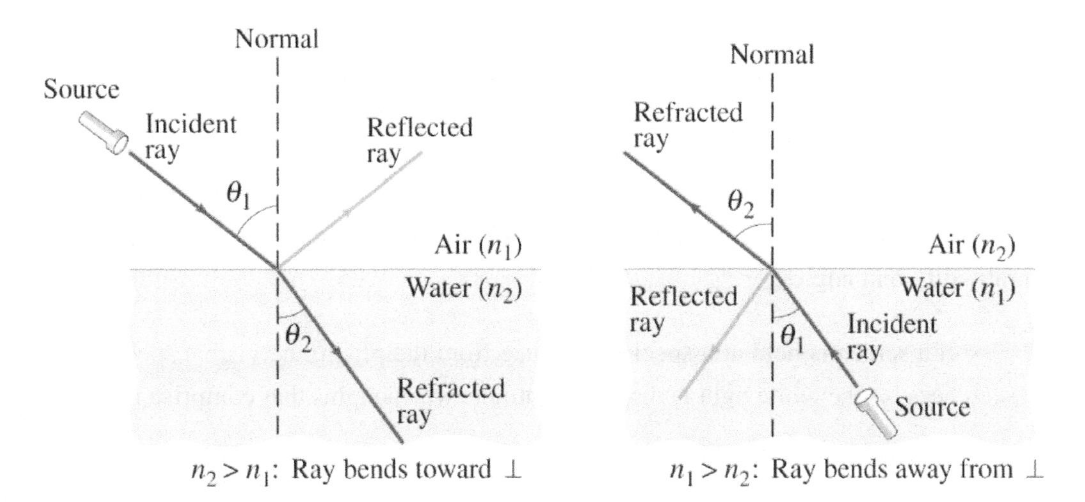

Given the angle of incidence, the angle of refraction is computed by Snell's Law.

$$n_1 \sin(\theta_1) = n_2 \sin(\theta_2)$$

Mirages are also produced by refraction. Hot surfaces warm up the air above them, producing air with a slightly lower index of refraction than the ambient air.

When a light ray passes through this air, it curves the light such that the observer sees an image that is a form of reflection off the ground.

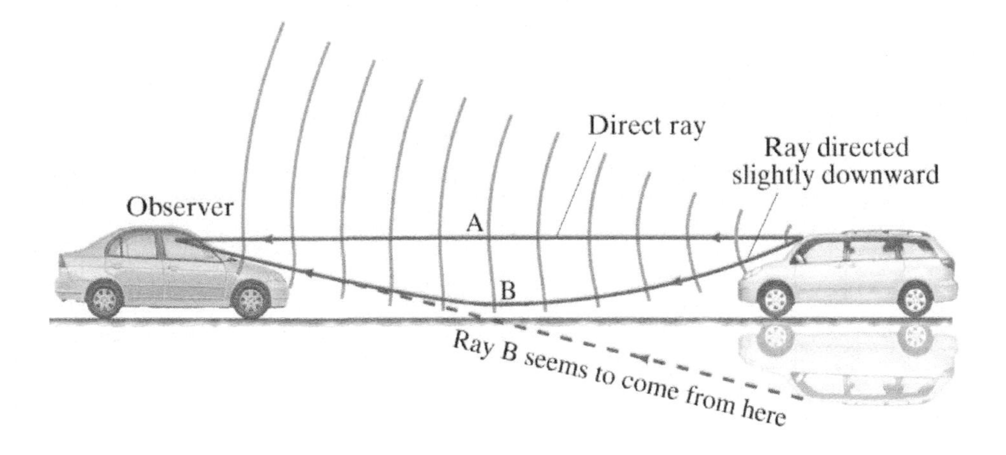

Dispersion

Dispersion is the phenomenon that separates light into its colors. All wavelengths of light refract upon crossing a boundary between two different mediums. However, different wavelengths of light refract at slightly different angles.

Lower wavelength light refracts the most. Thus, violet light refracts at a greater angle than blue light, which will refract at a greater angle than green light.

An example of dispersion is when white light passes through a prism. The diagram below represents a prism producing the visible spectrum of colors. White light is composed of the entire visible spectrum, in which each wavelength (color) refracts at slightly different angles.

If a screen is held at a specific distance from the prism, the visible spectrum can be seen because the white light is dispersed into the wavelengths that comprise it.

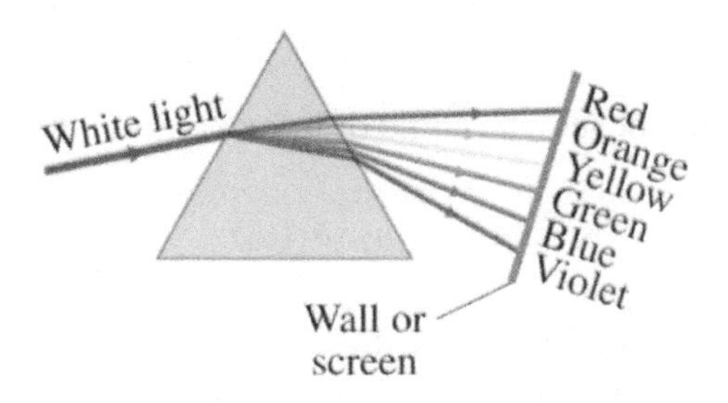

White light is dispersed by a prism into component colors based on wavelength

Atmospheric rainbows are created by dispersion. Light passing through small raindrops or mist refracts and disperses into the visible spectrum, like for a prism.

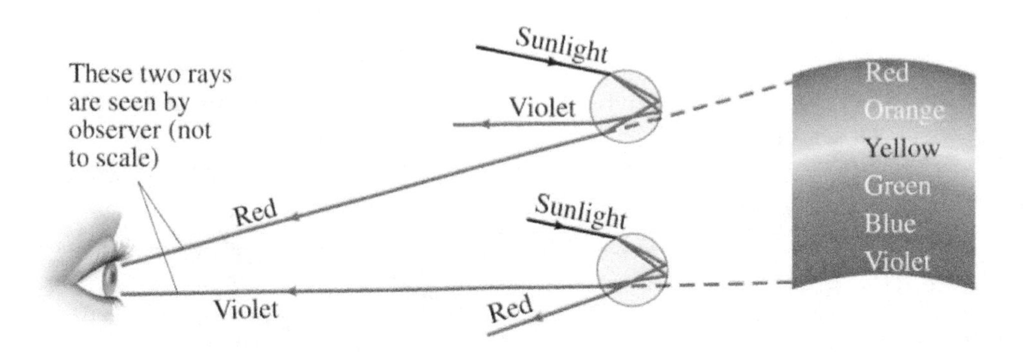

Conditions for Total Internal Reflection

The critical angle is defined as the angle of incidence for which an incident light ray makes an angle of refraction equal to 90°. The critical angle can be found by:

$$n_1 \sin(\theta_c) = n_2 \sin(90°)$$

The *total internal reflection* occurs when the angle of incidence is larger than the critical angle. No light is transmitted between the two mediums, and the light is only reflected in the original medium.

The diagram below illustrates various angles of incidence and their corresponding angles of diffraction. At points, I and J, the angle of incidence is lower than the critical angle, and the light is refracted through the surface (a portion is also partially reflected).

At point K, the critical angle is achieved, and the angle of refraction is exactly 90°.

At point L, the angle of incidence is higher than the critical angle. Therefore, all the light is reflected, resulting in total internal reflection.

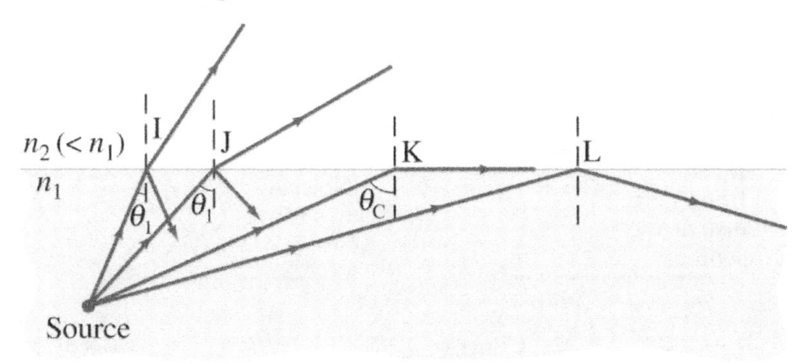

Total internal reflection is an important phenomenon because it is the principle behind the operation of fiber optics. In a fiber optic cable, light is transmitted along the fiber regardless of loops or bends in the cable, because the indices of refraction of the cable and outer sheath are such that the light always experiences total internal reflection.

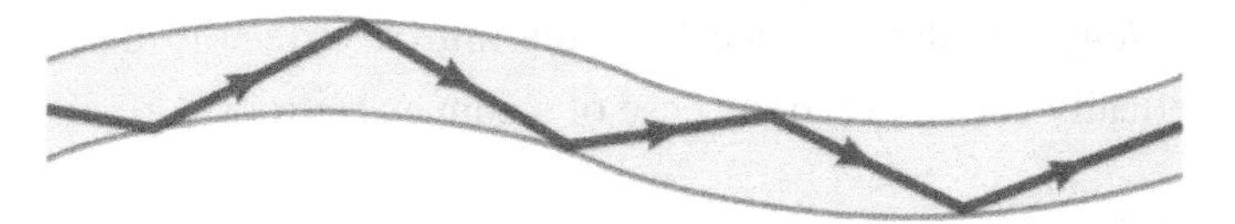

Schematic of total internal reflection for a fiber optic cable

Mirrors

Spherical mirrors allow users to see different images than what is possible with only flat mirrors.

Many parking garages have convex mirrors mounted near blind corners; they allow drivers to see a much wider image of what is around a corner than a flat mirror.

Similarly, many beauty mirrors are slightly concave and present a magnified image to the user. Both examples are types of spherical mirrors, which have distinctive properties depending on the curvature of the reflective surface.

Mirror curvature, radius, focal length

Mirror curvature can be concave or convex.

The *convex mirror* disperses light rays such that they are diverging.

The *concave mirror* focuses light rays such that they converge.

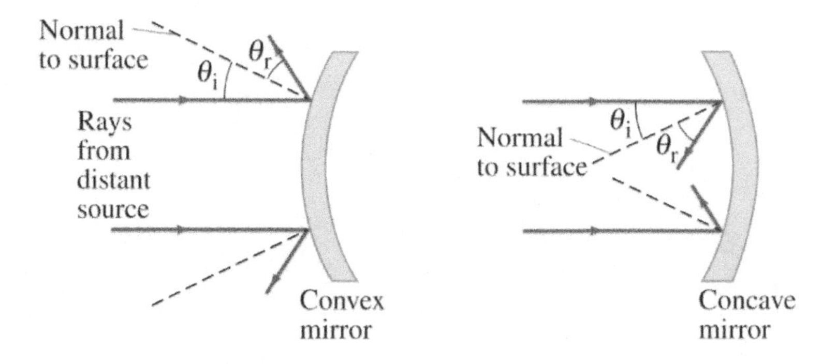

Both convex and concave spherical mirrors have defined radii of curvature and foci. The *radius of curvature* is the radius of the curve of the mirror; in spherical mirrors, the radius of curvature is constant throughout the surface of the mirror.

The *focal length* is the distance from the mirror surface to the where light rays converge. Theoretically converge, in the case of a convex mirror.

Concave mirrors have a *positive focal length.*

Convex mirrors have a *negative focal length.*

Both concave and convex mirrors have focal lengths equal to half the radius:

$$f = \frac{1}{2}R$$

where f is the focal length (m), and R is the radius of curvature (m).

The figure below displays the various features of a concave mirror. Point C is the *center of curvature* and is the point where the radius of curvature originates (center of a circle). Point F is the focus and demonstrates where all light rays focus if reflected off the mirror.

The distances f and r are the focal length and radius of curvature, which is the distances to points F and C, respectively. A convex mirror would have the same center of curvature, focal point, the radius of curvature and focal length; however, the light would be incident upon the outer edge of the mirror and disperse, rather than converge.

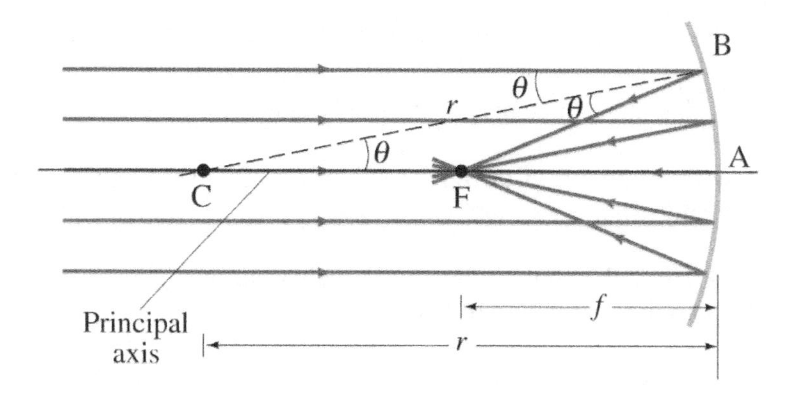

Mirror ray tracing

When looking at a mirror, it is essential to know the relationship of the object held up to the mirror and the image formed by the mirror. Depending upon the object distance from the mirror, the mirror, and the characteristics of the mirror, the formed image can be smaller, larger, virtual, real, erect, or inverted.

The image formed by a mirror is classified by size in relation to the original object, whether the image is inverted or not, and if the formed image is real or virtual.

The size of the image formed by a mirror can either be equal to, larger or smaller than the size of the object held up to the mirror.

Additionally, the image may be erect like the object held up to the mirror, or the image may be inverted and appear to be upside down.

For example, a person's face reflected on the inside of a shiny spoon appears inverted and smaller than the face.

Finally, the formed image may be real or virtual. Real images are images that are formed when light rays converge upon a point in a real location, whereas virtual images form from light rays that diverge.

A virtual image will be at the point where the light rays appear to diverge from (this point is not there, hence "virtual image").

Identify the image formed by a mirror by using *ray diagrams*.

Ray diagrams are tools that allow the user to visualize the image formed, depending upon the object location with respect to the mirror.

When drawing ray diagrams, three rays are typically drawn:

1. A ray from the top of the object height parallel to the axis—after reflection off the mirror, it passes through the focal point.

2. A ray from the top of the object height through the focal point—after reflection, it is parallel to the axis.

3. A ray perpendicular to the mirror passes through the object height and the center of curvature.

The figure below uses a ray diagram drawn for a concave mirror.

Diagrams below show the paths of rays 1, 2, and 3, respectively.

After the rays have been drawn, the convergence point is the location of the image.

The image height and orientation are then determined by drawing a line from the center axis to the convergence point.

The image formed by the mirror is inverted, larger, and real (the image is formed by the convergence of light rays on a real point).

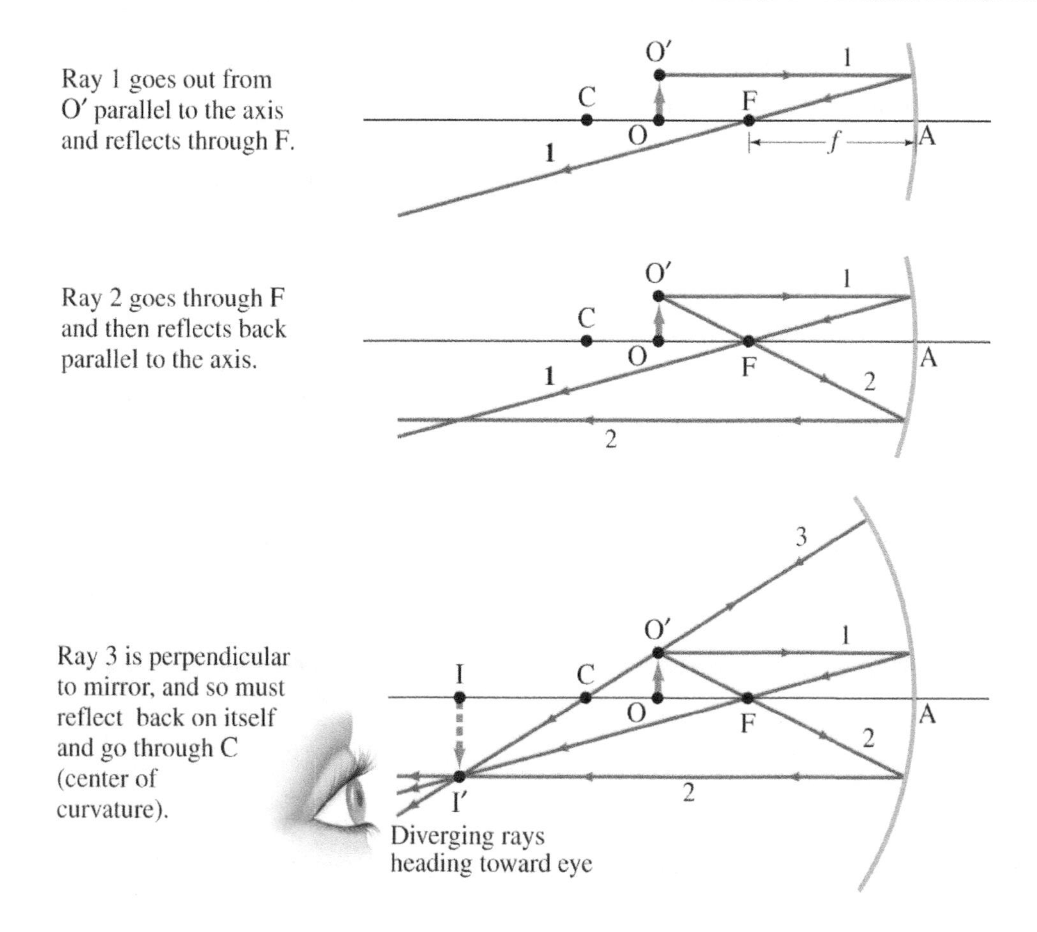

The location of the object determines the image. The illustration below depicts the image formed by an object located *in front of* the focal point of a concave mirror.

The formed image is erect, larger, and virtual (the image is formed by the convergence of light rays on a virtual point)

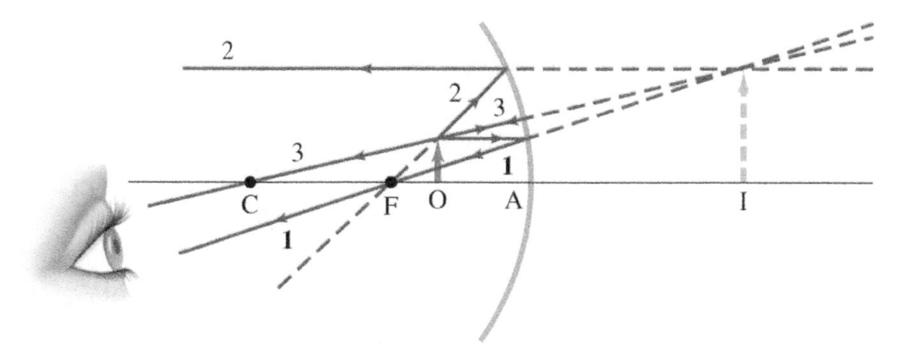

A concave mirror produces a larger, erect, virtual image when the object is placed within the focal point.

Concave mirrors can produce both real and virtual images, depending on where the object is located at in relation to the mirror.

If the object is outside the radius of the sphere of the mirror, the image is real.

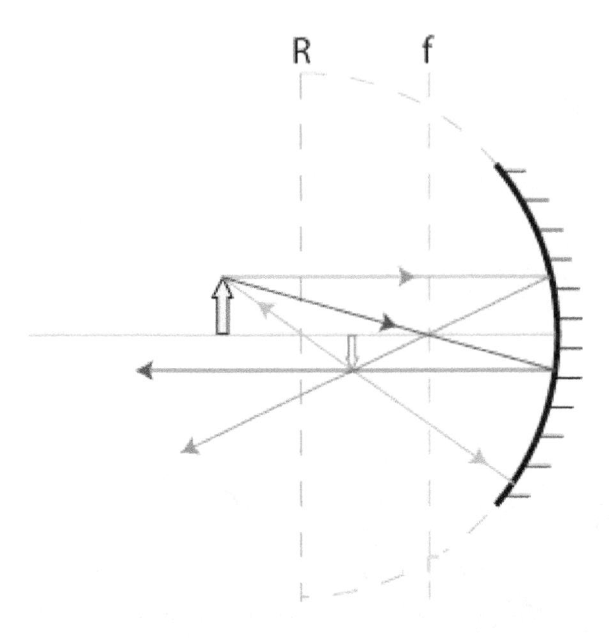

A concave mirror produces a real image if the object is beyond the radius of curvature

If the object is inside the radius of the sphere of the mirror, the image is virtual.

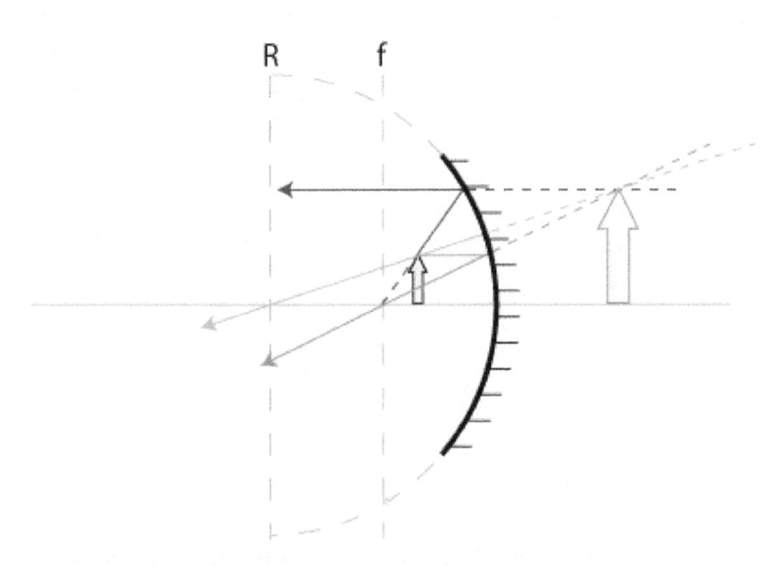

A concave mirror produces a virtual image if the object is inside the radius of curvature

For convex mirrors, only a virtual image is created because the curve of the mirrors is different.

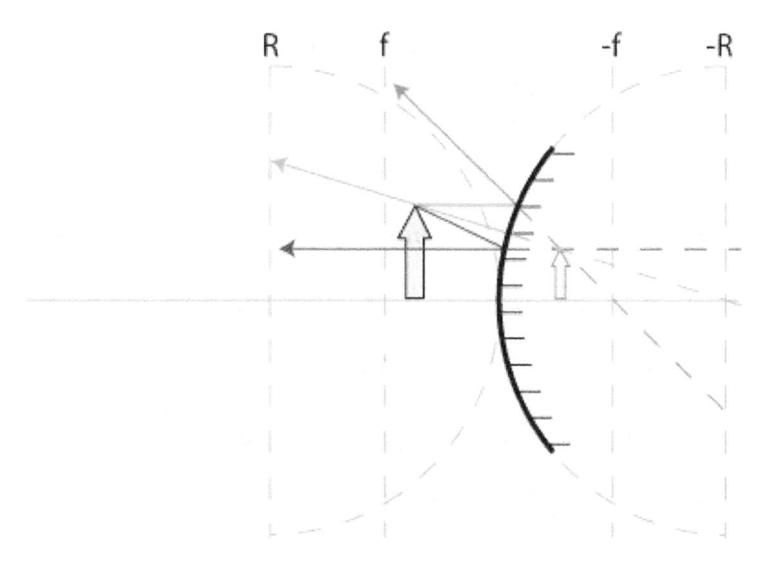

A convex mirror produces a virtual image

For mirrors:

1. First, draw a parallel line from the object; as it bounces off the mirror, it intersects the focal point.

   Now, which focal point to intersect? The left or right?

   For concave mirrors, it focuses the ray to the left focal point.

   For convex mirrors, which cannot focus, it diverges the ray; an extrapolation to the right focal point is needed.

2. Draw a line that intersects the R point on the principal axis (left, or right?). The ray drawn should bounce back its original path, and not be reflected elsewhere. By analysis of the mirror, this is decided.

   Now that two rays are drawn, an intersection can be made. Use this intersection as a guide to draw the last ray.

3. The last ray should first intersect the focal point, then bounce off the mirror parallel to the principal axis.

Which focal point to intersect? Is extrapolation necessary? There is only one combination for the ray to fit the intersection already made by the previous two rays.

Important, draw the parallel line first and force it to cross the intersection already made by the previous two rays.

The object-image relationship between concave and convex mirrors

Object Distance	Concave Mirror	Convex Mirror
Before C	inverted, smaller, real	erect, smaller, virtual
At C	inverted, equal size, real	erect, smaller, virtual
Between C and F	inverted, larger, real	erect, smaller, virtual
At F	no image	erect, smaller, virtual
Before F	erect, larger, virtual	erect, smaller, virtual

F represents the focal point, and C is the radius of curvature

Thin Lenses

A lens is a device used to bend and focus light using the refractive properties of the lens material.

The lenses in a microscope bend and focus light such that objects become magnified. Eyeglasses correct vision problems by focusing the light in front of eyes so that those who wear glasses can see without blurry vision. Regardless of the application, lenses are an important tool in many devices and scientific instruments.

Many similarities exist between lenses and mirrors regarding image formation; however, there are a few differences to note.

A convex lens is like a concave mirror (both are referred to as converging) except for the following issues:

- Real images are on the opposite side of the lens as the object because light travels through the lens and focus on a screen behind the lens.
- Virtual images are on the same side of the lens as the object because light cannot focus in front of a lens and be cast on a screen.
- Concave lenses are the same as convex mirrors (both are referred to as diverging) except for in the instance:
- Virtual images formed by the lens are on the same side of the lens as the object; light cannot focus in front of a lens and be cast on a screen.

Converging and diverging lenses, focal length

Most lens problems approximate the lens as a *thin lens*. Thin lenses are those whose thickness is small compared to their radius of curvature.

A thin lens may be either converging (convex) or diverging (concave).

The figure below depicts a converging lens.

Like with mirrors, the light which passes through the lens converges on the focal point of the lens (two focal points, one on each side).

The radius of curvature is the same as that of a spherical mirror, except the lens is double-sided, and it has two radii of curvature (most of the time, these are equal).

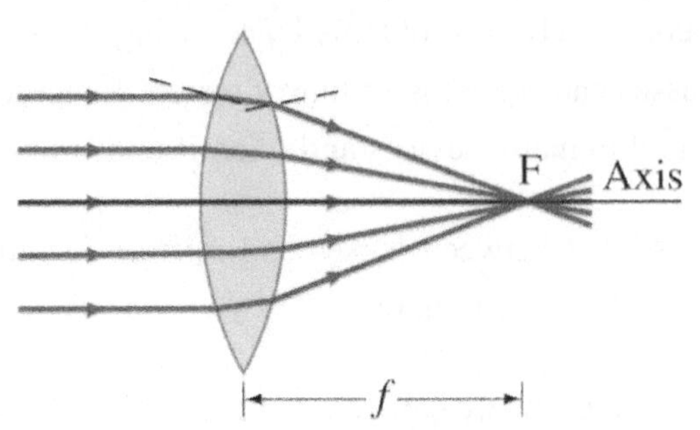

A diverging lens (concave) makes parallel light diverge; the focal point is the point where the diverging rays would converge if projected back.

Like the converging lens, the diverging lens has two radii of curvature (most times, these are equal).

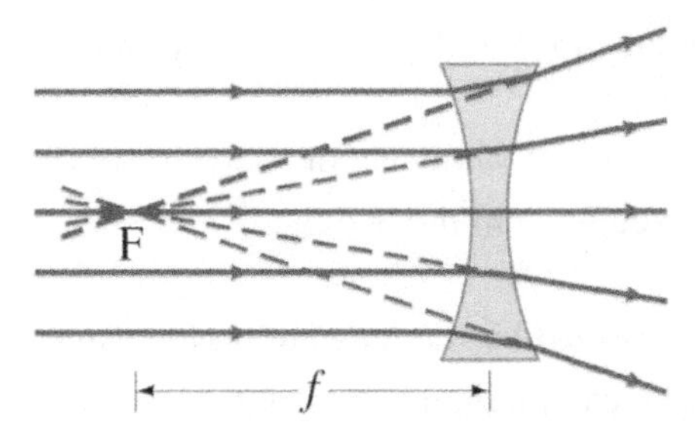

Ray tracing thin lens

Ray tracing for thin lenses is like that for mirrors. There are three key rays:

1. A ray from the top of the object height parallel to the axis—after transmission through the lens, it passes through the focal point.

2. A ray from the top of the object height through the focal point—after transmission, it is parallel to the axis.

3. A ray from the top of the object height through the center of the lens.

For example, the figure below demonstrates ray tracing through a converging lens.

Diagrams below show the paths of rays 1, 2, and 3, respectively. Notice that after the rays have been drawn in, the convergence point is the location of the image.

The image height and orientation are determined by drawing a line from the center axis to the convergence point.

In this case, the image formed by the lens is inverted and real.

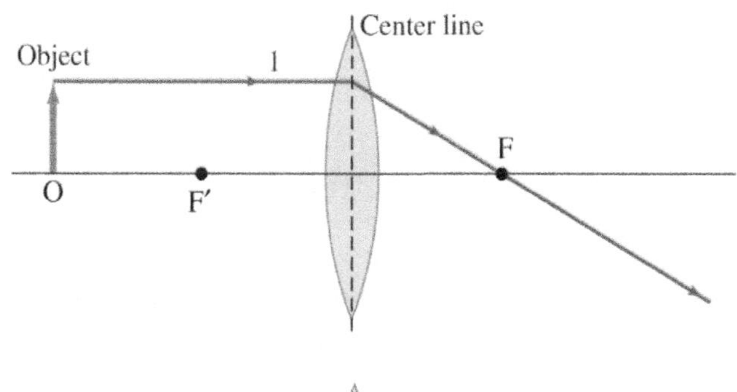

Ray 1 leaves one point on object going parallel to the axis, then refracts through focal point behind the lens.

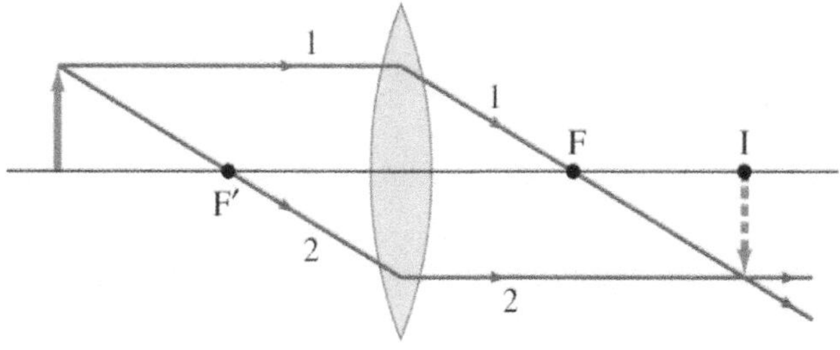

Ray 2 passes through F′ in front of the lens; therefore it is parallel to the axis behind the lens.

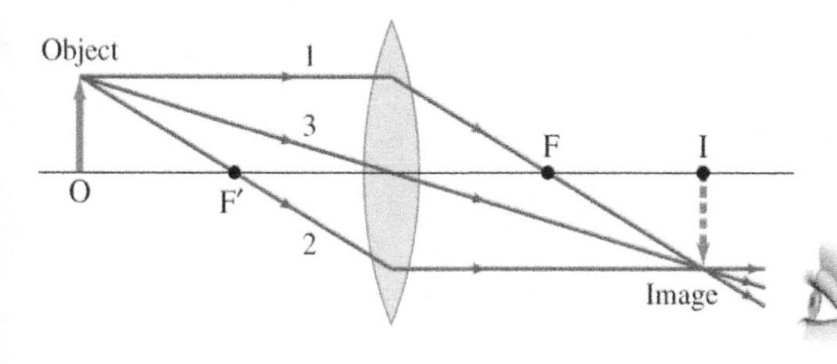

Ray 3 passes straight through the center of the lens (assumed very thin).

For a diverging lens, the same three rays can be used; the image is upright and virtual.

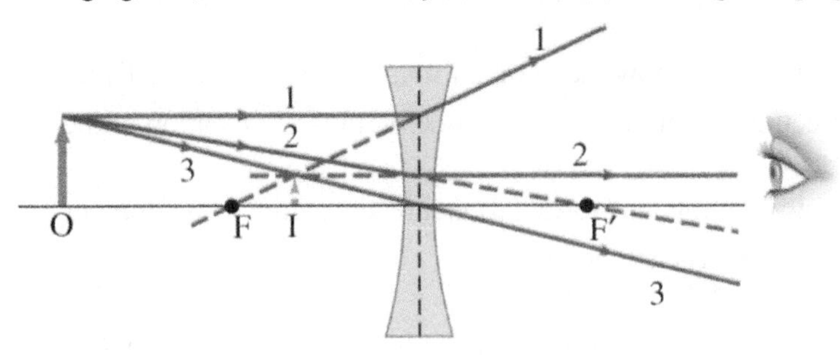

As stated previously, lenses are similar to mirrors.

A convex lens (like a concave mirror) can create either a real or a virtual image, depending on where the real object is positioned.

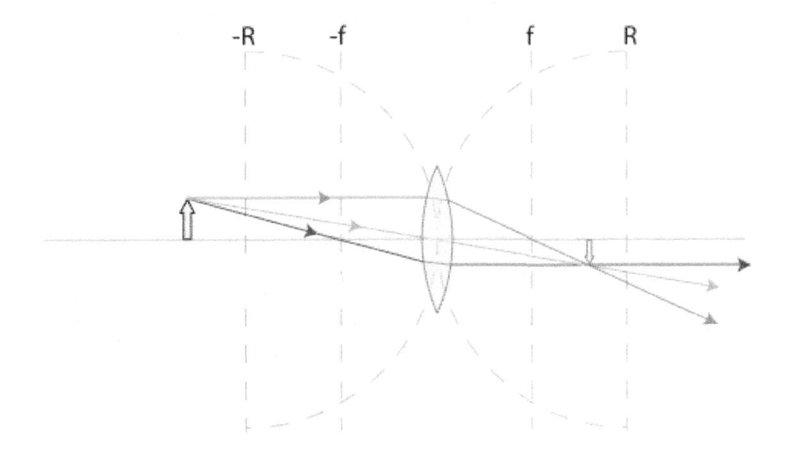

A convex lens produces real images if the object is beyond the focal length

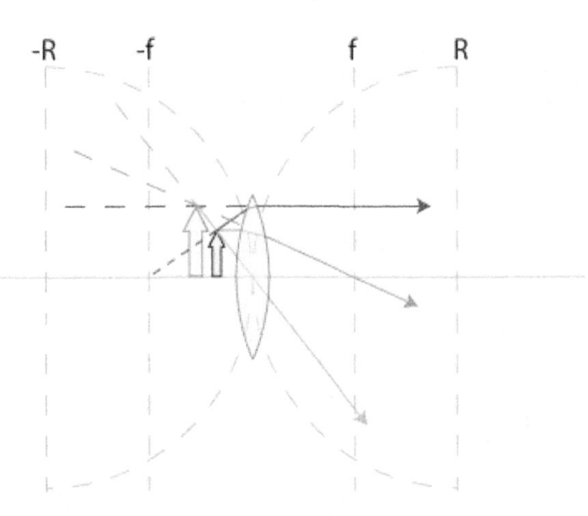

A convex lens produces a virtual image if the object is within the focal length

A concave lens acts similar to a convex mirror, and only produce virtual images.

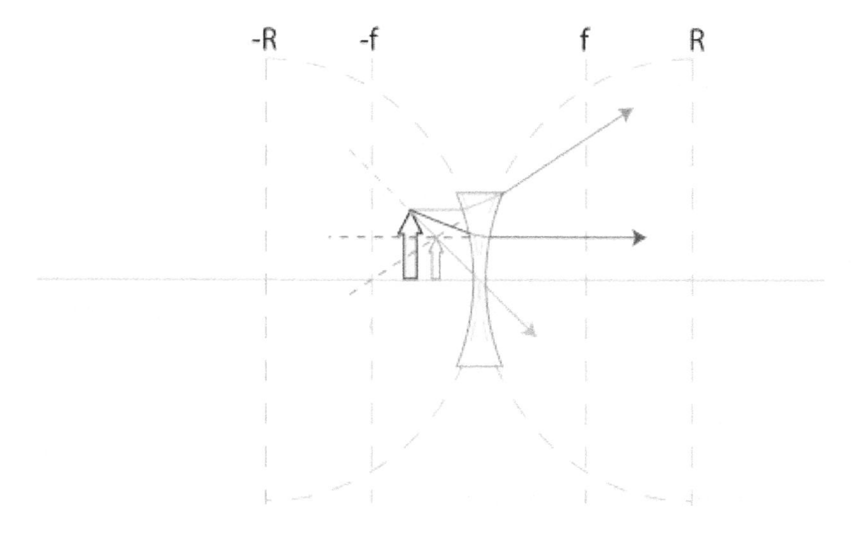

A concave lens produces a virtual image if the object is within the focal length

For lenses, use the method like drawing rays for mirrors:

1. Draw the parallel → focal point ray. It should make sense which focal point the ray should hit or extrapolate, given the converging or diverging nature of the lens.

2. Draw a ray intersecting the center of the lens.

3. Using the intersection already made by the previous two rays as a guide, draw the focal point → parallel ray.

   Draw the parallel line first and force it to pass through the intersection already made by the previous two rays.

The formula (1/p) + (1/q) = 1/f with sign conventions

To determine the distance from either the mirror or lens that an image forms, the thin lens equation can be used:

$$\frac{1}{d_o} + \frac{1}{d_i} = \frac{1}{f}$$

where d_o is the distance of the object away from the lens or mirror (m), d_i is the distance of the image away from the lens or mirror (m), and f is the focal length of the lens or mirror (m).

Although the thin lens equation is relatively simple, the proper sign conventions must be followed for the types of mirrors and lenses.

For a lens, several differences are depending on whether the lens is converging or diverging. The focal length is written positive for converging lenses and negative for diverging.

The object distance is considered positive when the object is on the same side as the light entering the lens; otherwise, it is negative. Finally, the image distance is positive if the image is on the opposite side from the light entering the lens; otherwise, it is negative.

For mirrors, the sign conventions are different. The focal length is written positive for concave mirrors and negative for convex mirrors.

The object distance is positive when the object is in front of the mirror, and negative if the object is behind the mirror (i.e., virtual object, this is a rare case).

The image distance is positive if the image is in front of the mirror (real image), and negative if the image is behind the mirror (virtual image).

When solving problems using the thin lens equation, be sure to perform as many steps as possible.

Draw a ray diagram and place the image where the principal rays intersect.

Solve for unknowns by following the sign conventions and check that the answers are consistent with the ray diagram.

Magnification and lens power formula

The magnification of an object, in relation to image height, can also be obtained for lenses and mirrors. The *magnification equation* is expressed as:

$$M = \frac{h_i}{h_0} = -\frac{d_i}{d_0}$$

where M is magnification, h_i is the height of the image (m), h_0 is the height of the object (m), d_o is the distance of the object away from the lens or mirror (m), and d_i is the distance of the image away from the lens or mirror (m).

Like the lens equation, the magnification equation has specific sign conventions for the type of mirrors and lenses. The magnification is expressed as positive for all types of lenses when the image is upright with respect to the object, and magnification is expressed as negative when the image is inverted.

Magnification is expressed as positive for all types of mirrors when the image is upright with respect to the object, and negative when the image is inverted.

The *power of a lens* is not related to its magnification but rather is expressed as the inverse of its focal length, where lens power has units of diopters (D), which are equal to m^{-1}.

The power of a lens is positive if it is converging, and negative if it is diverging:

$$P = \frac{1}{f}$$

where P is the lens power (D).

Lensmaker's equation

The Lensmaker's equation relates the radii of curvature of a lens and the produced focal length:

$$\frac{1}{f} = (n-1)\cdot\left(\frac{1}{R_1} + \frac{1}{R_2}\right)$$

where f is the focal length of the lens (m), n is the index of refraction of the lens material, R_1 is the radius of curvature of the first lens face (m), and R_2 is the radius of curvature of the second lens face (m).

Combination of Lenses

Combinations of lenses are used to multiply the magnification of objects.

This is demonstrated in light microscopes and telescopes. These devices contain multiple lenses to allow the user to magnify objects beyond what a single lens can produce. This is possible because the real image formed by a lens can be used as the object for another lens.

The magnification by multiple lenses is the product of all the individual magnifications and is expressed as:

$$M_{total} = M_1 M_2$$

where M_{total} is the total magnification, M_1 is the magnification of the first lens, and M_2 is the magnification of the second lens.

The simplest form of a combination of lenses is when two lenses are in contact.

In this case, the total power of the combination of lenses is the sum of the two lens's powers:

$$P_{total} = P_1 + P_2$$

or

$$\frac{1}{f_{total}} = \frac{1}{f_1} + \frac{1}{f_2}$$

If some distance separates the two lenses, then the total power of the combination is expressed as:

$$\frac{1}{f_{total}} = \frac{1}{f_1} + \frac{1}{f_2} - \frac{d}{f_1 f_2}$$

where d is the distance between the two lenses (m)

Lens Aberration

In real life, no lens is ideal. The light rays tend to converge in slightly different spots. This phenomenon is an *aberration*, and there are several types.

Spherical lenses and mirrors often suffer from *spherical aberration*. This occurs because the curvature of the lens/mirror is often large, and the rays striking the outer edges refract/reflect more and therefore bend more than the rays striking the center portion of the lens/mirror. This can be seen in the converging lens figure shown below.

Spherical aberration is reduced by utilizing only the center portion of the lens or mirror.

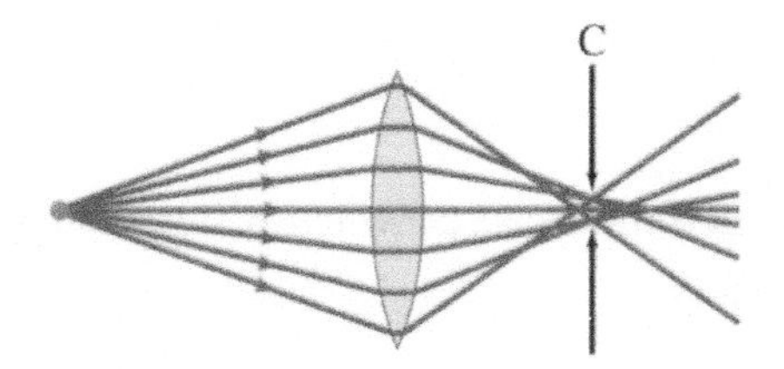

Spherical aberrations may be reduced by using a mirror or lens

Chromatic aberration occurs in lenses and is when the light of different wavelengths focuses at different points.

Since different wavelengths of light refract at slightly different angles, this is the phenomenon of dispersion.

When light shines through a lens, this results in different focus points, depending upon the wavelength of light. For instance, observe the figure below.

Blue light is refracted more than red light so that blue light focuses at an earlier point than the red light.

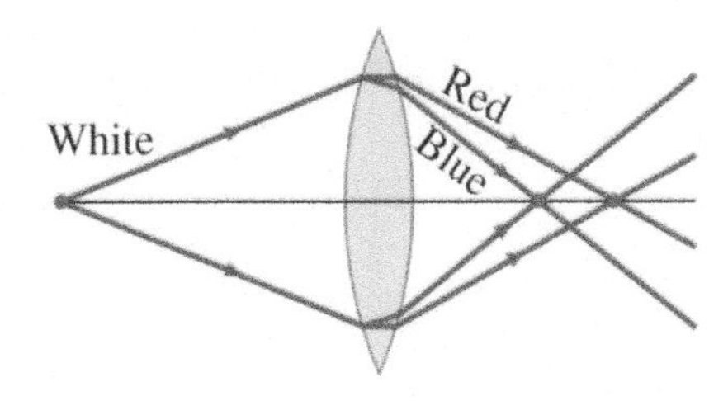

Chromatic aberrations focus light of different wavelength to produce colors

A solution for chromatic aberration is a device called an *achromatic doublet.*

A second lens made of different materials is attached to the first lens.

When light passes through the chromatic aberration, it is corrected in the second lens, which focuses all the wavelengths to one point.

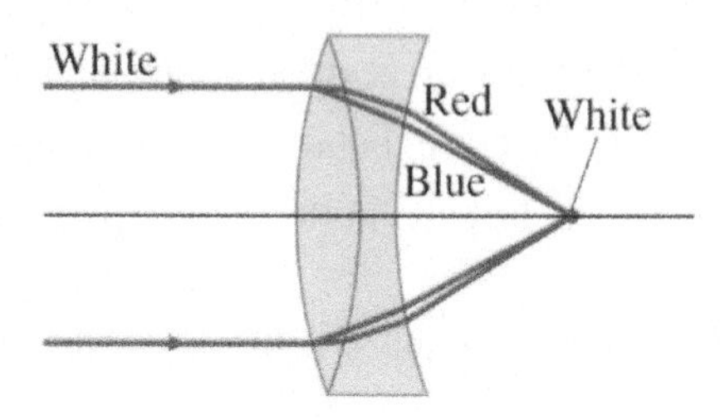

Chromatic aberrations are corrected for with a second lens

Distortion is an aberration caused when different parts of a lens have slightly different magnification factors.

The two diagrams below illustrate two types of distortion.

The diagram below on the left represents *barrel distortion*, where the magnification decreases with distance from the axis.

The diagram below on the right represents *pincushion distortion*, where magnification increases with distance from the optical axis.

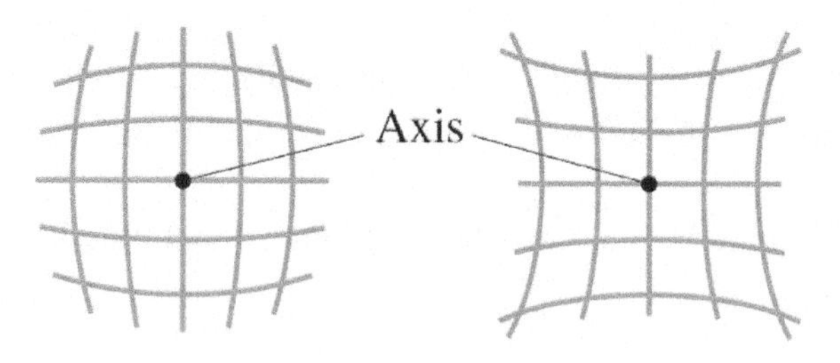

Barrell distortion *Pincushion distortion*

Optical Instruments

Like cameras, the human eye manipulates electromagnetic waves using an adjustable lens and an iris.

The figure below is a cross-section of a human eye, with the iris and lens labeled.

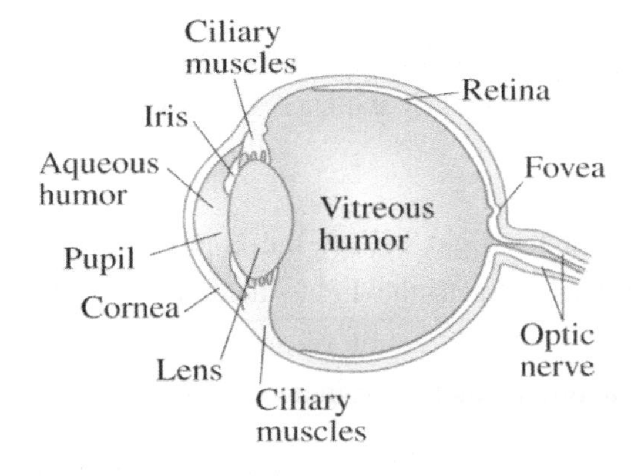

Anatomy of the human eye for vision perception

The eye works by making small adjustments of the lens shape to allow objects at different distances to be focused on the retina. Most of the refraction of the light occurs at the surface of the cornea, while the lens only makes fine-tuned adjustments.

The diagram below illustrates the lens system of the human eye, adjusting to compensate for object distance.

When the object is far away (far away objects have rays assumed to be parallel), the lens of the eye adjusts so that its radius of curvature creates a focal point at the retina.

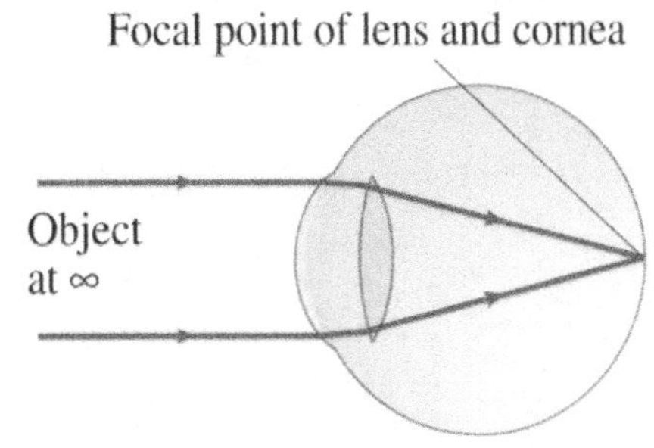

Parallel light rays are focused on the retina of the eye

The near point is the closest distance at which the eye can focus clearly. The typical distance is about 25 cm. The far point is the farthest distance at which an object can be seen clearly. The typical distance is at infinity.

Nearsightedness (*myopia*) occurs when a person's far point is too close.

Farsightedness (*hyperopia*) occurs when a person's near point is too far away.

Nearsightedness can be corrected with a diverging lens. In the diagram on the left, an eye with myopia refracts light too much and causes the focal point to be located before the retina.

In the diagram on the right below, the correction lens is shown for myopia (nearsightedness). The lens refracts the light such that the light diverges before the cornea. Consequently, the higher refraction of the eye is compensated for, and the focal point occurs at the retina, producing a clear image.

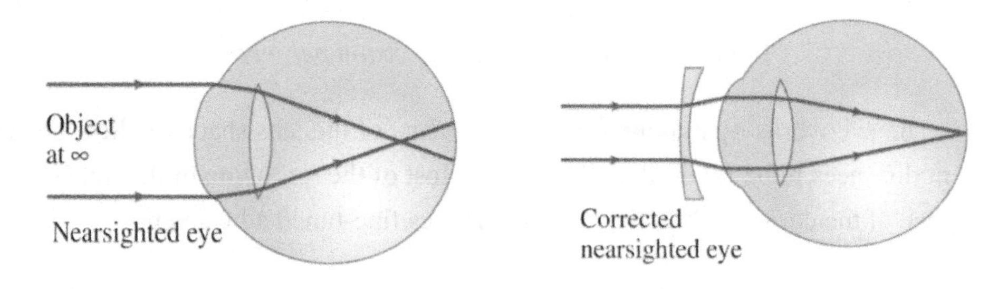

Farsightedness (*hyperopia*) is corrected with a converging lens. Hyperopia occurs when the eye refracts the incoming light too little. The focal point occurs behind the retina, as in the diagram on the left below.

Farsightedness can be corrected with a converging lens, which brings the light rays to a more parallel orientation before entering the eye (diagram on the right below). The lower refractive index of the eye is compensated, and the focal point is at the retina.

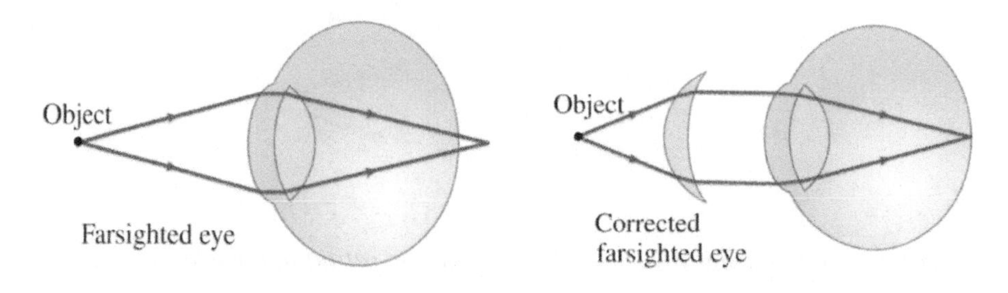

Magnifying glass

A magnifying glass is a converging lens. It allows the user to focus on objects closer than the near point so that they create a larger image on the retina.

The figure below demonstrates the use of a magnifying glass.

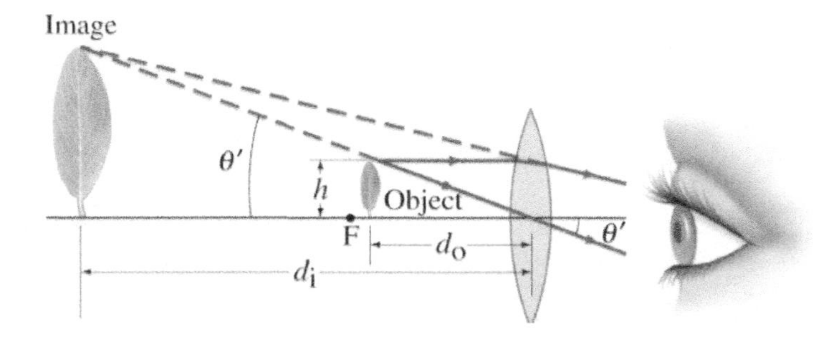

Its angular magnification describes the power of a magnifying glass:

$$M = \frac{\theta'}{\theta}$$

$$\theta \approx \frac{h}{f}$$

$$\theta' \approx \frac{25cm}{h}$$

If the eye is relaxed (N is the near point distance and f the focal length), magnification is given as:

$$M = \frac{\theta'}{\theta} = \frac{h/f}{h/N} = \frac{N}{f}$$

$$M = \frac{25cm}{f}$$

If the eye is focused at the near point, magnification is given as:

$$M = \frac{N}{f} + 1$$

$$M = \frac{25cm}{f} + 1$$

Refracting telescope

A refracting telescope consists of two lenses at opposite ends of a long tube.

The *objective lens* is closest to the object.

The *eyepiece* is closest to the eye.

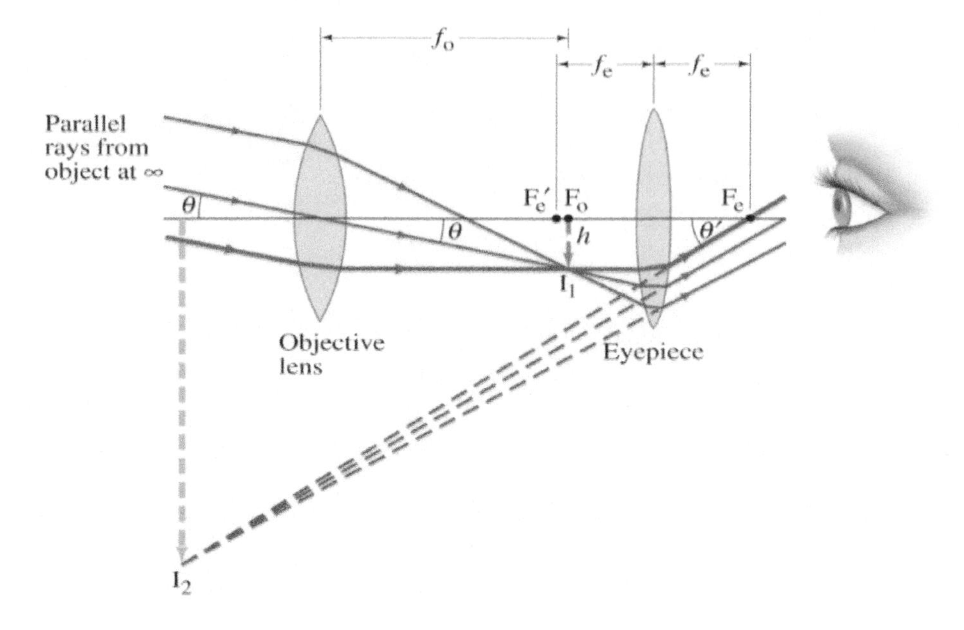

The magnification is given by:

$$M = \frac{\theta'}{\theta} = \frac{(h/f_e)}{(h/f_o)}$$

$$M = -\frac{f_o}{f_e}$$

Chapter Summary

Light, Refraction, and Reflection

- Light paths are rays.
- Visible spectrum of light ranges approximately from 400 nm to 750 nm.
- Two sources of light are coherent if they have the same frequency and maintain the same phase relationship.
- The index of refraction varies with wavelength, leading to dispersion.
- There are boundaries between different materials, as well as between regions of the same material that is under different conditions (e.g., temperature or pressure), which cause a change in the index of refraction.
- Index of refraction: $n = \frac{c}{v}$
- Law of refraction (Snell's Law): $n_1 \sin(\theta_1) = n_2 \sin(\theta_2)$
- The angle of reflection equals the angle of incidence.
- Total internal reflection occurs when the angle of incidence is greater than the critical angle:

$$\sin(\theta_C) = \frac{n_2}{n_1}\sin(90°) = \frac{n_2}{n_1}$$

Diffraction—Single-slit and Double-slit

- Wavelength of light in a medium with an index of refraction n: $\lambda_n = \frac{\lambda}{n}$
- Wavelength can be measured precisely with a spectroscope.
- Young's double-slit experiment demonstrated interference:

  constructive interference occurs when:
  $d \sin(\theta) = m\lambda$, $m = 0, 1, 2, 3, \ldots$

  and destructive interference when:
  $d \sin(\theta) = (m + \frac{1}{2})\lambda$, $m = 0, 1, 2, 3, \ldots$

- A *diffraction grating* has many small slits or lines, and the same condition for constructive interference.

- Light bends around obstacles and openings in its path, producing diffraction patterns.
- Light passing through a narrow slit produces a bright central maximum of width:

$$\sin(\theta) = \frac{\lambda}{D}$$

- Interference can occur between reflections from the front and back surfaces of a thin film.

Polarization

- Light whose electric fields are all in the same plane is plane-polarized.
- The intensity of plane-polarized light is reduced after it passes through another polarizer:

$$I = I_o \cos^2(\theta)$$

- Light can also be polarized by reflection; it is entirely polarized when the reflection angle is the polarization angle: $\tan(\theta_p) = n$

Mirrors

- Plane mirror: image is virtual, upright, and same size as the object.
- A real image means light passes through it.
- A virtual image means light does not pass through it.
- A spherical mirror can be concave or convex.
- The focal length of the mirror: $f = \frac{r}{2}$
- Mirror equation: $\frac{1}{d_o} + \frac{1}{d_i} = \frac{1}{f}$
- Magnification: $m = \frac{h_i}{h_o} = -\frac{d_i}{d_o}$

Lenses

- A converging lens focuses incoming parallel rays to a point, while a diverging lens spreads incoming rays so that they appear to come from a point.
- Power of a lens: $P = \frac{1}{f}$
- The thin lens equation is given by: $\frac{1}{f} - \frac{1}{d_i} = \frac{1}{d_o}$ or $\frac{1}{d_o} + \frac{1}{d_i} = \frac{1}{f}$

Lens Aberrations

- Spherical aberration occurs when rays far from axis do not go through the focal point.
- Chromatic aberration occurs when different wavelengths have different focal points.

Optical Devices

- The resolution of optical devices is limited by diffraction.
- Nearsighted vision is corrected by the diverging lens, farsighted by converging.
- Simple magnifier: object at the focal point

$$M = \frac{\theta'}{\theta} = \frac{h/f}{h/N} = \frac{N}{f}$$

- Magnification:

$$M = \frac{\theta'}{\theta} = \frac{(h/f_e)}{(h/f_o)} = -\frac{f_o}{f_e}$$

Practice Questions

1. Which color of the visible spectrum has photons with the most energy?

A. Violet **B.** Green **C.** Orange **D.** Red **E.** Yellow

2. An index of refraction of less than one for a medium implies that:

A. the speed of light in the medium is lower than the speed of light in a vacuum
B. refraction is not possible
C. the speed of light in the medium is the same as the speed of light in a vacuum
D. the speed of light in the medium is higher than the speed of light in a vacuum
E. reflection is not possible

3. The colors on an oil slick are caused by reflection and:

A. refraction
B. polarization
C. diffraction
D. interference
E. diffusion

4. Which of the following describes the best scenario if a person wants to start a fire using sunlight and a mirror?

A. Use a concave mirror with the object to be ignited positioned at the center of the curvature of the mirror
B. Use a concave mirror with the object to be ignited positioned halfway between the mirror and its center of curvature
C. Use a plane mirror
D. Use a convex mirror
E. A fire cannot be started using a mirror since mirrors form only virtual images

5. A lens has a focal length of 2 m. What lens could be combined with it to get a combination with a focal length of 3 m?

A. A lens of power 1/6 diopters
B. A lens of power 6 diopters
C. A lens of power of –6 diopters
D. A lens of power of –1/6 diopters
E. A lens of power 3 diopters

6. An incident ray traveling in the air makes an angle of 30° with the surface of a medium with an index $n = 1.73$. What is the angle that the refracted ray makes with the surface? (Use the index of refraction for air $n = 1$)

A. 30° **B.** $90\ sin^{-1}(0.5)$ **C.** 60° **D.** $sin^{-1}(0.5)$ **E.** 90°

7. A mirage is produced because:

A. light travels faster through the air than through water
B. images of water are reflected from the sky
C. warm air has a higher index of refraction than cool air
D. warm air has a lower index of refraction than cool air
E. water from oceans and lakes is highly reflective

8. An object is viewed at various distances using a concave mirror with a focal length 14 m. What happens when a candle is placed at the focus?

A. Light rays end up parallel going to infinity
B. Light rays meet at the focus
C. An image is formed 7 m in front of the mirror
D. An image is formed 7 m behind the mirror
E. Light rays end up parallel going out 7 m on either side

9. A material which can rotate the direction of polarization of linearly polarized light is said to be:

A. diffraction limited
B. dichroic
C. circularly polarized
D. birefringent
E. optically active

10. When light enters a material of a higher index of refraction, its speed:

A. increases
B. stays the same
C. first increases then decrease
D. decreases
E. first decreases then increase

11. What is the correct order of the electromagnetic spectrum from the shortest to the longest wavelength?

A. Radio waves → X-rays → Ultraviolet radiation → Visible light → Infrared radiation → Microwaves → Gamma rays
B. Gamma rays → X-rays → Visible light → Ultraviolet radiation → Infrared radiation → Microwaves → Radio waves
C. Gamma rays → X-rays → Ultraviolet radiation → Visible light → Infrared radiation → Microwaves → Radio waves
D. Visible light → Infrared radiation → Microwaves → Radio waves → Gamma rays → X-rays → Ultraviolet radiation
E. Gamma rays → X-rays → Infrared radiation → Visible light → Ultraviolet radiation → Microwaves → Radio waves

12. The angle of incidence can vary between zero and:

A. 2π radians
B. π radians
C. $\pi/2$ radians
D. 1 radian
E. $3\pi/2$ radians

13. An amateur astronomer grinds a double-convex lens whose surfaces have radii of curvature of 40 cm and 60 cm. What is the focal length of this lens in the air? (Use index of refraction for glass $n = 1.54$)

A. 44 cm
B. 88 cm
C. 132 cm
D. 22 cm
E. 1.32 cm

14. When light reflects from a stationary surface, there is a change in its:

I. frequency II. speed III. wavelength

A. I and II only
B. I and III only
C. III only
D. I, II and III
E. none of the above

15. If a distant galaxy is moving away from the Earth at 4,300 km/s, how do the detected frequency (f_{det}) and λ of the visible light detected on Earth compare to the f and λ of the light emitted by the galaxy?

A. The f_{det} is lower, and the λ shifts towards the red end of the visible spectrum
B. The f_{det} is lower, and the λ shifts towards the blue end of the visible spectrum
C. The f_{det} is the same, but the λ shifts towards the red end of the visible spectrum
D. The f_{det} is the same, but the λ shifts towards the blue end of the visible spectrum
E. The f_{det} and the λ are the same as the sources

Solutions

1. A is correct.

Violet light has the highest frequency of the visible light colors with a range of 668 to 789 THz. Because the frequency is related to photon energy by:

$$E = hf$$

Violet light is also the most energetic of the visible light spectrum.

2. D is correct.

The index of refraction for a given material is expressed as the speed of light in a vacuum divided by the speed of light in that material:

$$n = c / v$$

If the index of refraction is less than one, it implies that the speed of light in the material is greater than the speed of light in vacuum. However, this is never the case because the speed of light in a vacuum cannot be exceeded.

3. D is correct.

The colors observed on an oil slick pool are caused by reflection and thin-film interference. This is the process by which the incoming light is reflected off the top and bottom layer of the oil slick and producing reflected waves in and out of phase. This phase change determines the interference (constructive or destructive) and colors form as a result.

4. B is correct.

$$f = \tfrac{1}{2}R$$

where R is the radius of curvature

5. D is correct.

The first lens has a power:

$$P_1 = 1 / f$$

$$P_1 = \tfrac{1}{2}\ \text{D}$$

For a combination of total power:

$P_{tot} = 1 / f_{tot}$

$P_{tot} = 1/3$ D

Thus,

$P_2 = P_{tot} - P_1$

$P_2 = 1/3$ D – 1/2 D

$P_2 = -1/6$ D

6. A is correct.

$n_1 \sin \theta_1 = n_1 \sin \theta_2$

$(1) \sin (90° - 30°) = (1.73) \sin \theta_2$

$\sin (60°) = 1.73 \sin \theta_2$

$\sin \theta_2 = \sin (60°) / 1.73$

$\sin \theta_2 = 0.5$

$\theta_2 = \sin^{-1} (0.5)$

$\theta_2 = 30°$

7. D is correct.

A mirage is produced due to the higher index of refraction of cold air compared to warm air. As light from the sky goes through the denser cold air and approaches the warm air (which is less dense and therefore has a lower refraction index) near the ground (generated by a hot surface of asphalt or sand), the light bends away from the warm air. The reflection of the sky can be observed on the ground.

8. A is correct.

Mirror equation:

$1 / f = 1 / d_o + 1 / d_i$

If the object is placed at the focus:

$d_o = f$

$1 / f = 1 / f + 1 / d_i$

$1 / f - 1 / f = 1 / d_i$

$0 = 1 / d_i$

Only if $d_i = \infty$ is this true.

This implies that no image is formed. Thus the light rays neither converge nor diverge and travel parallel to infinity.

9. E is correct.

An *optically active* material rotates the direction of polarization of linearly polarized light.

10. D is correct.

When light enters a material of a higher index of refraction, its speed *decreases*.

11. C is correct.

The correct order of the electromagnetic spectrum from shortest to longest wavelength is Gamma rays → X-rays → Ultraviolet radiation → Visible light → Infrared radiation → Microwaves → Radio waves

12. C is correct. The angle of incidence ranges from 0° to 90°.

Convert degrees to radians:

$0° = 0$ radians

$(90° / 1)\cdot(\pi / 180°) = \pi/2$ radians

13. A is correct.

Double convex lens:

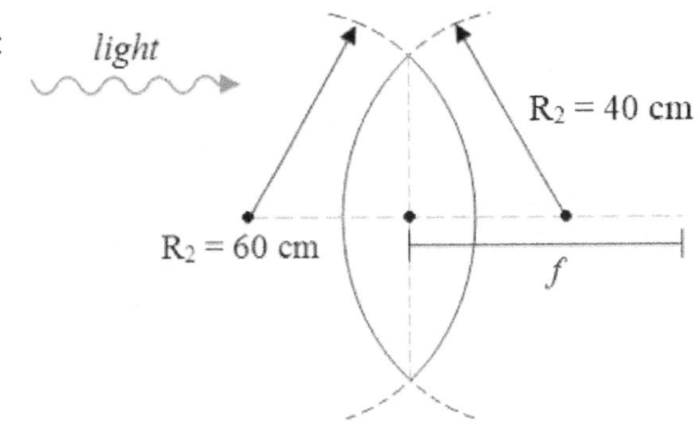

Lensmaker's formula:

$1 / f = (n - 1)\cdot(1 / R_1 - 1 / R_2)$

$1 / f = (1.54 - 1)\cdot(1 / 40 \text{ cm} - 1 / {-60} \text{ cm})$

$1 / f = -0.0225 \text{ cm}^{-1}$

$f = 44 \text{ cm}$

If light passes through the center of the radii of curvature (as it does to R_2) before the curve itself in the lens, then that R is negative by convention.

14. E is correct.

Light does not experience a change in frequency, wavelength, or speed when it reflects from a stationary surface.

15. A is correct.

The Doppler effect is qualitatively similar for both light and sound waves.

If the source and the observer move towards each other, the f_{det} is higher than the f_{source}.

If the source and the observer move away, the f_{det} is lower than the f_{source}.

Since the galaxy is moving away from the Earth, the f_{det} is lower.

The speed of light through space is constant ($c = \lambda f$).

A lower f_{det} means a longer λ_{det}, so the λ_{det} is longer than the λ_{source}.

The λ has been shifted towards the red end of the visible spectrum because red light is the visible light with the longest λ.

Chapter 8

Electric Circuits

CIRCUITS

- **Batteries, Electromotive Force, Voltage**
- **Current**
- **Resistance**
- **Power**
- **Kirchhoff's Laws**
- **Capacitance**
- **Inductors**
- **DC Circuits**

ALTERNATING CURRENTS AND REACTIVE CIRCUITS

- **LCR circuits**
- **Transformers and Transmission of Power Transformers**
- **Measuring Devices**

CIRCUITS

An *electric circuit* is a series of connections between a voltage source, circuit elements, and conducting wires, which result in a flow of current.

The voltage source is the energy input and is necessary for continuing the flow of energy.

The circuit elements can use some of the energy from the voltage source as heat or light.

The current in a circuit always flows from high potential to low potential through the external circuit.

A complete circuit is one where current can flow in a complete loop.

The top image shows a complete circuit. The battery provides a potential difference, which causes current to flow from its positive terminal, through the light bulb, then to its negative terminal, completing the loop.

The bottom figure is a schematic circuit drawing of the circuit shown above it. Note that the schematic drawing looks different than the photograph of the physical circuit that it represents.

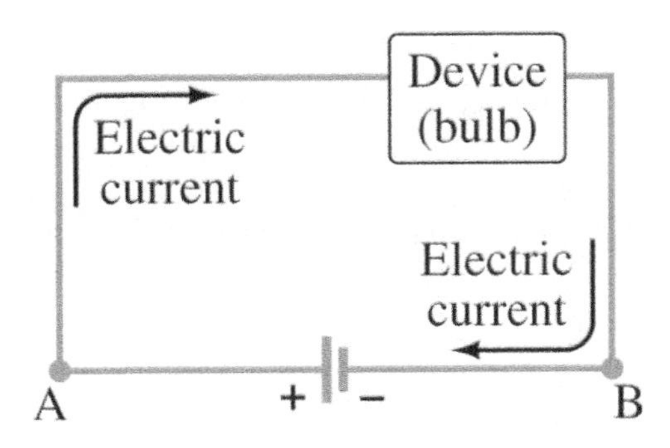

Batteries, Electromotive Force, Voltage

The electric potential difference is important for solving circuit problems.

The *electromotive force* (EMF) describes the measure of energy that causes the flow of current in a circuit. The EMF is defined as the potential difference between two points in a circuit. It is the energy per unit of charge given by an energy source. This value is the same as the amount of work done on one unit of electric charge, given in volts (V).

Batteries are physical devices that produce an electric potential difference and are often used to provide the EMF in electric circuits. These devices operate by transforming chemical energy into electrical energy.

Chemical reactions within the battery cell create a potential difference between the two terminals.

This potential difference can be maintained, even if a current is flowing until, eventually, the chemical reaction has exhausted itself, and the battery can no longer maintain a potential difference.

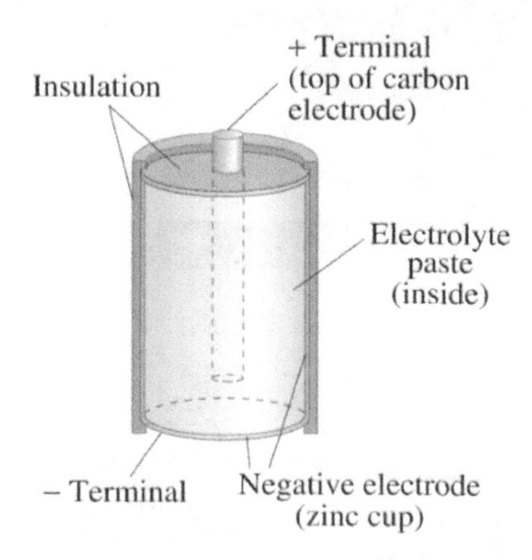

Component of a battery for stored voltage

Internal resistance of a battery

Although batteries are often depicted as having no resistance, all non-ideal batteries will have some internal resistance. The internal resistance of a battery is like a resistor right next to the battery, connected in series.

If the battery in a circuit has no internal resistance (ideal case), the potential difference across the battery equals the EMF. If the battery does have an internal resistance, the potential difference across the battery equals the EMF, minus the voltage drop due to internal resistance.

The *terminal voltage* is given by the equation:

$$V_{ab} = V_0 - IR_{battery}$$

where V_{ab} is the terminal voltage of the battery (V), V_0 is the voltage before internal resistance (V), I is the current through the battery (A), and $R_{battery}$ is the internal resistance of the battery (Ω).

The figure below depicts a non-ideal battery with internal resistance and resulting terminal voltage.

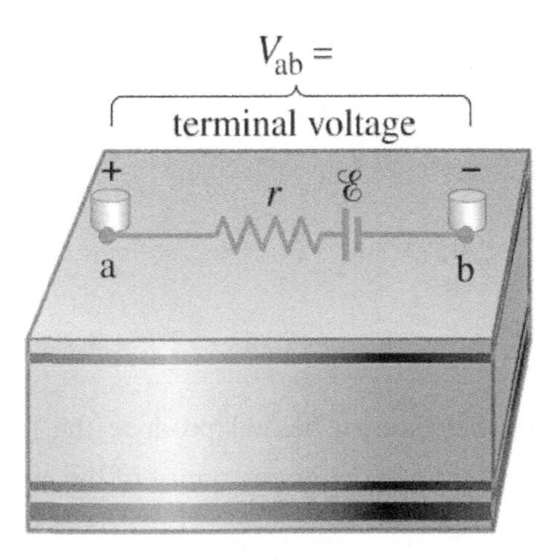

Battery with internal resistance and terminal voltage

Batteries in series and parallel

When two or more batteries are in series, the total potential difference provided is the sum of the EMF provided by each battery. This is only the case if the terminals are aligned such that the batteries' terminals go from positive to negative, or vice versa.

The figure below shows two batteries in series (note that these batteries have internal resistance and the voltage shown is the terminal voltage). The total voltage provided by the two batteries is expressed as:

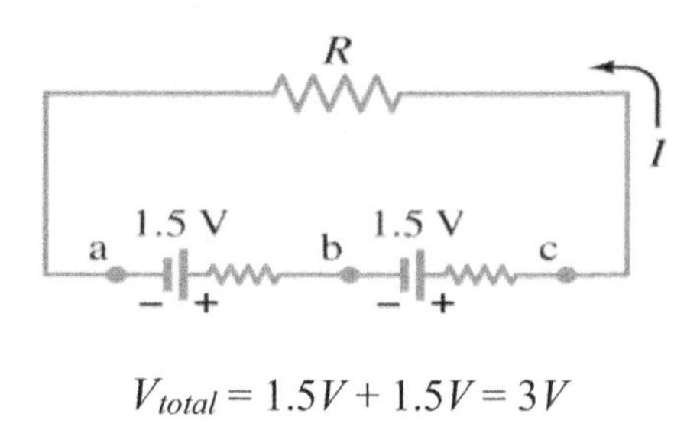

$$V_{total} = 1.5V + 1.5V = 3V$$

When the batteries are in series but aligned such that the positive terminal goes to positive (or negative to negative), the total voltage provided is the difference between the two batteries. In this scenario, the battery with less voltage is being charged:

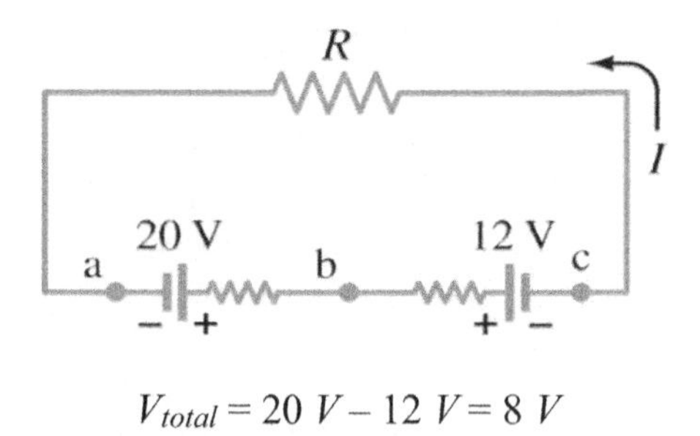

$$V_{total} = 20\ V - 12\ V = 8\ V$$

Batteries in parallel (shown below) produce the same voltage but can support larger current draws depending upon the resistance of the circuit.

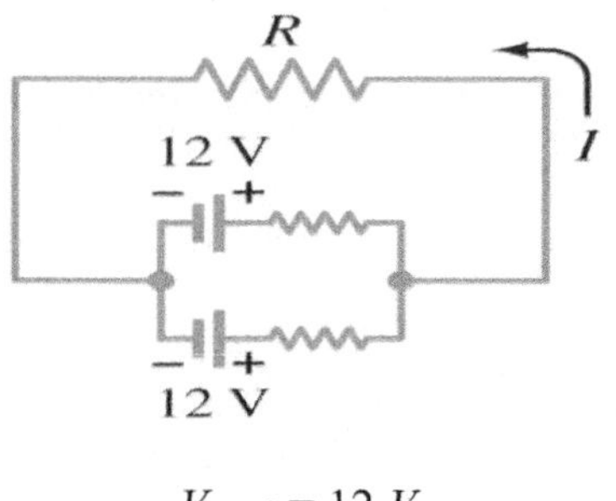

$$V_{total} = 12\ V$$

Current

Current results from the flow of electrons through a conductor. Electrons in a conductor are loosely bound to the nuclei and have large, random speeds dependent upon the temperature of the conductor. Metals are well-known conductors because one electron from each atom is loosely bound and free to move through the metal lattice.

When an electric potential difference is applied across a metal or any conducting material, it creates an electrical field that passes through the conductor at near the speed of light. The free electrons in the conductor are attracted against the line direction of the electric field due to Coulomb attraction forces.

As the electrons pass through the conduction material, these electrons acquire an average drift velocity, which, although considerably smaller than the thermal velocity of the electrons, creates the electric current through the conductor.

The diagram below shows a conductor with an applied potential difference and the resulting electric field.

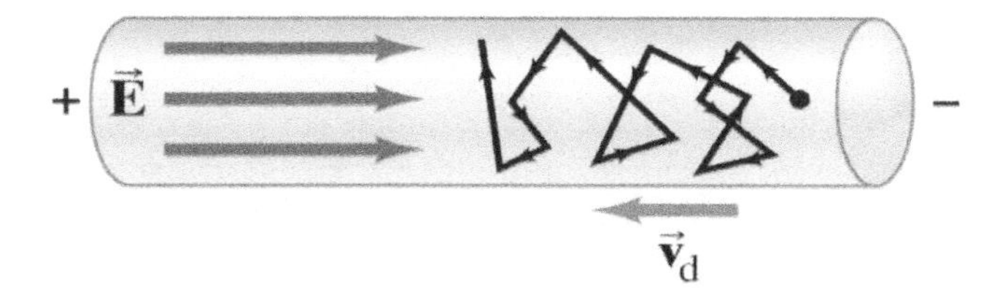

The current through the conductor is calculated as the number of electrons drifting through a unit's volume at the drift velocity:

$$I = nq_eAv_d$$

where I is the current (A), n is the number of electrons per unit volume (electrons/m^3), A is the area of the conductor (m^2), v_d is the drift velocity (m/s) and q_e is the charge of an electron (1.6×10^{-19} C).

More generally, the electric current is defined as the rate at which charge flows through a conductor.

Current I is expressed as:

$$I = \frac{\Delta q}{\Delta t}$$

There are two types of current: direct current (DC) and alternating current (AC).

A *direct current* (DC) is when a constant potential difference is applied to the circuit, and the current moves only in one direction. Direct current is characteristic of electronic devices, batteries, and solar cells.

Most generated electricity is *alternating current* (AC) in which the potential difference is cycled between high and low, creating an oscillatory current flow. The alternating current is transmitted from the power source to the consumer over high voltage lines and "stepped down" for use in homes and businesses.

Resistance

Resistivity ($\rho = RA/l$)

Electrical resistance is the loss of current energy (i.e., electron flow) through a material.

There are two sources of electrical resistance:

collisions with other electrons in current, or

collisions with other charges in the material.

The collisions are factors that account for resistance. Resistance is determined by the resistivity of a material (a measure of conductance) and other factors. In most electronic circuit applications, wires transmit current.

The resistance of a wire depends upon the length, the cross-sectional area, the temperature, and the material:

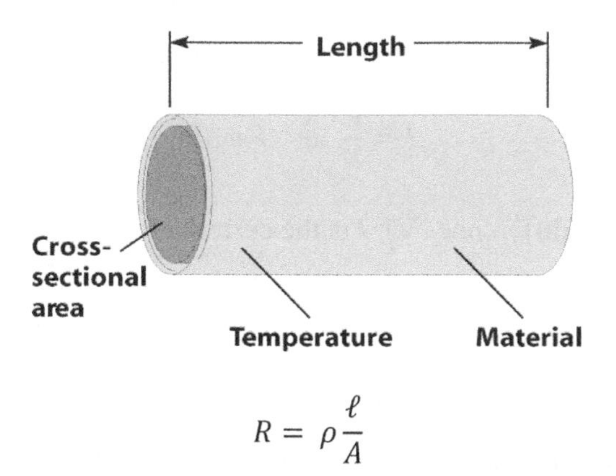

$$R = \rho \frac{\ell}{A}$$

where R is the resistance (Ohms or Ω), ρ is the resistivity ($\Omega \cdot m$), ℓ is the length of the wire (m), and A is the cross-sectional area of the wire (m^2).

The resistivity is directly proportional to the resistance. Therefore, the greater the resistivity, the higher the resistance of the material. Resistance is inversely proportional to the area (inverse square to the radius and diameter).

The greater the diameter of the wire, the lower is the resistance; there is a larger area for electrons to flow.

From the resistance equation above, a wire of low resistance will be short, of large diameter, and made from a material that has low resistivity. Extension cords are made of wires with thicker diameters to keep the resistance low. The resistance of a wire is also dependent upon the temperature of the material.

For any given material, the resistivity increases with temperature:

$$\rho_T = \rho_0[1 + \alpha(T - T_0)]$$

where ρ_T is the resistivity at a temperature T ($\Omega \cdot m$), ρ_0 is the standard resistivity at a standard temperature of T_0 ($\Omega \cdot m$), α temperature coefficient of resistivity (K^{-1}) and T is the temperature (K)

Ohm's Law (I = V/R)

Ohm's Law gives the relationship between the potential difference across resistance and the current through the resistance.

Ohm's Law states that, in electrical circuits, the current through a conductor between two points is directly proportional to the potential difference (or voltage) across the two points. Current is inversely proportional to the resistance between them:

$$I = \frac{V}{R} \quad \text{or} \quad V = IR$$

where V is the potential difference (V), I is the current (A), and R is the resistance (Ω).

Resistors in series

In a circuit, *resistors* are arranged in series or parallel.

For series circuit (shown below), the resistors have current pass through the first resistor, then into the second, into the third and so on:

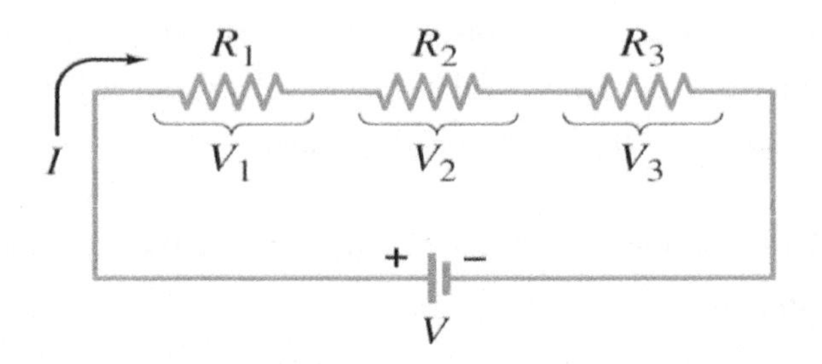

For series circuits, the total resistance of the circuit is the sum of all the resistances of the components. Adding elements to the circuit increases the total resistance and decreases the total current.

The equivalent resistance of resistors in series is found by:

$$R_{\text{equivalent}} = R_1 + R_2 + R_3 + \ldots$$

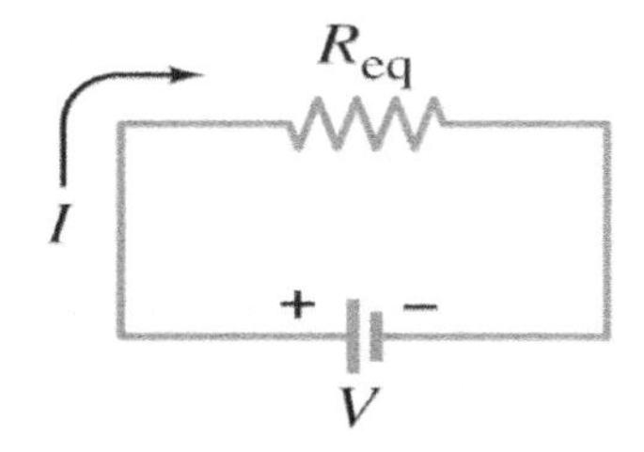

In a series circuit, the current through each resistor is the same ($I_{\text{series}} = I_1 = I_2 = I_3$); however, the voltage drops across each resistor. The voltage drop among resistors in series is split according to the resistance—a greater resistance equals a greater voltage drop (V = IR).

This can also be expanded for more than one resistor, by splitting it into components:

$$V = V_1 + V_2 + V_3 = IR_1 + IR_2 + IR_3$$

Resistors in parallel

Parallel resistors are arranged such that all resistors experience the same voltage drop across them but have different currents according to their resistances.

In a parallel circuit, there is more than one available path for the current.

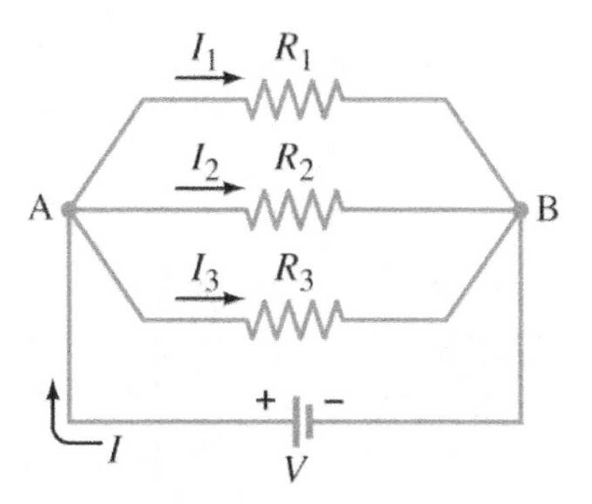

Parallel circuit offers more than one route for the current to flow

The equivalent resistance of resistors in parallel is:

$$\frac{1}{R_{equivalent}} = \frac{1}{R_1} + \frac{1}{R_2} + \frac{1}{R_3} + \cdots$$

As stated earlier, for parallel resistors, the voltage across each resistor is the same ($V_{\text{parallel}} = V_1 = V_2 = V_3$); however, the current across each resistor is not.

For parallel resistors, the total current is the sum of currents across each resistor:

$$\frac{V}{R_{\text{eq}}} = \frac{V}{R_1} + \frac{V}{R_2} + \frac{V}{R_3}$$

$$I = \frac{V}{R_{\text{eq}}}$$

Most typical circuits (e.g., household circuits) are combinations of series and parallel circuits. Generally, electrical outlets in the same room are in series, but the electrical outlets in separate rooms are in parallel. This is the reason why a blown fuse may cut the power to an entire room, but not the entire house.

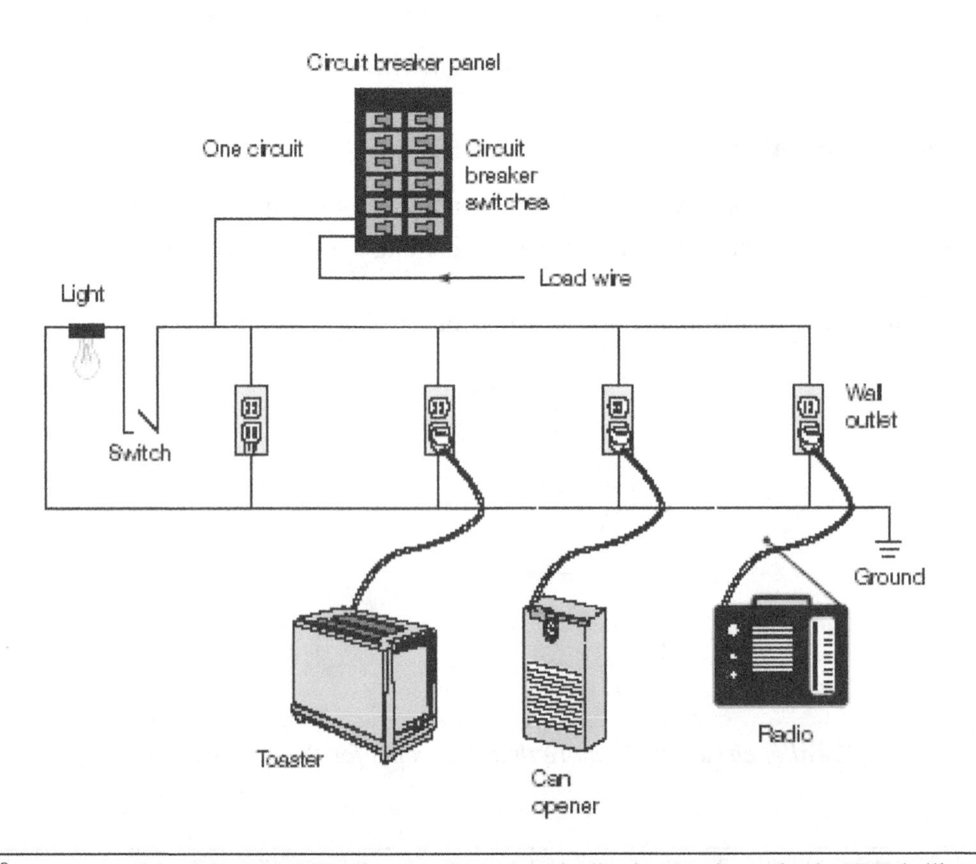

Power

The *power* (P) in a circuit is the rate at which energy converts from electrical energy into some other form, such as heat or mechanical energy.

The SI unit for power is the watt (W), which is equal to joules per second (J/s).

For a component in a DC circuit, the power is given by the equation:

$$P = IV$$

where P is the power (W), I is the current (A or ampere), and V is the voltage (V).

Using Ohm's Law, power in an electrical circuit is expressed as:

$$P = I^2R$$

Kirchhoff's Laws

Kirchhoff's Laws are useful to solve for the unknown components within a circuit.

Kirchhoff's Current Law

Kirchhoff's Current Law is an expression of the conservation of charge. By the law of the conservation of charge, a charge neither is created nor destroyed.

Therefore, in a circuit, the sum of all currents entering a junction must be equal to the sum of all currents exiting the junction.

$$I_{entering,total} = I_{exiting,total}$$

Kirchhoff's Voltage Law (Kirchhoff's Loop Rule)

Kirchhoff's Voltage Law is also an expression of the conservation of energy.

The voltage law states that the sum of all voltage drops, across any resistive elements within a closed-circuit loop, must equal the EMF in the same loop.

$$\sum V_{drop} = EMF_{total}$$

For any complete loop (i.e., traveling around a circuit loop and returning to the point of origin), the total change in voltage is 0. Choose a starting point on the circuit and proceed around the circuit in a complete loop (returning to the point of origin). Evaluate the boosts and drops in voltage as each circuit element is passed.

Moving across a battery is the same as "up" a battery when approaching in the direction of the EMF (from the – to + terminal).

Moving across a battery is the same as "down" a battery if approaching against the EMF (from + to – terminal).

Moving across a resistor is the same as "down" a resistor by the amount IR, where I is the current through the resistor, and R is the resistance in Ω when approaching in the same direction as the current.

Moving across a resistor is the same as "up" a resistor by the amount IR, where I is the current through the resistor, and R is the resistance in Ω when approaching in the opposite direction as the current.

When solving problems that use of Kirchhoff's rules, follow these steps:

1. Label each current.
2. Identify unknowns.
3. Apply junction and loop rules; as many independent equations will be needed as there are unknowns.
4. Solve the equations, being careful with signs.

For example, apply Kirchhoff's Circuit Laws to the circuit below:

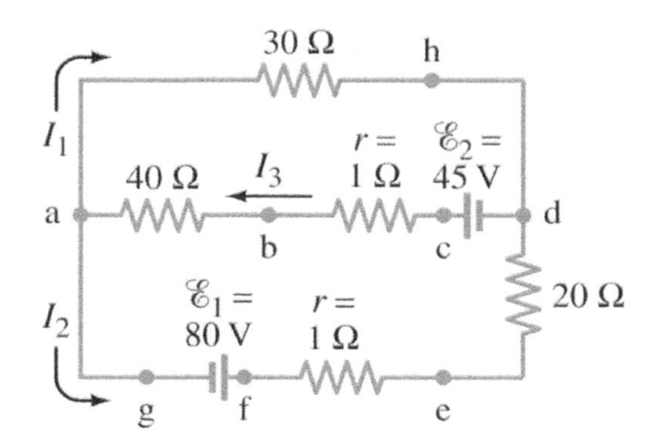

Apply Kirchhoff's Current Law:

$$I_1 + I_2 = I_3$$

Apply Kirchhoff's Voltage Law:

Loop CBAHD: $0\text{ V} = -I_3(1\ \Omega) - I_3(40\ \Omega) - I_1(30\ \Omega) + 45\text{ V}$

Loop FEDCBAG: $0\text{ V} = -I_2(1\ \Omega) - I_2(20\ \Omega) + 45V - I_3(1\ \Omega) - I_3(40\ \Omega) + 80\text{ V}$

Loop FEDHAG: $0\text{ V} = -I_2(1\ \Omega) - I_2(20\ \Omega) + I_1(30\ \Omega) + 80\text{ V}$

Capacitance

A *capacitor* is a device used to store electrical charge. A basic capacitor usually consists of two conductors that are close to each other but are not touching. An insulator of a dielectric separates the two conducting plates or are separated by air.

In the diagram on the left, the parallel plate capacitor has no insulator. Only air is between the plates.

The diagram on the right depicts a variation of the parallel plate capacitor in which there is an insulating material between the coiled plates.

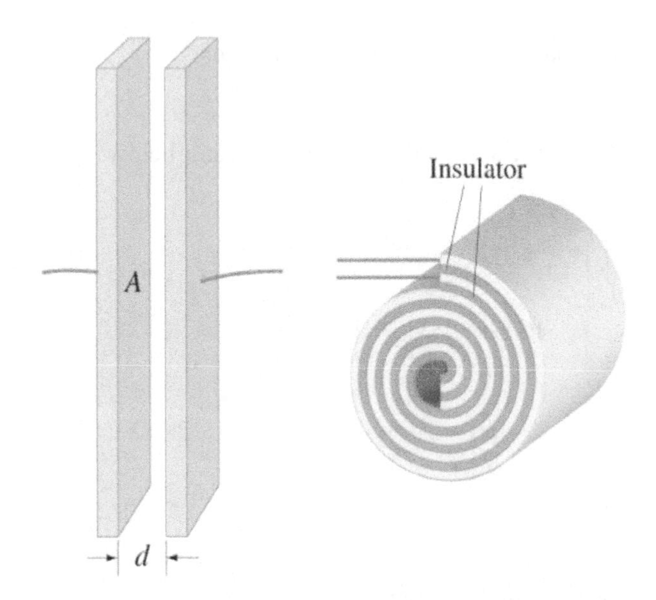

A capacitor's ability to store charge is *capacitance*, measured in Farads (F).

The capacitance does not depend on the voltage. Capacitance is a function of the geometry and materials of the capacitor.

Capacitance is expressed by:

$$C = \frac{Q}{V}$$

where C is the capacitance (F), Q is the charge (J or joules), and V is the voltage (V).

Parallel-plate capacitor

The simplest form of a capacitor is that of the parallel-plate capacitor. It consists, as its name implies, of two conductors in the form of plates, which are parallel and separated by air or a dielectric material. In the diagrams below, a parallel-plate capacitor is connected to a battery. The equivalent circuit diagram is on the right.

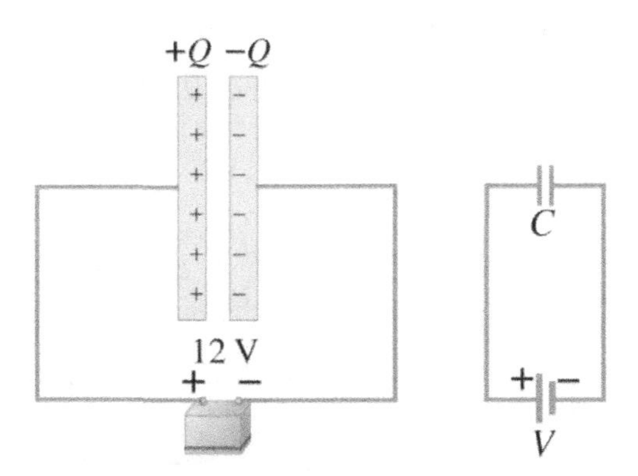

The capacitance of a parallel plate capacitor is expressed as:

$$C = \frac{Q}{V} = \frac{\epsilon_0 A}{d}$$

where ϵ_0 is the permittivity of space (8.854×10^{-12} F/m), k is the relative permittivity of the dielectric material; A is the area of the plates, not the sum of the areas (m^2) and d is the plate separation (m).

The voltage (V) across the capacitor is the product of the electric field between the plates and the plate separation:

$$V = E \times d$$

where E is the electric field between the capacitor ($V \cdot m^{-1}$).

Dielectrics

A *dielectric* is a non-conductive material that does not readily allow a current to pass through the material. Inserting a dielectric between the plates of a capacitor increases the capacitance. The dielectric allows more charge to be stored on the plates of the capacitor.

The diagrams below represent an air-filled parallel plate capacitor (left and center) and a dielectric-filled parallel plate capacitor (right).

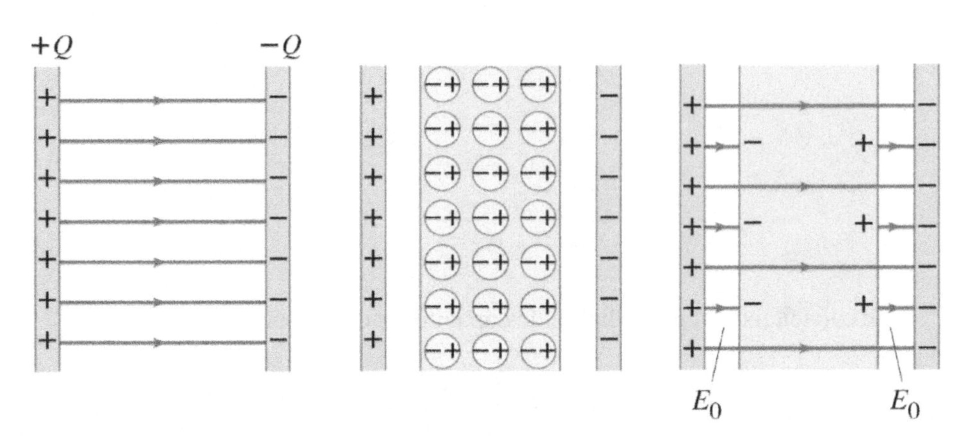

The diagram on the left shows an air-filled parallel plate capacitor. The amount of charge Q stored is limited by the strength of the electric field between the plates.

In the center diagram, if too much charge is stored between the plates, a strong electric field develops. This stored charge ionizes the air and causes an arc of charge to jump across the plates.

In the diagram on the right, a dielectric is placed between the plates. The dielectric material becomes polarized (insulator in an electric field) and inhibits the formation of a strong electric field between the plates. Accordingly, more charge (and thus a higher capacitance) can be stored on the plates before the dielectric material breaks down and allows current to pass through it.

The capacitance of a parallel plate capacitor with a dielectric is expressed as:

$$C = \frac{Q}{V} = \frac{k\,\epsilon_0\,A}{d}$$

where C is the capacitance of the capacitor, A is the area of the capacitor, k is the dielectric constant, ϵ_0 is the permittivity of free space.

Energy of charged capacitor

A charged capacitor stores electric energy; the energy stored is equal to the work done to charge the capacitor.

$$U = \frac{1}{2}QV = \frac{1}{2}CV^2 = \frac{1}{2}\frac{Q^2}{C}$$

where U is the potential energy of the charged capacitor (J), Q is the charge stored (magnitude of either $+Q$ or $-Q$ on one plate) (C or coulombs), and C is capacitance (F).

The *energy density* of the capacitor is defined as the energy of the electric field per unit volume. Energy density can be expressed as:

$$\text{energy density} = \frac{PE}{volume} = \frac{1}{2}k \in_0 E^2$$

Capacitors in series

Like resistors, capacitors can be connected in series or parallel in a circuit.

For capacitors in series, the equivalent capacitance is found by the sum of the reciprocals of the individual capacitance:

$$\frac{1}{C_{\text{eq}}} = \frac{1}{C_1} + \frac{1}{C_2} + \frac{1}{C_3}$$

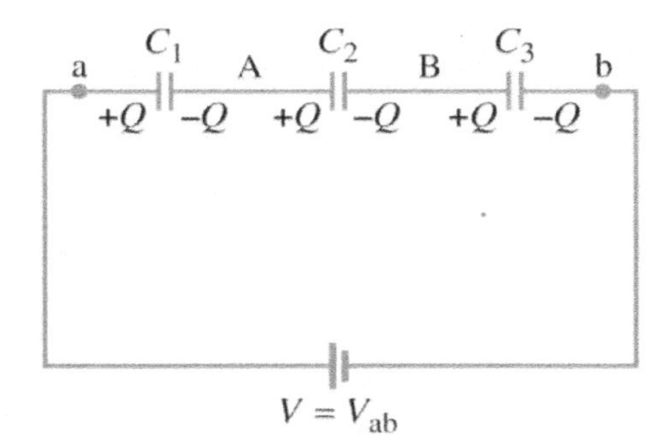

All capacitors in series have the same charge Q, but the voltage V across each capacitor is different.

$$V_1 \neq V_2 \neq V_3$$

$$Q_1 = Q_2 = Q_3$$

Capacitors in parallel

Capacitors may also be connected in parallel, as depicted below:

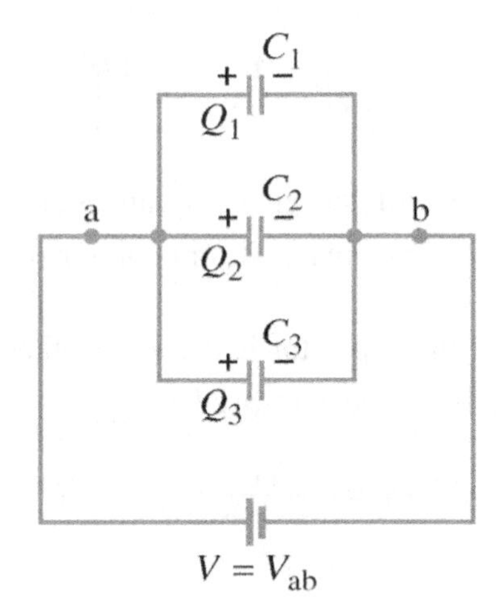

In this example above, all the capacitors have the same voltage.

The equivalent capacitance is the sum of the individual capacitances.

$$V_1 = V_2 = V_3$$

$$C_{equivalent} = C_1 + C_2 + C_3$$

Capacitors in parallel hold different charges:

$$Q_1 \neq Q_2 \neq Q_3$$

Inductors

Inductors are circuit devices that resist changes in current, through the application of Faraday's Law.

Typical inductors consist of coils of wire wrapped around a different core material. Inductors operate by producing a back EMF against a change in a current (according to Lenz's Law and Faraday's Law).

The unit for the inductor is the *Henry* (H), and the back EMF produced is:

$$EMF = -L\frac{\Delta I}{\Delta t}$$

where *EMF* is the opposing voltage produced (*V*), *L* is the inductance of the inductor (H), and $\Delta I/\Delta t$ is the change in current over the change in time (A/s).

The magnetic field produced by an inductor stores energy. This is similar to the electric field of a capacitor.

The energy stored in this magnetic field is expressed as:

$$U = \frac{1}{2} \times \frac{B^2}{\mu_0}$$

where *U* is the stored energy (J), *B* is the magnetic field (T or Tesla), and μ_0 is the permittivity of free space (1.2566×10^{-6} m kg/s^2 A^2)

Resistor-Capacitance (RC) Circuits

A simple resistor-capacitance (RC) circuit consists of a resistor and capacitor in series, connected to a potential difference. The capacitor begins to charge when the switch is closed, and the circuit forms a complete loop.

Below is a diagram of an RC circuit with the switch open.

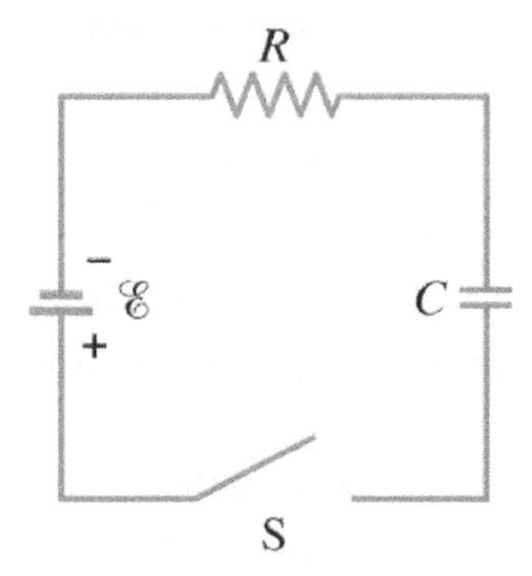

When the switch is closed, the capacitor begins to charge over time. The voltage across the capacitor increases with time and is expressed as:

$$V_C = V(1 - e^{-t/RC})$$

where V_C is the voltage across the capacitor (V), V is the potential difference across the circuit (V), t is time (s), R is resistance (Ω), and C is the capacitance (Farad).

The charge follows a similar curve:

$$Q_C = CV_b(1 - e^{-t/RC})$$

where $V_b = V_R + V_C$. V_R is the voltage across the resistor (V), and Q_C is the charge across the capacitor (C).

The current through the circuit is the same everywhere (series circuit):

$$I = \frac{V_b}{R}(e)^{-t/RC}$$

where I is the current through the circuit (A).

This curve has a characteristic time constant:

$$\tau = RC$$

where τ is the time constant (s).

The graphs below display the voltage across the charging capacitor and current across the circuit.

The diagram below illustrates the growth curve of voltage over time. Notice that the capacitor reaches 63% voltage in one time constant and has ~100% voltage after three-time constants.

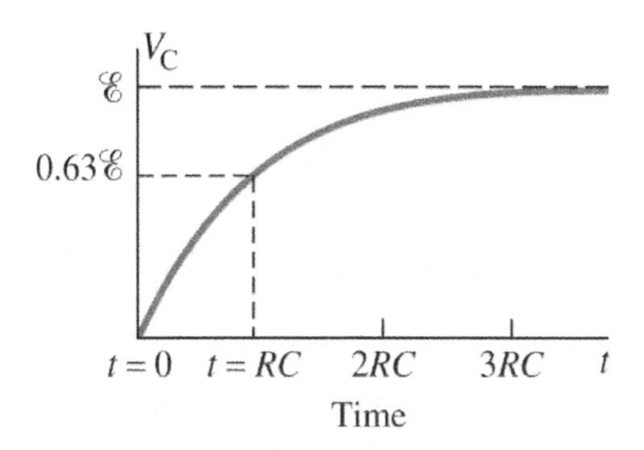

Current vs. time graph plotting charge developed on capacitor

The diagram below displays the decay in current over time. Notice that after one time constant, the current is 37% of its original value (t = 0 seconds), and after three-time constants, the current is nearly zero amps.

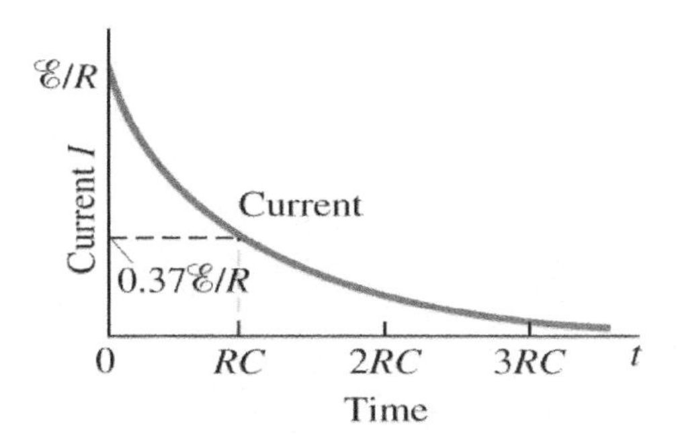

Current vs. time graph plotting current decay

During the discharge of a capacitor, the capacitor acts as a battery and drives the current flow. The current flow decreases with time as the capacitor discharges.

The capacitor voltage decays and is expressed as:

$$V_C = V_0(e)^{-t\,/\,RC}$$

Where V_0 is the charge voltage across the capacitor (V).

The charge through the discharging capacitor follows a similar decay:

$$Q = CV_0(e)^{-t\,/\,RC}$$

The current in the circuit when the capacitor is discharging is expressed as:

$$I = \frac{V_0}{R}(e)^{-t\,/\,RC}$$

When the capacitor is discharging, the voltage, charge, and current all follow a similar decay curve.

The figure below shows the voltage decay curve of a discharging capacitor. Although this curve represents the voltage, both the charge and current graphs would have exponential decay curves that follow this pattern.

For one time constant, the voltage, charge, and current would be at 37% of the max value ($t = 0$ seconds).

By three-time constants, the voltage, charge, and current would decay to nearly zero.

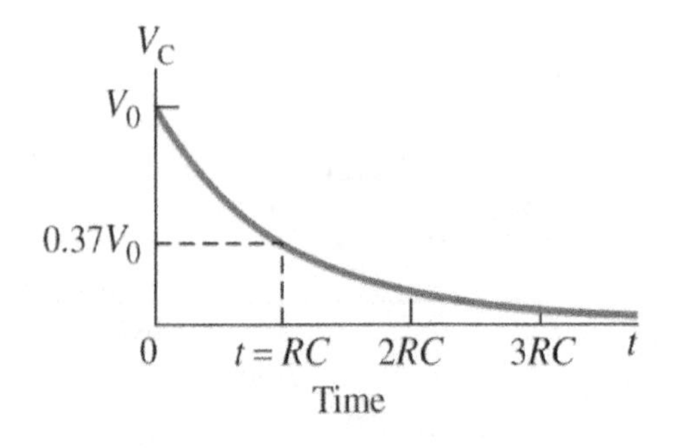

Charge voltage vs. time graph of decay across a capacitor

LR Circuits

A simple *LR circuit* consists of an inductor in series with a resistor connected to a potential difference. The inductor resists change in current through the circuit. Therefore, a gradual, rather than abrupt rise in current occurs when the loop is closed.

After a period, the current remains steady and is equal to that if the circuit were a resistor alone (assuming ideal inductor with no resistance).

Below is a figure of an LR circuit with a switch that can either connect or disconnect the circuit to a potential difference.

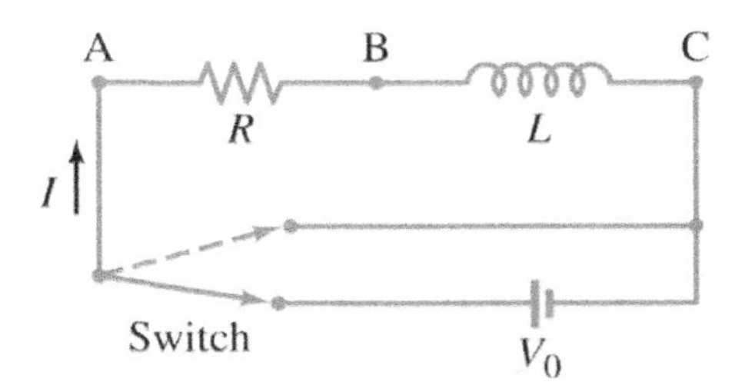

Schematic of a LR circuit with inductor L, resistor R, and circuit switch

The switch is initially open, so there is no potential difference, and no current flows through the circuit. When the switch is closed, the voltage across the capacitor is:

$$V_L = V_b e^{-tR/L}$$

where V_L is the voltage across the inductor (V), R is the resistance of the resistor (Ω), and L is the inductance of the inductor (H or Henry).

The voltage (V) across the resistor is expressed as:

$$V_b = V_L + V_R \text{ and } V_R$$

The current (I) across the circuit is expressed as:

$$I = \left(\frac{V_b}{R}\right)(1 - e^{-tR/L})$$

where I is the current across the circuit (A).

The time constant of an LR circuit is:

$$\tau = \frac{L}{R}$$

where τ is the time constant (s).

The growth curve of the current through the circuit is seen below. After one time constant, the circuit has reached 63% of its maximum current.

Although not displayed, the maximum current is reached in approximately three-time constants (like the RC circuit).

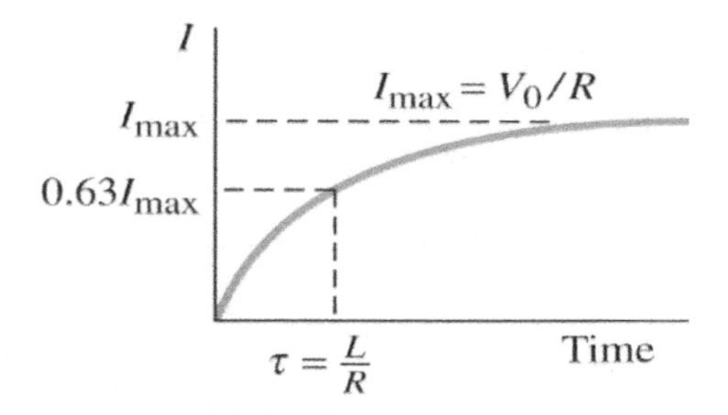

If the switch is opened, such that the potential difference is no longer across the circuit, but the circuit remains closed, the current decays according to:

$$I = I_{max}e^{-t/\tau}$$

In the graph below, the current decays from its maximum value.

After one time constant, the current is at 37% of its maximum.

After three-time constants, the circuit has approximately zero current.

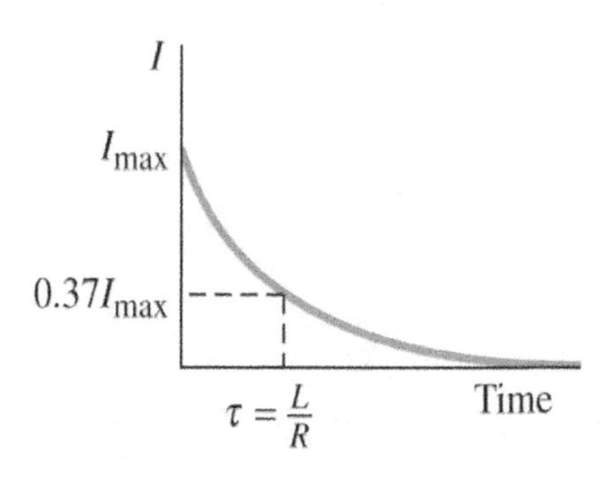

Current vs. time graph with current decay plotted as a line

ALTERNATING CURRENTS AND REACTIVE CIRCUITS

As mentioned earlier in this chapter, the current provided by a battery flows steadily in one direction. This is how *direct current* (DC) is named. However, the power supplied by an electrical generation plant is always delivered in alternating current (AC).

The figures below demonstrate DC vs. AC systems.

The diagram shows a DC system in which the current is held constant over time.

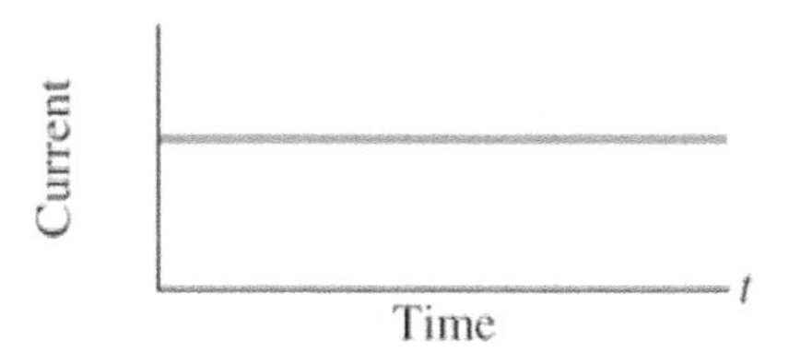

In the diagram below, the current varies sinusoidally and results in *alternating current*. An alternating current varies as a function of a square wave or a triangular wave. In most typical applications, the current is supplied as a sinusoidal function.

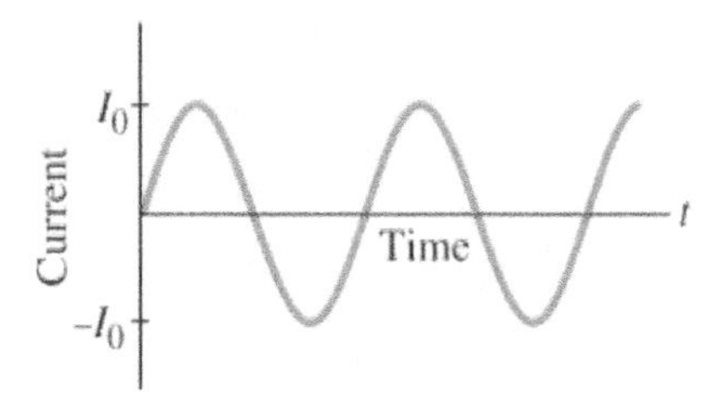

The sinusoidal variation in current is a function of the frequency:

$$I = I_0 \sin \omega t = I_0 \sin 2\pi f t$$

where I_0 is the max current (amplitude) given in (A), ω is the frequency (rad/s), and f is the frequency (Hz).

Like current, the voltage varies in a sinusoidal wave when graphed against time.

$$V = V_0 \sin 2\pi f t = V_0 \sin \omega t$$

where V_0 is the max voltage (amplitude) given in (V).

The power delivered is calculated by multiplying the current squared by voltage:

$$P = IV = I(IR) = I^2R = I_0^2R \sin^2 \omega t$$

Usually, the average power is of most interest:

$$\bar{P} = \frac{1}{2} I_0^2 R = \frac{1}{2}\frac{V_0^2}{R}$$

where $\bar{P}$ is the average power (W or watts), and R is the resistance (Ω or ohms).

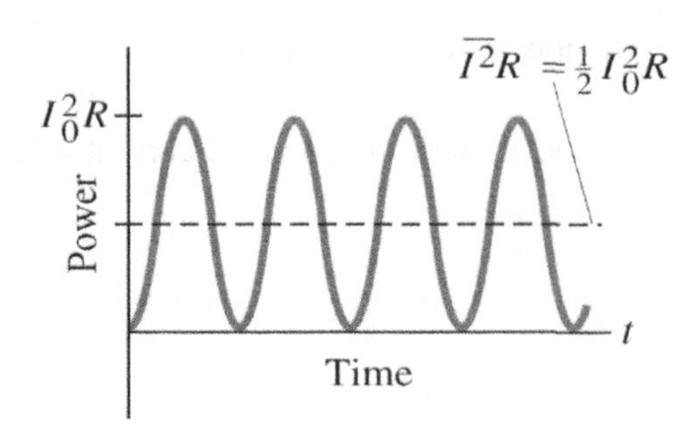

The current and voltage both have average values of zero.

Thus, current and voltage are expressed as root-mean-square values:

$$I_{rms} = \sqrt{\overline{I^2}} = \frac{I_0}{\sqrt{2}}$$

$$I_{rms} = 0.707\, I_0$$

$$V_{rms} = \sqrt{\overline{V^2}} = \frac{V_0}{\sqrt{2}}$$

$$V_{rms} = 0.707\, V_0$$

where I_{rms} is the root-mean-square current (A), and V_{rms} is root-mean-square voltage (V).

Ohm's Law is written using root-mean-square values:

$$V_{\text{rms}} = I_{\text{rms}} R$$

The average power can also be found using root-mean-square values:

$$P_{\text{avg}} = I_{\text{rms}} V_{\text{rms}}$$

$$P_{\text{avg}} = I^2_{\text{rms}} R$$

LRC Circuits

Resistors, capacitors, and inductors have different phase relationships between current and voltage when placed in an AC circuit as compared to a DC circuit. Below is a diagram of a simple AC circuit with an inductor, resistor, and capacitor (LRC circuit):

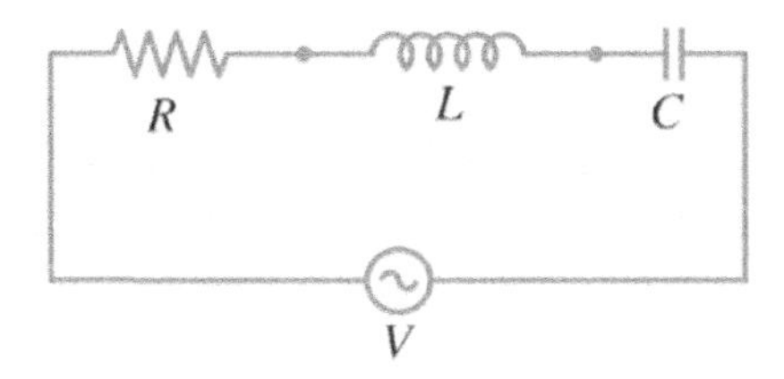

Unlike DC circuits, the current vs. voltage through the different elements of an LRC circuit is time dependent. As such, the unique characteristics of each element must be known before the entire circuit can be analyzed.

The current through a resistor in an AC circuit is in phase with the voltage, meaning both change signs at the same time.

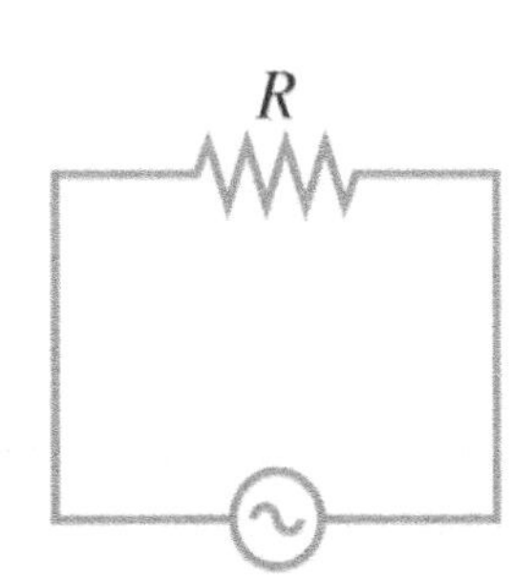

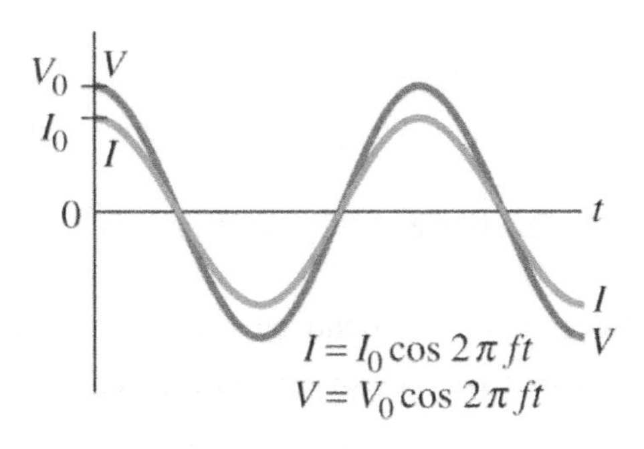

The phase relationship between current and voltage in inductors is different.

Inductors resist changes in current, so the current lags the voltage by 90°.

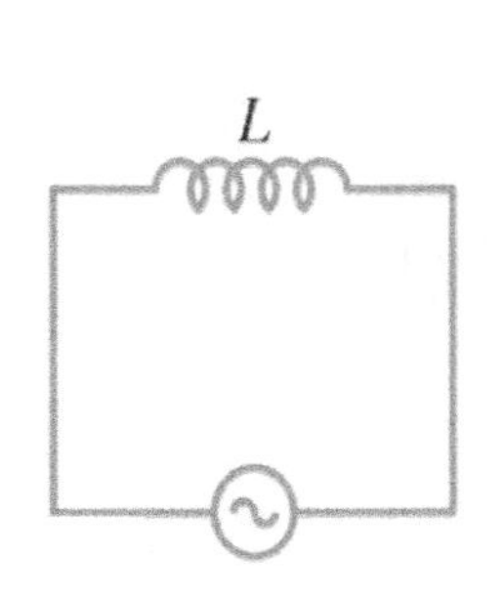

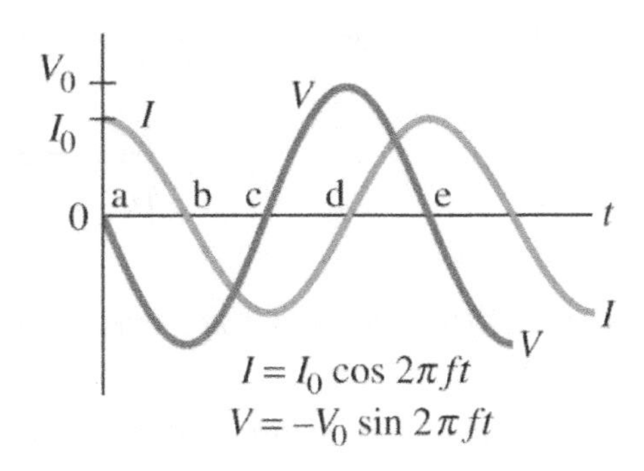

Conversely, the capacitor's current leads the voltage by 90°.

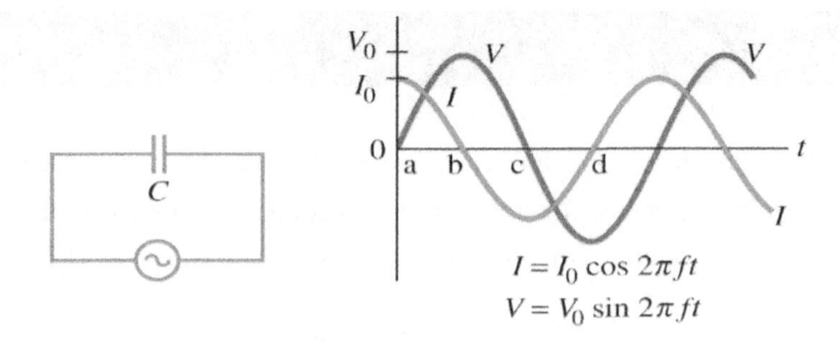

Because both the inductor and capacitor have phase differences with respect to the voltage of their effective resistance (ratio of voltage to current), the root-mean-square current is calculated differently.

The *reactance* is the effective resistance of an inductor or capacitor and is given by:

$$X_L = \omega L = 2\pi f L$$

where X_L is the reactance of the inductor (Ω).

The reactance of the capacitor (X_C) is given by:

$$X_c = \frac{1}{\omega C} = \frac{1}{2\pi f C}$$

where X_c is the reactance of the capacitor (Ω).

Both equations depend on frequency. From the ratio of voltage to current, the effective resistance (i.e., *impedance*) of the circuit is given by:

$$Z = \sqrt{R^2 + (X_L - X_C)^2}$$

where Z is the impedance (Ω or ohms).

The root-mean-square current in an AC circuit is:

$$I_{rms} = \frac{V_{rms}}{Z} = \frac{V_{rms}}{\sqrt{R^2 + (2\pi f L - \frac{1}{2\pi f C})^2}}$$

I_{rms} depends on the frequency and is at a maximum when $X_c = X_L$.

The frequency at which I_{rms} is maximum:

$$f_0 = \frac{1}{2\pi}\sqrt{\frac{1}{LC}}$$

where f_0 is the resonant frequency of the circuit (Hz).

The f_0 is the *resonant frequency*.

The figure below illustrates the current peak at the resonant frequency (f_0) for large resistances and small resistances:

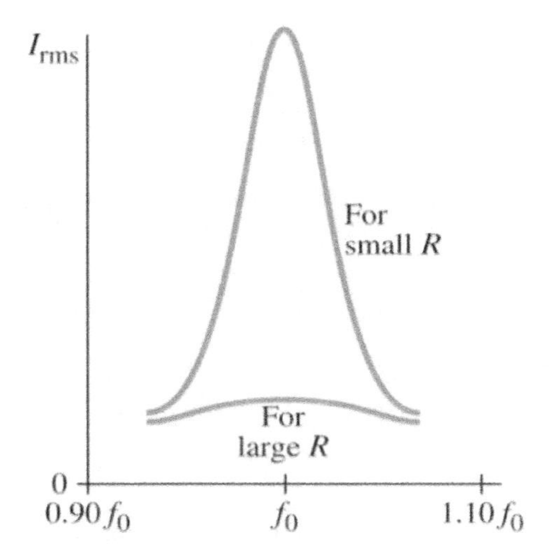

The current I_{rms} peak at the resonant frequency for large vs. small resistance

Transformers and Transmission of Power

Power is transmitted in alternating current (AC) for several reasons. A significant factor is the reduced power loss through the power lines when transporting electricity. If electricity were transported at high current, the result is a large Ohmic power loss through the power lines. These power losses are reduced by transporting electricity at high voltages rather than higher currents. According to the power loss equation:

$$P = I^2R$$

How can the voltage be manipulated such that the electricity is transported at high voltage, and then reduced to lower voltages for the consumer? This is possible through the electrical devices known as *transformers*. Transformers work only if the current is changing; this is an important reason why electricity is transmitted as AC.

A transformer operates according to Faraday's Law of electromagnetic induction (1831) that steps-up or steps-down the voltage through a line. This principle permits power companies to reduce power loss during the delivery of electricity. The transformer consists of an iron core around which an input wire and output wire are wrapped. In a step-up transformer (i.e., increase voltage), the input wire has a few primary coils around the iron core. The output wire has many secondary coils.

When alternating current travels through the input coil, the changing current (sinusoidal variation in current) induces a magnetic field in the iron core. The magnetic field loops through the iron core and induces a voltage in the output wire wrapped around the core.

The output voltage is proportional to the number of coils wrapped around the core.

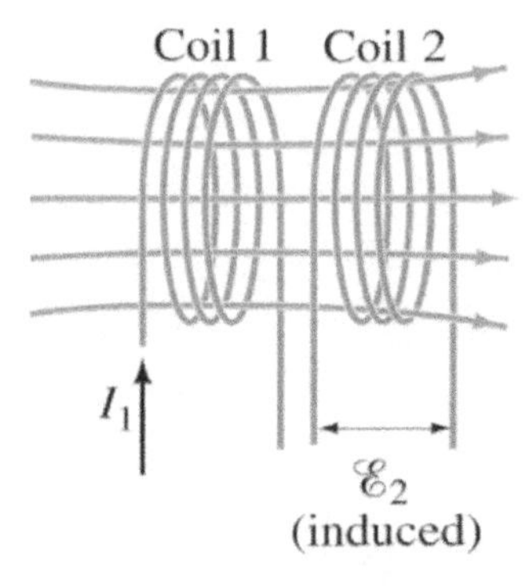

A transformer showing the input (coil 1) and output coil (coil 2)

The figure below shows a basic power delivery system. The power station generates electricity, which is sent through a step-up transformer to reduce the power loss during transmission. The AC electricity runs the high voltage transmission line to a step-down transformer near the consumer. The electricity is then sent to homes on low voltage lines, where a final step-down transformer lowers the voltage for the end-user.

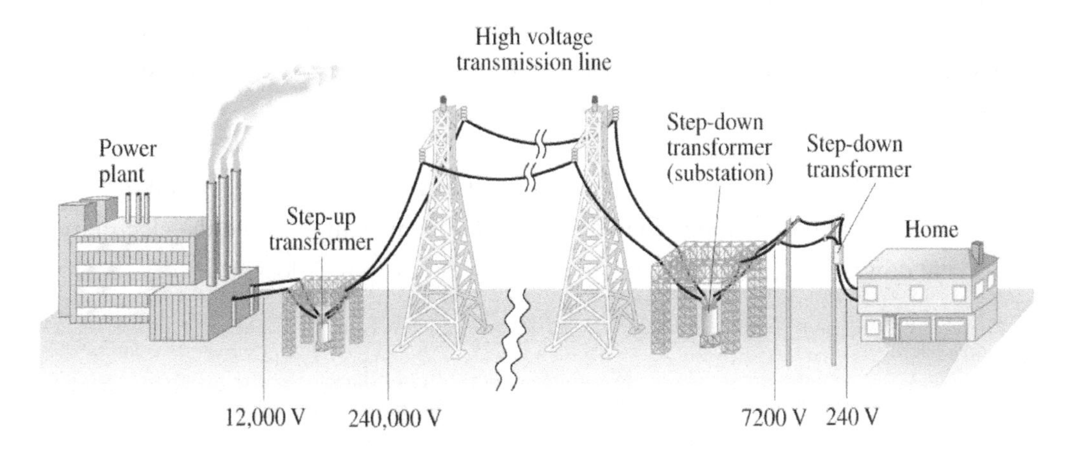

The ratio of the voltages in the primary and secondary coils of a transformer is equal to the ratio of the number of turns in each coil:

$$\frac{V_S}{V_P} = \frac{N_S}{N_P}$$

where V_S is the voltage through the secondary coil (V), V_P is the voltage through the primary coil (V), N_S is the number of turns in the secondary coil and N_P is the number of turns in the primary coil.

Power is determined by calculating the product of voltage (V) and the current (I). The input power in the primary coil must equal the output power in the secondary coil.

This relationship between voltage and current is expressed by:

$$V_P I_P = V_s I_s$$

where I_p is the current through the primary coil (A) and I_s is the current through the secondary coil (A).

Energy must be conserved. Therefore, in an ideal system, the ratio of the currents must be the inverse of the ratio of turns:

$$\frac{I_P}{I_S} = \frac{N_S}{N_P}$$

The diagram below is of a step-up transformer. These types of transformer are used at power-generating stations to increase the voltage of the power lines and to reduce Ohmic losses in the lines.

Notice that the primary coil has far fewer turns than the secondary coil. Accordingly, the voltage through the secondary coil is much higher, and the electricity can be sent down power lines at a higher voltage with less power loss.

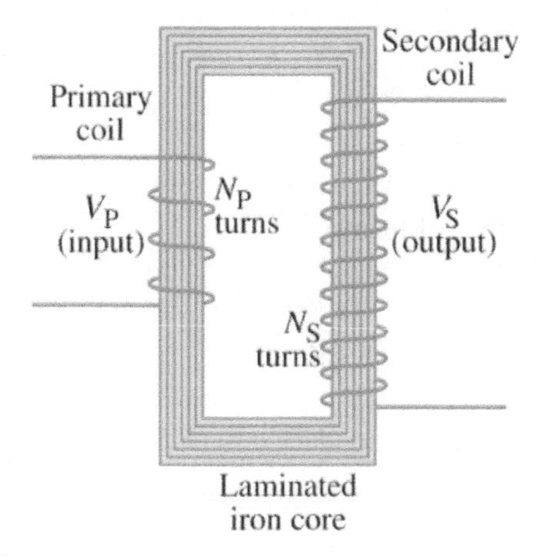

A step-up transformer: the primary coil has fewer turns than the secondary coil.

Measuring Devices

Electric meters measure various aspects of an electric circuit. There are many types of electric meters. However, the ammeter and the voltmeter are the two that are most used.

Ammeters

An *ammeter* is a device used to measure the current through a circuit. Ammeters usually have two connecting electrodes that are placed alongside the point in the circuit where current is to be measured. The current passes through the ammeter and the results are displayed by the ammeter.

An ammeter must have almost no internal resistance. The current in the circuit passes through the ammeter, so the ammeter needs to have low resistance as not to affect the reading for the current.

A *galvanometer* is an older but accurate type of ammeter. The figure below displays a typical galvanometer setup. Two electrodes are placed at a point in the circuit, and the current flows through them.

As the current passes through the coil, it creates a magnetic field whose direction depends upon the direction of the current flow. The permanent magnets on the side of the coil produce a torque about the coil, which then rotates to display the measured current against a calibrated dial.

The symbol for an ammeter in a circuit diagram is shown below:

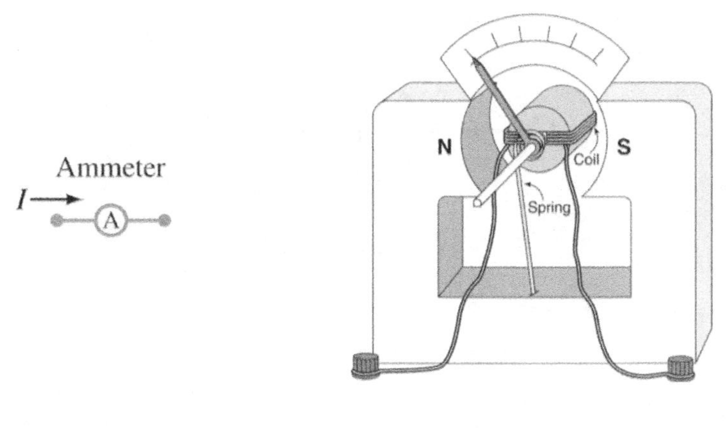

Galvanometer

Voltmeters

Voltmeters measure the voltage through two points within a circuit.

Like ammeters, voltmeters have two connecting electrodes, which are placed on either side of the element to be measured.

However, voltmeters are designed such that they have infinitely high resistance, theoretically pushing all the current through the circuit element and not letting any current flow through the voltmeter.

The figure below depicts a voltmeter connected across opposite sides of a resistor to measure the voltage through it.

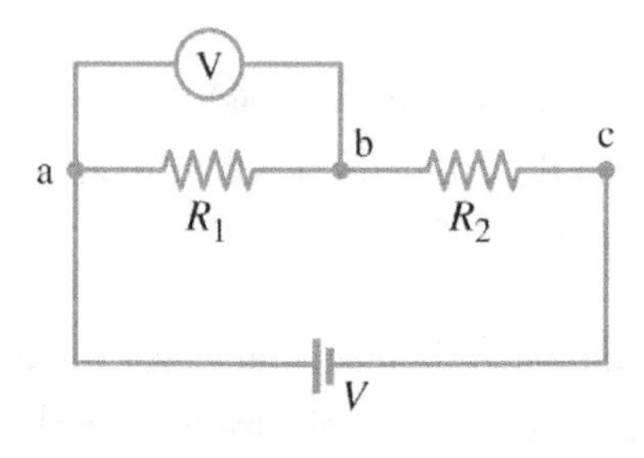

Voltmeter

Ohmmeter

An *ohmmeter* is a device used to measure electrical resistance.

Unlike ammeters and voltmeters, the ohmmeter has a battery that passes current through resistive elements to measure their resistance.

The figure below depicts an ohmmeter symbol:

The symbol indicating an ohmmeter

Chapter Summary

Electric Potential

- Electric potential is potential energy per unit charge: $V_a = \frac{PE_a}{q}$
- A source of an electromotive force (*EMF*) transforms energy from some other form of electrical energy.
- A battery is a source of constant potential difference.
- An electric potential of a point charge:

$$V = k\frac{Q}{r} \quad \text{[single point charge V = 0 at r = ∞]} = \frac{1}{4\,\pi\,\epsilon_0}\frac{Q}{r}$$

- Kirchhoff's rules:

  The sum of the currents entering a junction equals the sum of the currents leaving it.

  The total potential difference around a closed loop is zero.

Magnetic Fields and Flux

- Magnetic flux: $\Phi_B = B_{\perp}A = BA\cos(\theta)$
- Changing magnetic flux induces EMF: $\mathcal{E} = -N\frac{\Delta\,\Phi_B}{\Delta t}$
- An induced EMF produces current that opposes the original flux change.
- A changing magnetic field produces an electric field.
- Energy density stored in a magnetic field:

$$u = \text{energy density} = \frac{1}{2} \times \frac{B^2}{\mu_0}$$

Transformers

- Transformers use induction to change voltage: $\frac{V_S}{V_P} = \frac{N_S}{N_P}$
- Mutual inductance: $\mathcal{E}_2 = -M\frac{\Delta I_1}{\Delta t}$, $\mathcal{E}_1 = -M\frac{\Delta I_2}{\Delta t}$

Circuits

- Electric current is the rate of flow of electric charge.
- Conventional current is in the direction that positive charge would flow.
- A direct current is constant.
- An alternating current varies sinusoidally:

$$I = \frac{V}{R} = \frac{V_0}{R}\sin\omega t = I_0 \sin\omega t$$

- Power in an electric circuit: $P = IV$
- An RC circuit has a characteristic time constant: $\tau = RC$

Resistance

- Resistance is the ratio of voltage to current: $V = IR$
- Ohmic materials have constant resistance, independent of voltage.
- Resistance is determined by shape and material: $R = \rho\frac{\ell}{A}$ (ρ is the resistivity)
- A battery is a source of EMF in parallel with internal resistance.
- Resistors in series: $R_{eq} = R_1 + R_2 + R_3$
- Resistors in parallel: $\frac{1}{R_{eq}} = \frac{1}{R_1} + \frac{1}{R_2} + \frac{1}{R_3}$

Capacitance

- Capacitors are non-touching conductors carrying an equal and opposite charge.
- Capacitance is given by the equation: $Q = CV$
- The capacitance of a parallel-plate capacitor is: $C = \epsilon_0 \frac{A}{d}$
- A dielectric is an insulator (usually used between the plates of a parallel-plate capacitor)
- Dielectric constant gives a ratio of the total field to an external field.
- Energy density in an electric field:

  $$\text{Energy density} = \frac{PE}{volume} = \frac{1}{2} \epsilon_0 E^2$$

- Capacitors in parallel:
  - $V = C_{eq} V = C_1 V + C_2 V + C_3 V = (C_1 + C_2 + C_3)V$
  - $C_{eq} = C_1 + C_2 + C_3$
- Capacitors in series:
  - $\frac{Q}{C_{eq}} = \frac{Q}{C_1} + \frac{Q}{C_2} + \frac{Q}{C_3} = Q\left(\frac{1}{C_1} + \frac{1}{C_2} + \frac{1}{C_3}\right)$
  - $\frac{1}{C_{eq}} = \frac{1}{C_1} + \frac{1}{C_2} + \frac{1}{C_3}$

Average RMS

- Current: $I_{rms} = \sqrt{\overline{I^2}} = \frac{I_0}{\sqrt{2}} = 0.707\, I_0$
- Voltage: $V_{rms} = \sqrt{\overline{V^2}} = \frac{V_0}{\sqrt{2}} = 0.707\, V_0$
- $I = \frac{\Delta Q}{\Delta t} = neAv_d$
- LRC series circuit: $z = \sqrt{R^2 + (X_L - X_c)^2}$

Practice Questions

1. Two parallel circular plates with radii 7 mm carrying equal-magnitude surface charge densities of ± 3 μC/m^2 are separated by a distance of 1 mm. How much stored energy do the plates have? (Use dielectric constant $k = 1$ and electric permittivity $\mathcal{E}_0 = 8.854 \times 10^{-12}$ F/m)

A. 226 nJ **B.** 17 nJ **C.** 127 nJ **D.** 78 nJ **E.** 33 nJ

2. The potential difference between the plates of a parallel plate capacitor is 75 V. The magnitude of the charge on each plate is 3.5 μC. What is the capacitance of the capacitor?

A. 116×10^{-6} F
B. 37×10^{-6} F
C. 0.7×10^{-6} F
D. 478×10^{-6} F
E. 4.7×10^{-8} F

3. An electron is released from rest at a distance of 5 cm from a proton. How fast will the electron be moving when it is 2 cm from the proton? (Use Coulomb's constant $k = 9 \times 10^9$ N·m^2/C^2, the mass of an electron = 9.1×10^{-31} kg, the charge of an electron = -1.6×10^{-19} C and the charge of a proton = 1.6×10^{-19} C)

A. 92 m/s
B. 123 m/s
C. 147 m/s
D. 1.3×10^3 m/s
E. 4.8×10^5 m/s

4. A current of 5 A flows through an electrical device for 12 seconds. How many electrons flow through this device during this time? (Use the charge of an electron = -1.6×10^{-19} C)

A. 5.2×10^{18} electrons
B. 3.8×10^{20} electrons
C. 6.3×10^{19} electrons
D. 1.2×10^8 electrons
E. 9.1×10^{14} electrons

5. The distance of 1 mm separates two parallel plates. If the potential difference between them is 3 V, what is the magnitude of their surface charge densities? (Use the electric permittivity $\mathcal{E}_0 = 8.854 \times 10^{-12}$ F/m)

A. 64×10^{-9} C/m^2
B. 33×10^{-9} C/m^2
C. 27×10^{-9} C/m^2
D. 16×10^{-9} C/m^2
E. 76×10^{-9} C/m^2

6. A 3 Ω resistor is connected in parallel with a 6 Ω resistor; both resistors are connected in series with a 4 Ω resistor, and all three resistors are connected to an 18 V battery as shown. If 3 Ω resistor burnt out and exhibits infinite resistance, which of the following is correct?

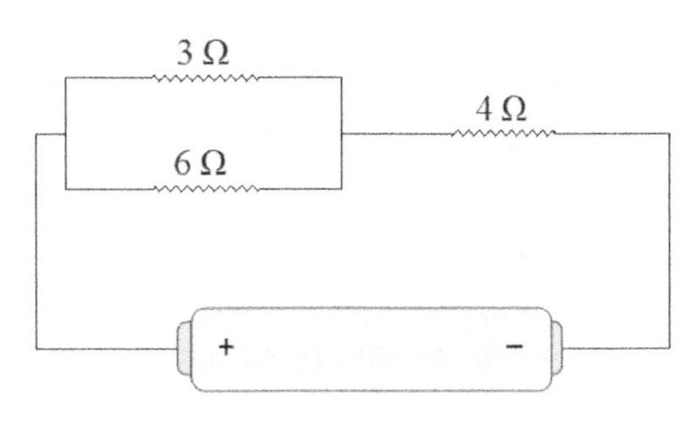

A. The power dissipated in the circuit increases
B. The current provided by the battery remains the same
C. The current in the 6 Ω resistor decreases
D. The current in the 4 Ω resistor decreases to zero
E. The current in the 6 Ω resistor increases

7. Each plate of a parallel-plate air capacitor has an area of 0.005 m^2, and the separation of the plates is 0.08 mm. What is the potential difference across the capacitor when an electric field of 5.6×10^6 V/m is present between the plates?

A. 367 V **B.** 578 V **C.** 448 V **D.** 227 V **E.** 834 V

8. What is the current in a wire if a total of 2.3×10^{13} electrons pass a given point in a wire in 15 s? (Use the charge of an electron = -1.6×10^{-19} C)

A. 0.25 μA **B.** 3.2 μA **C.** 7.1 μA **D.** 1.3 μA **E.** 0.88 μA

9. The resistivity of gold is 2.44×10^{-8} Ω·m at a temperature of 20 °C. A gold wire, 0.6 mm in diameter and 48 cm long, carries a current of 340 mA. What is the number of electrons per second passing a given cross-section of the wire? (Use the charge of an electron = -1.6×10^{-19} C)

A. 2.1×10^{18} electrons
B. 2.8×10^{14} electrons
C. 1.2×10^{22} electrons
D. 2.4×10^{17} electrons
E. 6.3×10^{15} electrons

10. When the frequency of the AC voltage across a capacitor is doubled, the capacitive reactance of that capacitor will:

A. become zero
B. decrease to ½ its original value
C. increase to 4 times its original value
D. decrease to ¼ its original values
E. increase to 2 times its original value

11. A charged particle of mass 0.006 kg is subjected to a 6 T magnetic field, which acts at a right angle to its motion. The particle moves in a circle of radius 0.1 m at a speed of 3 m/s. What is the magnitude of the charge on the particle?

A. 3 C **B.** 30 C **C.** 3.6 C **D.** 0.03 C **E.** 1.8 C

12. When a current flows through a metal wire, the moving charges are:

I. protons II. neutrons III. electrons

A. I only **B.** II only **C.** III only **D.** I and II only **E.** I and III only

13. A kilowatt-hour is equivalent to:

A. 3.6×10^6 J/s
B. 3.6×10^6 J
C. 3.6×10^3 W
D. 3.6×10^3 J
E. 3.6×10^3 J/s

14. For the graph shown, what physical quantity does the slope of the graph represent?

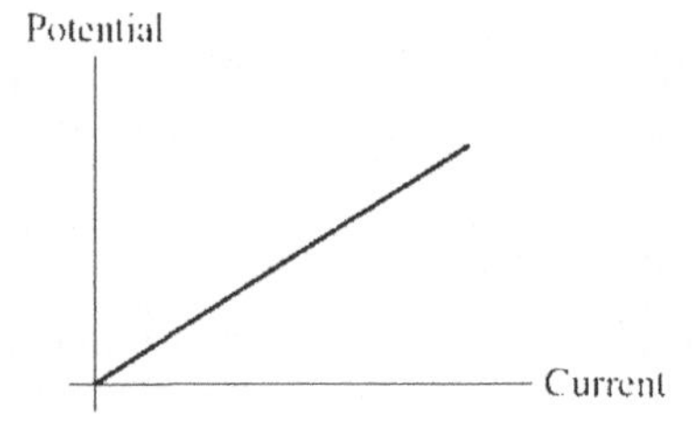

A. 1 / Voltage
B. Resistance
C. Power
D. Charge
E. 1 / Resistance

15. What is the voltage across a 15 Ω resistor that has 5 A current passing through it?

A. 3 V **B.** 5 V **C.** 15 V **D.** 30 V **E.** 75 V

Solutions

1. D is correct.

Find the area:

$A = \pi r^2$

$A = \pi(7 \times 10^{-3}\ \text{m})^2$

$A = 1.54 \times 10^{-4}\ \text{m}^2$

Find the capacitance:

$C = A\mathcal{E}_0 k / d$

$C = [(1.54 \times 10^{-4}\ \text{m}^2)\cdot(8.854 \times 10^{-12}\ \text{F/m})\cdot(1)] / (1 \times 10^{-3}\ \text{m})$

$C = 1.36 \times 10^{-12}\ \text{F}$

Find the charge:

$\sigma = 3 \times 10^{-6}\ \text{C/m}^2$

$\sigma = Q / A$

$Q = \sigma A$

$Q = (3 \times 10^{-6}\ \text{C/m}^2)\cdot(1.54 \times 10^{-4}\ \text{m}^2)$

$Q = 4.6 \times 10^{-10}\ \text{C}$

Find potential energy:

$U = ½Q^2 / C$

$U = ½(4.6 \times 10^{-10}\ \text{C})^2 / (1.36 \times 10^{-12}\ \text{F})$

$U = 78 \times 10^{-9}\ \text{J}$

2. E is correct.

Use the relationship:

$Q = CV$

$C = Q / V$

$C = (3.5 \times 10^{-6}\ \text{C}) / (75\ \text{V})$

$C = 4.7 \times 10^{-8}\ \text{F}$

3. B is correct.

Use the energy relationship:

$PE_{before} = PE_{after} + KE$

Electrostatic Potential Energy:

$PE = kQq / r$

Solve:

$kQq / r_1 = kQq / r_2 + ½mv^2$

$½mv^2 = kQq(1 / r_1 – 1 / r_2)$

$v^2 = (2kQq / m)·(1 / r_1 – 1 / r_2)$

$v = \sqrt{[(2kQq / m)·(1 / r_1 – 1 / r_2)]}$

$v = \sqrt{[(2)·(9 \times 10^9\ N·m^2/C^2)·(1.6 \times 10^{-19}\ C)·(1.6 \times 10^{-19}\ C) / (9.1 \times 10^{-31}\ kg)]·[(1 / 0.02\ m) – (1 / 0.05\ m)]}$

$v = 123$ m/s

Note: only the magnitude of the charge is used.

4. B is correct.

1 amp = 1 C/s

# of electrons = (5 C/s)·(12 s / 1)·(1 electron / 1.6×10^{-19} C)

# of electrons = 3.8×10^{20} electrons

5. C is correct.

$Q = A\mathcal{E}_o V / d$

$Q / A = \mathcal{E}_o V / d$

$Q / A = (8.854 \times 10^{-12}\ F/m)·(3\ V) / (1 \times 10^3\ m)$

$Q / A = 27 \times 10^{-9}\ C/m^2$

6. E is correct.

The equivalent resistance of the 3 Ω and 6 Ω resistors is:

$1 / R_{eq} = 1 / (3\ \Omega) + 1 / (6\ \Omega)$

$R_{eq} = 2\ \Omega$

The voltage across the equivalent resistor (i.e., the 6 Ω resistor) is given by the voltage divider relationship:

$V_6 = 18\ \text{V}\ (2\ \Omega) / (2\ \Omega\ + 4\ \Omega) = 6\ \text{V}$

The current through the 6 Ω resistor is:

$I_6 = V_6 / 6\ \Omega = 1\ \text{A}$

After the 3 Ω resistor burns out, the voltage across the 6 Ω resistor is found using the voltage divider relationship:

$V_6' = 18\ \text{V}\ (6\ \Omega) / (6\ \Omega\ +\ 4\ \Omega) = 10.8\ \text{V}$

Now the current through the 6 Ω resistor is:

$I_6' = V_6' / 6\ \Omega = 1.8\ \text{A}$

The current has increased.

7. C is correct.

Parallel-plate capacitor voltage:

$V = Ed$

$V = (5.6 \times 10^6\ \text{V/m})\cdot(0.08 \times 10^{-3}\ \text{m})$

$V = 448\ \text{V}$

8. A is correct.

1 amp = 1 C/s

$I = (2.3 \times 10^{13}\ \text{electrons} / 1)\cdot(1.6 \times 10^{-19}\ \text{C} / 1\ \text{electron})\cdot(1 / 15\ \text{s})$

$I = 0.25 \times 10^{-6}\ \text{amps}$

$I = 0.25\ \mu\text{A}$

9. A is correct.

Use the definition of one ampere:

$$1 \text{ amp} = 1 \text{ C/s}$$

Let the current expressed in electrons per second be denoted by I_e.

Solve:

$$I_e = (340 \times 10^{-3} \text{ C} / 1 \text{ s})\cdot(1 \text{ electron} / 1.6 \times 10^{-19} \text{ C})$$

$$I_e = 2.1 \times 10^{18} \text{ electrons/s}$$

10. B is correct.

Capacitive Reactance formula:

$$X_c = 1 / (2\pi Cf)$$

If f is doubled:

$$X_{c2} = 1 / [2\pi C(2f)]$$

$$X_{c2} = ½[1 / (2\pi Cf)]$$

$$X_{c2} = ½X_c$$

Capacitive reactance is halved.

11. D is correct.

Find the forces acting on the particle:

$$F_{\text{centripetal}} = F_{\text{magnetic}}$$

$$ma_{\text{centripetal}} = qvB$$

$$mv^2 / r = qvB$$

$$q = mv / Br$$

$$q = [(0.006 \text{ kg})\cdot(3 \text{ m/s})] / [(6 \text{ T})\cdot(0.1 \text{ m})]$$

$$q = 0.03 \text{ C}$$

12. C is correct.

In a solid, the locations of the nuclei are fixed; only the electrons move.

13. B is correct.

1 kW = 1,000 W

1,000 W = 1,000 J/s

1 hr = (60 min)·(60 sec/min)

1 hr = 3,600 s

kW·hr = (1,000 J/s)·(3,600 s)

kW·hr = 3.6×10^3 J

14. B is correct.

$V = IR$

$R = V / I$

15. E is correct.

Voltage = current × resistance

$V = IR$

$V = (5 \text{ A})\cdot(15\ \Omega)$

$V = 75 \text{ V}$

Chapter 9

Electrostatics and Electromagnetism

ELECTROSTATICS

- **Charges, Electrons, Protons, Conservation of Charge**
- **Conductors, Insulators**
- **Coulomb's Law**
- **Electric field *E***
- **Potential Difference, Electric Potential at Point in Space**
- **Equipotential Lines**
- **Electric Dipole**
- **Electrostatic Induction**
- **Gauss's Law**

MAGNETISM

- **Magnetic Fields and Poles**
- **Magnetic Field Force**
- **Faraday's Law**
- **Torque on Current-Carrying Wire**

ELECTROMAGNETIC RADIATION (LIGHT)

- **Properties of Electromagnetic Radiation**
- **Classification of Electromagnetic Spectrum, Photon Energy**

Charges, Electrons and Protons, Conservation of Charge

What is charge?

Like mass, the *charge* is an intrinsic property of all matter. Every particle in the universe has a mass and a charge measured in coulombs (C). Unlike mass, however, three distinct types of charges exist: positive charge, negative charge, and neutral charge. (Neutral charge does NOT mean that the particle has no charge, the net charge is neutral.)

Where does charge come from?

Today, it is known that the basic unit of matter is the atom, which has electrons orbiting a nucleus of protons and neutrons. These subatomic particles, specifically the electron and proton, are the basis of all macroscopically observed charge phenomenon. The electron is negatively charged and has a measured charge of -1.60×10^{-19} coulombs.

Conversely, the proton is positively charged and has an equal but opposite charge of $+1.60 \times 10^{-19}$ coulombs. For macroscopic objects that contain a charge, it is essential to note that only the electrons contribute to the charge because they are mobile. Protons are part of the core nuclei of the atom and are not mobile.

The charge on an atom is either due to an excess of electrons (if the observed charge is negative) or a shortage of electrons (if the observed charge is positive).

Negative charge: # electrons > # protons

Positive charge: # electrons < # protons

Neutral charge: # electrons = # protons

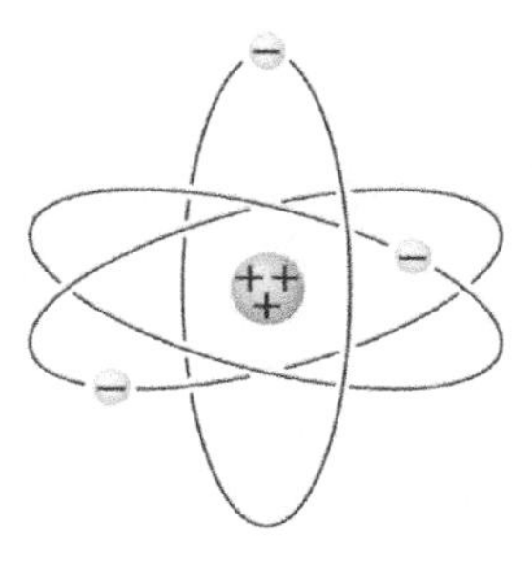

Schematic of an atom with a core of protons and neutron and orbiting electrons

What are the properties of charge?

Charges have several unique properties that govern the laws of nature. Most importantly, charges exhibit force upon each other according to the types of charge. The same charges produce repulsive forces, and opposite charges produce attractive forces.

The figures below show a ruler that is negatively charged (i.e., excess of electrons) and a glass rod that is positively charged (i.e., deficiency of electrons).

In the top diagram, two negatively charged rulers are brought close together and repel due to their like charges.

In the middle diagram, two positively charged rulers are brought close together and repel due to their like charges.

In the bottom diagram, the charged ruler and the glass rod attract each other due to their opposite charges.

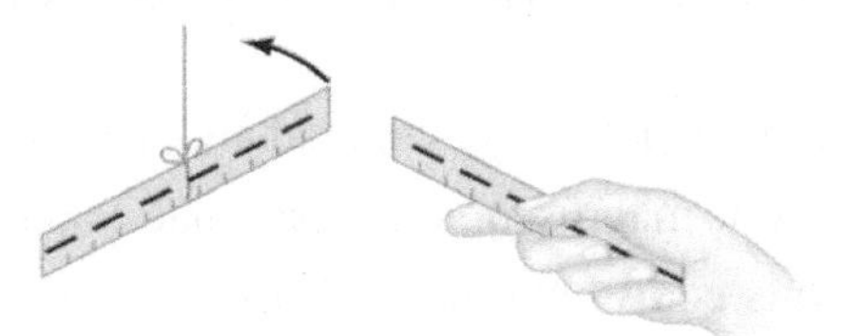

Two charged plastic rulers repel

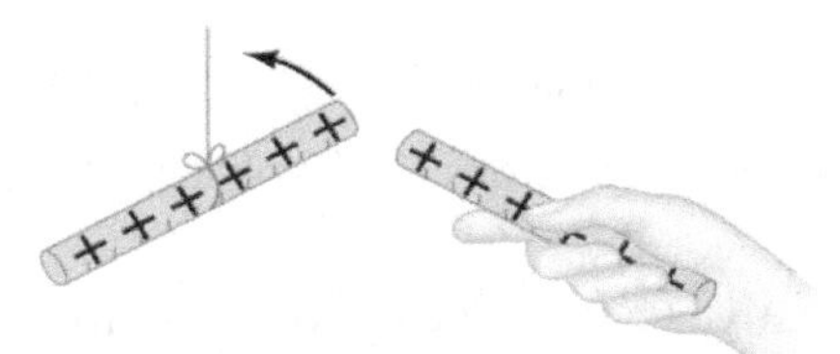

Two charged glass rods repel

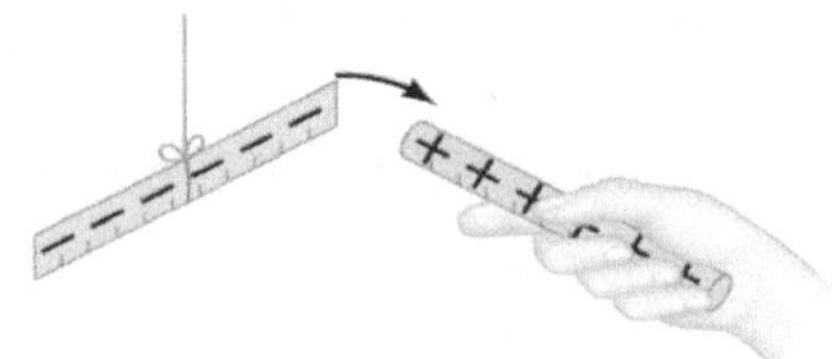

Charged glass rod attracts charged plastic ruler

The charge is always conserved. Like mass and energy, the charge cannot be created nor destroyed. The charge can only be transferred from one source to another.

If a charged object (e.g., the ruler in the above example) is placed on a table for some time, and later found to be neutrally charged, the charge was not destroyed, but instead transferred to the surroundings (e.g., air and table).

Another unique property of charge is that all charge is quantized.

The magnitude of the electron's charge (or proton) is the *fundamental charge* and is the smallest unit of charge that exists.

All macroscopically charged objects have a net charge equal to an integer multiple of the fundamental charge.

The fundamental charge can be expressed as:

$$q = ne$$

where q is the net charge (C), n is the integer multiple (i.e., excess electrons or protons), and e is the magnitude of the fundamental charge (1.60×10^{-19} C).

The quantization of charge enables the calculation for the excess or deficiency of electrons in an object that contributes to the object's net charge.

For example, if the negatively charged ruler from the earlier example is measured to have one coulomb of charge, then the number of excess electrons in the ruler is:

$$n = \frac{q}{e}$$

$$n = \frac{1.00 \text{ Coulomb}}{1.60 \times 10^{-19} \text{ Coulomb}}$$

$$n = 6.25 \times 10^{18} \text{ electrons}$$

Conductors, Insulators

Conductivity is a measure of a material's ability to transmit a charge.

Conductors are materials that have a high conductivity. These types of materials allow electrical charge (electrons) to "flow" through the material.

Copper is a good conductor. In most wires for electrical equipment, the electricity is transferred via copper wires because copper allows the electrical charges to flow easily.

Insulators are materials that do not transmit charge well and are poor conductors. Conventional insulators are wood, glass, and paper.

In the diagram on the left, one sphere has been charged positively (i.e., a deficit of electrons), and the other sphere is neutral.

In the middle diagram, a metal nail is placed on top of both spheres. The metal nail acts as a conductor and allows electrons to flow through it (from the neutral to the positive sphere) until the charge in each sphere is equal.

In the diagram on the right, a piece of wood is placed across the spheres. The piece of wood is an insulator and does not allow the flow of electrons from the neutral sphere to the positive sphere.

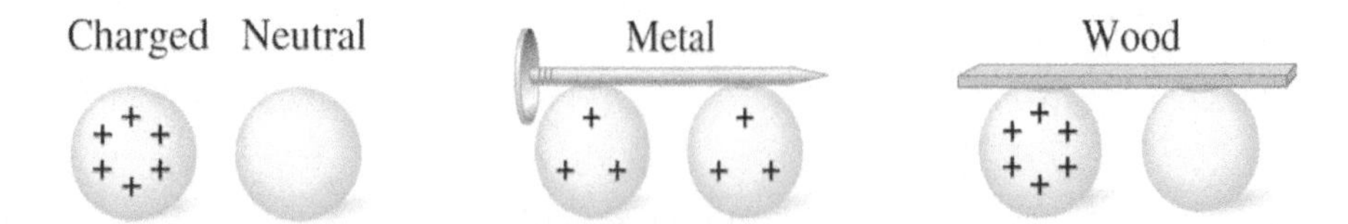

It is the electrons (i.e., negatively charged particles) that move and not the protons (i.e., positively charged particles).

Coulomb's Law

As mentioned earlier, charges either repel or attract one another, depending on if the charges are similar (repel) or opposite (attract).

This resulting force is the *electrostatic force* (Coulomb force) and can be found using *Coulomb's Law*, which describes the interaction between two charged particles:

$$F = \frac{1}{4\pi \epsilon_0} \frac{q_1 q_2}{r^2}$$

which simplifies to:

$$F = k \frac{q_1 q_2}{r^2}$$

where F is the electrostatic force between the two charges (N), q is the magnitudes of the charges (C), r is the distance between them (m), ϵ_0 is the permittivity of free space (8.854×10^{-12} $C^2/N{\cdot}m^2$), and k is Coulomb's constant with a value of 9×10^9 $N{\cdot}m^2/C^2$.

Coulomb's Law strictly applies only to point charges. If two objects have a net charge, then the objects must be approximated as point charges to calculate the force each exerts on the other.

The electrostatic force is always along the line connecting the charges.

If the charges have the same sign, the force is repulsive.

If the charges have opposite signs, the force is attractive.

To find the electrostatic force on more than two point charges, use the superposition principle and sum the force from each charge in all axial directions.

For example, what is the net force on Q_3 in the example below?

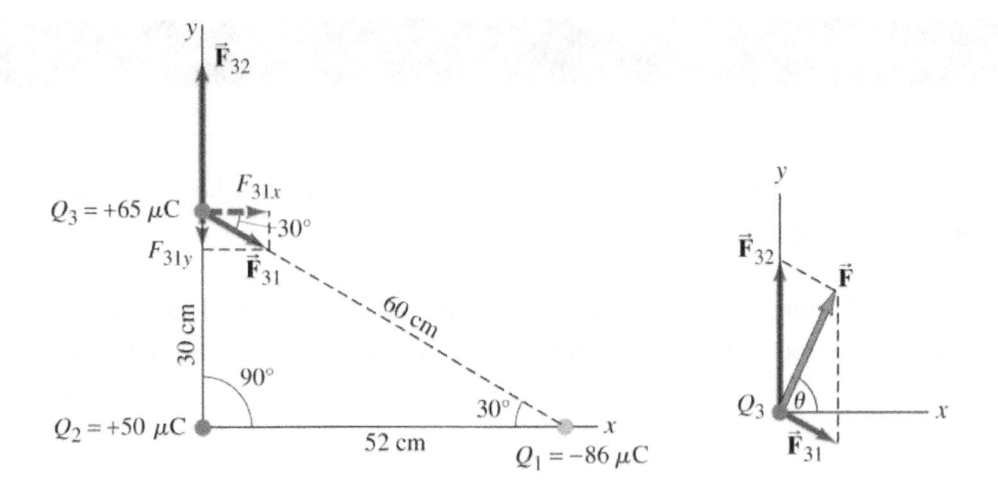

$$F_{32} = k\frac{q_1 q_2}{r^2} = \frac{(9 \times 10^9 \text{Nm}^2)\cdot(65 \times 10^{-6}\text{C})\cdot(50 \times 10^{-6}\text{C})}{(0.3\ \text{m})^2} = 325\ \text{N}$$

$$F_{31y} = \sin\theta\, k\frac{q_1 q_2}{r^2} = \left|\frac{\sin 30\,(9 \times 10^9 \text{Nm}^2)\cdot(65 \times 10^{-6}\text{C})\cdot(-86 \times 10^{-6}\text{C})}{(0.6\ \text{m})^2}\right| = 70\ \text{N}$$

$$F_{31x} = \cos\theta\, k\frac{q_1 q_2}{r^2} = \left|\frac{\cos 30\,(9 \times 10^9 \text{Nm}^2)\cdot(65 \times 10^{-6}\text{C})\cdot(-86 \times 10^{-6}\text{C})}{(0.6\ \text{m})^2}\right| = 121\ \text{N}$$

$$F_{net} = \sqrt{(121\ \text{N})^2 + (325\ N - 70\ \text{N})^2} = 282\ \text{N}$$

Electric Field *E*

Every charge is surrounded by *electric fields* (E), which emanate in all directions around the charge. These fields are directional, vector fields. The orientation of any field is in the direction that a positive charge is pushed when placed in the field.

Consequently, for positive charges, the electric field points outward in every direction, and for negative charges, the electric field points inward in every direction.

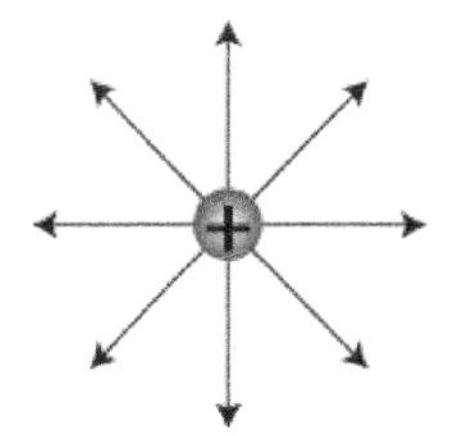

Electric field for (+) charge

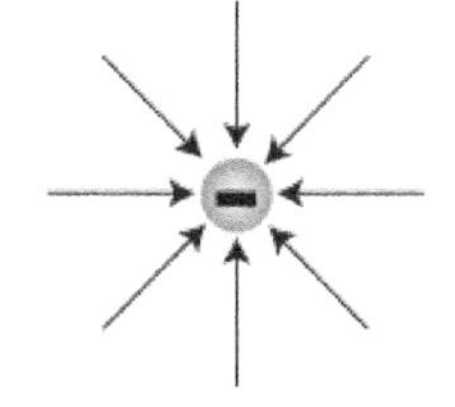

Electric field for (–) charge

The electric field (E) can be expressed as the force (F) on a charge, divided by the magnitude of the charge (q):

$$\vec{E} = \frac{\vec{F}}{q}$$

where (E) is the electric field measured in Newton/Coulombs (N/C).

Substituting in the electrostatic equation for force, the electric field from a single point charge (Q) is:

$$\vec{E} = \frac{\vec{F}}{q} = \frac{k\,q\,Q\,/\,r^2}{q} = k\frac{Q}{r^2}$$

These equations are rearranged to solve for the force (F) on a point charge (q) in an electric field (E):

$$\vec{F} = q\vec{E}$$

In the top figure below, an electric field *E* points in an arbitrary direction.

In the middle figure, if a positive charge *q* is placed in the same electric field, then the force *F* on the charge is in the direction of the electric field *E*.

In the middle figure, if a negative charge *q* is placed in the same electric field *E*, then the force *F* is in the opposite direction.

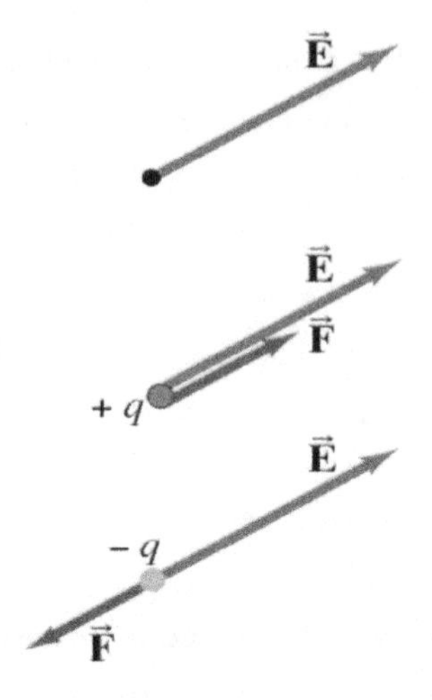

When solving electrostatics problems with electric forces and electric fields:

Draw a diagram of the entire situation.

Show all charges with signs, electric fields, and force.

Include the direction of electric fields and force.

Use Coulomb's Law and add the forces as vectors.

Like for all vectors, the superposition principle is used to calculate the net strength of multiple electric fields on a point charge:

$$\vec{E} = \overrightarrow{E_1} + \overrightarrow{E_2} + \ldots$$

The resultant vector (E) is calculated from the component vectors (E_1, E_2, ...)

Field lines

Like all field lines, the density of electric field lines denotes the strength of the field. For places where the field lines are close together, the electric field is stronger. If the electric field lines are spread out, the electric field is weaker.

The density of electric field lines determines the magnitude of the charge emitting from the field.

If the density of field lines is high, then the charge must have a higher magnitude.

If the density of the electric field is low, the charge must have a lower magnitude.

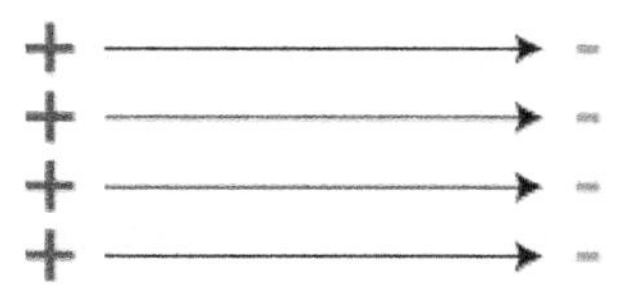

The electric field E points from the positive terminus to the negative terminus

Field due to the charge distribution

When two charges are brought close, their electric fields interact depending on the types of charges.

If two positive charges are brought close, the field lines repel.

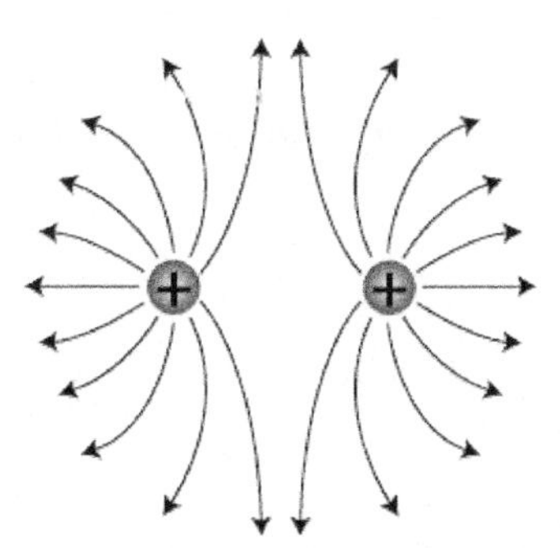

The electric field lines E repel upon approach for two positive point charges

If the two charges were negative, the field lines are the same as for two positive charges above, except that the direction of the field lines is reversed.

An *electric dipole* is created when two opposite charges are brought together.

In the diagram below, an electric dipole has the electric field lines originate from the positive charge and moves to the negative charge.

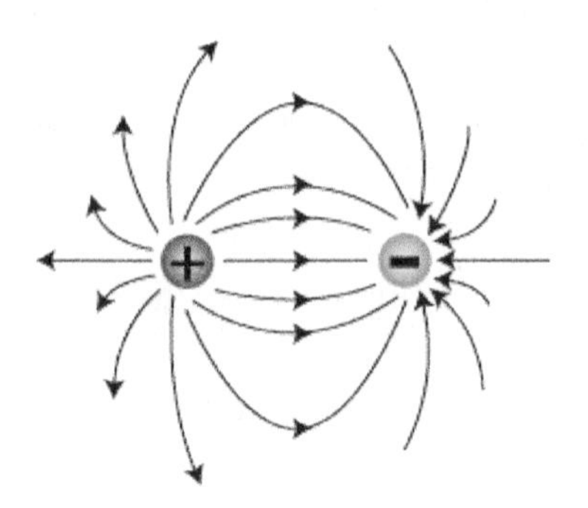

Electric field lines originate at the positive charge

The net charge on a conductor is on its surface, which radiates electric field lines.

For a charged cylinder, the electric field runs radially perpendicular to the cylinder and is zero (nonexistent) inside the cylinder. This effect is for all conducting objects; the net electric field is always from the surface and nonexistent within the object.

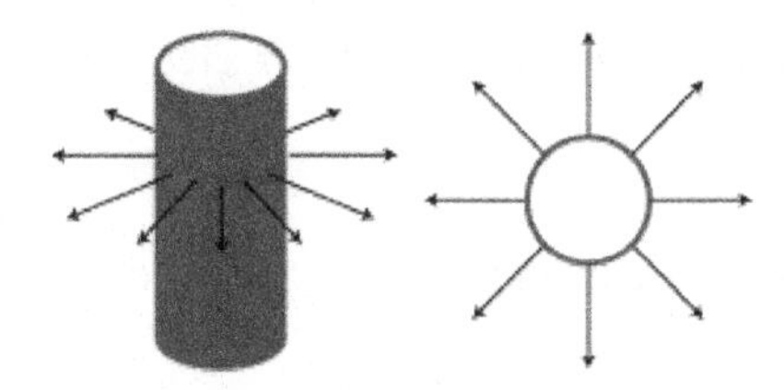

Electric field lines for a positive cylinder; sideview and cross sectional perspective

Summary of electric field lines:

1. Field lines indicate the direction of the field; the electric field is tangent to the line.

2. The magnitude of the field is proportional to the density of the lines.

3. Field lines start on positive charges and end on negative charges; the number is proportional to the magnitude of the charge.

4. The net electric field within a conductor is zero.

Potential Difference, Electric Potential at Point in Space

Because charges exert a force on each other in relation to the distance between the charges, charges must also have potential energy.

The *electric potential energy* has units of joules (J) and is written as:

$$U = \frac{k \cdot Q \cdot q}{r}$$

where U is the electric potential energy (J), Q and q are the point charges (C), r is the distance between the point charges (m), and k is Coulomb's constant with a value of 9×10^9 N·m²/C².

The electric potential energy of a single point charge is related to, but different from, its electric potential. The *electric potential* is the amount of energy per charge that something possesses.

The unit for electric potential is joules per coulomb (J/C) or volts (V) and is a scalar field associated with potential energy.

An electric potential is found using either of the following two equations.

$$V = \frac{U}{q} \ \ or \ \ V = \frac{kQ}{r}$$

where V is the electric potential (V), Q is the charge that is causing the potential (C), q is the charge experiencing the potential (the magnitude of q is small), U is the electrical potential energy possessed by q (J), k is Coulomb's constant and r is the distance between the two charges (always positive).

If multiple charges contribute to the electric potential, then the total electric potential is the sum of the potentials caused by the individual components.

Positive charges produce positive potentials.

Negative charges produce negative potentials.

The plot below on the left shows the potential due to a positive charge.

The plot below on the right shows the potential due to a negative charge.

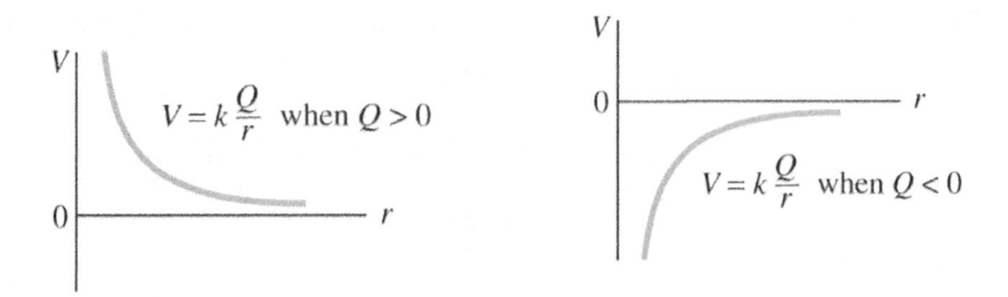

The potential difference is vital in producing forces and moving charges. The process is analogous to moving masses in gravitational fields.

ΔV gives the potential as the difference between the two potentials.

$$\Delta V = V_B - V_A$$

The potential difference is used for several applications. To determine the difference in potential between two plates of a capacitor or the positive and negative terminals of a battery. Using potentials instead of fields can make solving problems easier because the potential is a scalar quantity, whereas a field is a vector.

The Coulomb force is conservative. Therefore, the work in an electric field is conservative. As such, the work required to move a point charge in an electric field is only dependent upon the displacement of the charge, not the path taken by the point charge.

A straightforward method to calculate this work is to use electric potential differences, which can be written to relate it to work involved in positioning charges.

$$\Delta V = \frac{W}{q}$$

where ΔV is the electric potential difference (V), W is the work needed to move the point charge (J), and q is the magnitude of the moving charge (C).

Equipotential Lines

Equipotential lines are placed around a charge and mark where the electric potential is the same. Equipotential lines are always perpendicular to electric field lines.

In 3D, equipotential lines form a "surface" called an *equipotential surface*.

Any movement along an equipotential surface requires no work because the movement is perpendicular to the electric field.

The figure below depicts equipotential lines under different circumstances.

On the left is a single positive charge, along with its electric field lines and corresponding equipotential lines.

On the right, there is a dipole with its curved electric field lines and mirrored equipotential lines.

In both instances below, the equipotential lines are perpendicular to the electric field lines at every point.

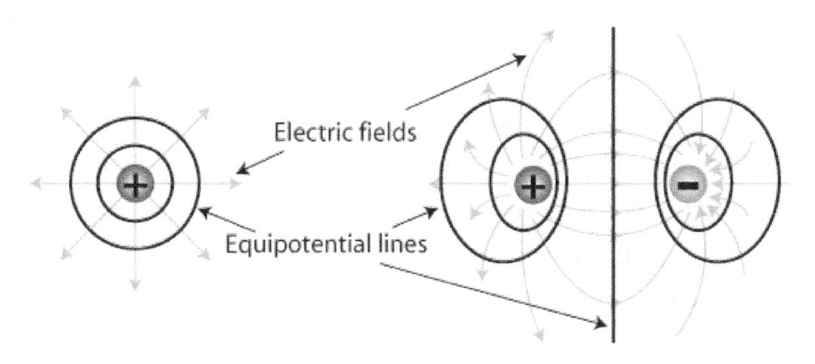

Equipotential lines are perpendicular to the electric field lines

Electric Dipole

Definition of dipole

A dipole is an interaction between a positive charge and a negative charge that is separated by some distance.

Behavior in an electric field

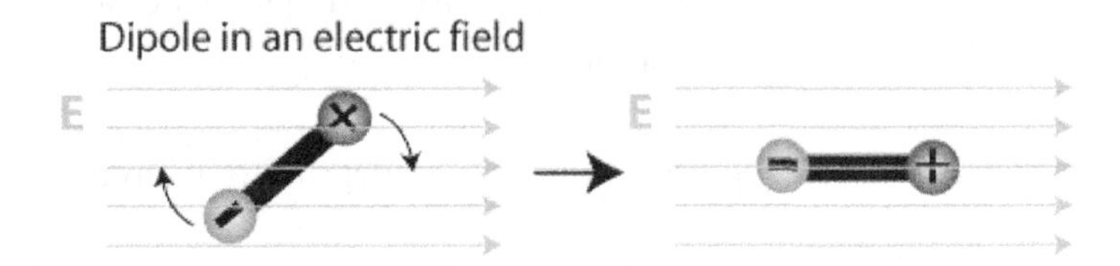

A dipole in an electric field aligns itself with the electric field. The positive end of the dipole is in the direction of the electric field (furthest from the source of the electric field). The dipole spins or rotates until it reaches such a position.

Many molecules exhibit electric dipoles and are *polar molecules*.

The water (H_2O) molecule shown below is polar. In water molecules, a dipole moment arises from the negative charge near the electronegative oxygen atom, and two positive charges near each hydrogen atom. The asymmetric distribution of charge creates an electric dipole, which plays a vital role in the physical and chemical properties of water.

A microwave heats food by heating the water in the food. This heating occurs because the oscillating electric field of a microwave (see the chapter on electromagnetic waves) rotates the water molecule such that the electric dipoles align with the electric field. The movement (kinetic energy) of water molecules creates heat, which increases the temperature of the food. Temperature is a measure of kinetic energy.

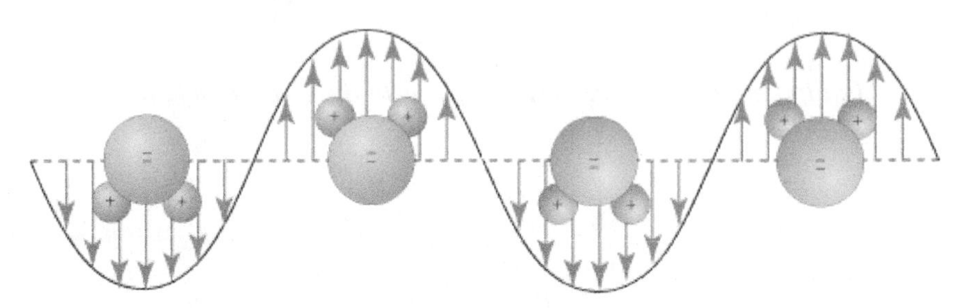

Water molecules have dipoles that align with the electric field

Electrostatic Induction

Electrostatic energy is a charge more or less fixed in a single place. However, this charge can be transferred to other objects in two ways.

Conduction is the transfer of energy where there is a point of direct contact between the two objects.

Induction is the transfer of energy where there is no direct contact between the two objects, and the transfer of energy is across an open space between two objects.

Electrostatic induction is where a charged object induces the movement or redistribution of charges in another object. Electrostatic induction occurs when an object is placed in or near an electric field.

Metal objects can be charged by conduction. In the diagram below, the metal rod B is initially neutral. When the neutral rod B touches a charged rod A, negatively charged electrons flow from the neutral rod B to positively charged rod A.

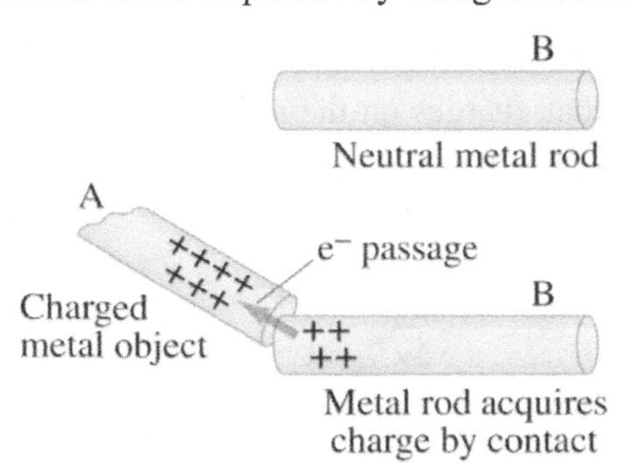

Induction in a neutral metal rod by a charged metal object

The rods can also be charged by induction.

In the figure below, the two rods never touch, and the neutral rod is grounded.

Electrons from the neutral rod can be transferred to the ground and leave the surface of their object.

In the middle diagram, when the negatively charged rod is brought near the neutral rod, the electric field attracts a positive charge. It causes the negatively charged electrons to be repelled and travel into the ground.

In the bottom diagram, if the connection to ground is cut (such that electrons cannot flow back into the system) and the negatively charged rod is removed, the neutral rod has a positive charge evenly distributed across its surface.

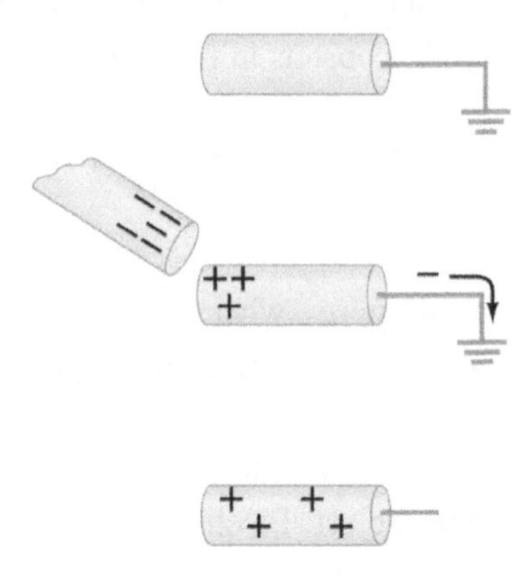

Insulators will not be charged by conduction or induction but experience a charge separation as *polarization*.

Unlike conductors, insulators do not have electrons that can move freely about the material. When a strong charge is brought near the insulator, the molecules become polarized and orient such that charges on the molecule align with the electric field (like electric dipoles in an electric field), as shown below.

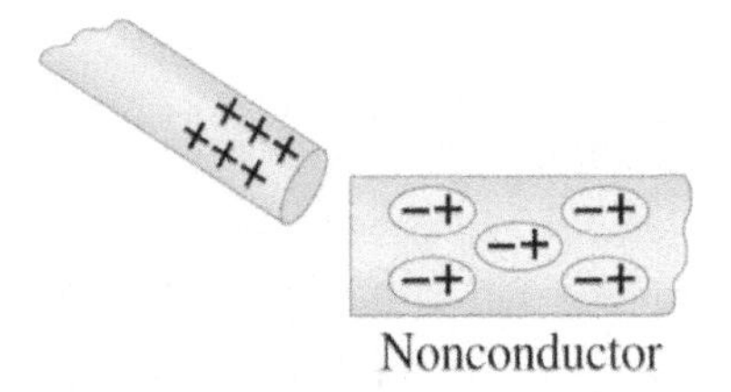

An *electroscope* detects charge in an object. Below is a schematic of a gold-leaf electroscope. The gold arms in the middle hang together if there is no charge, and the arms spread apart when a charge is present.

In the diagram on the left, an object is held near the metal knob on top.

In the diagram on the right, an object is held near the metal knob on top.

If electrons flow due to induction (no contact) or conduction (contact between objects), the gold leaves become similarly charged and repel.

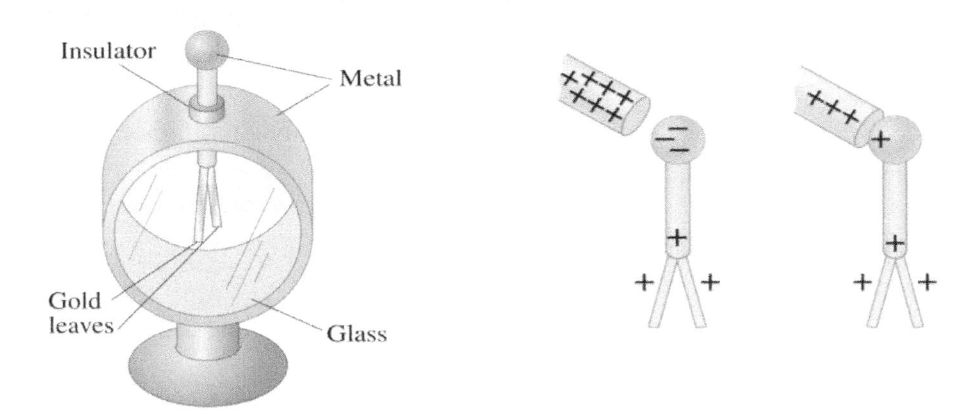

Electroscopes and the leaves that are used determines the sign of an unknown charge

The charged electroscope determines the sign of an unknown charge.

In the middle diagram below, if a negatively charged object is held near the knob and the leaves expand. Additional electrons are trying to get away from the like-charged object, causing an even stronger negative charge in both leaves.

In the diagram on the right, a positively charged object is held near the knob, and the leaves move toward each other. Electrons are moving into the knob to be close to the presence of the opposite charge. The leaves will not be as negatively charged as before, and there is less of a force driving the leaves apart.

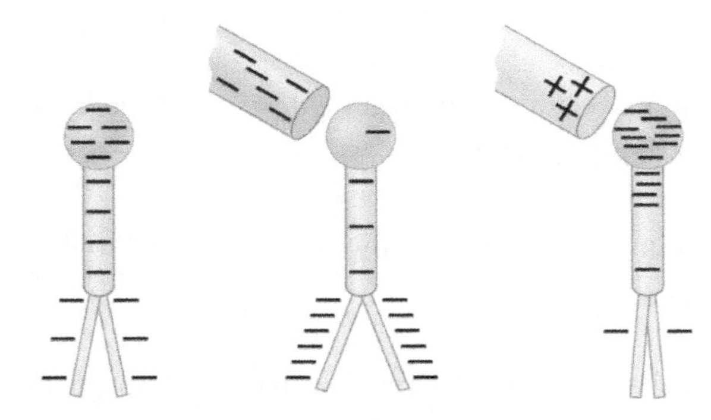

Electroscope determines the sign of an unknown charge

Gauss's Law

Gauss's Law relates the flow of a vector field through a surface to the behavior of the vector field inside the surface.

The vector field is an electric field and is expressed as:

$$\Phi_E = EA \cos(\theta)$$

where Φ_E is the electric flux (V·m), E is the electric field (N/C), A is the area that the field covers, and θ is the angle between the field and plane that runs perpendicular to the direction of field flow.

For an enclosed surface, the electric flux Φ_E is equal to the charge inside the enclosure (q) over the permittivity of free space (ε_0).

$$\Phi_E = \frac{q}{\varepsilon_0}$$

The net electric flux through an enclosed surface is dependent on the charge inside. If there is no charge inside, then the net electric flux through the enclosure is zero.

The *electric flux* through an area is proportional to the total number of field lines crossing the area.

In the top diagram below, an electric field passes through an area A that is at an angle with respect to the electric field.

In the middle diagram, the electric flux is proportional to the electric field that is perpendicular to that area.

In the bottom diagram, the electric field passing through the projection of the area is perpendicular to the electric field.

Regardless of the method, the two values will be equal.

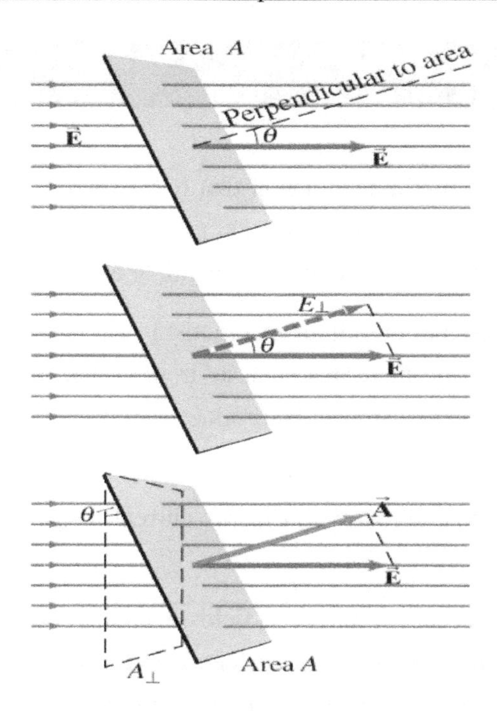

The flux through a closed surface of multiple electrical fields is found by:

$$\Phi_E = E_1 \Delta A_1 \cos(\theta_1) + E_2 \Delta A_2 \cos(\theta_2) + \ldots = \sum E \Delta A \cos(\theta) = \sum E \perp \Delta A$$

where $\sum$ represents the closed surface.

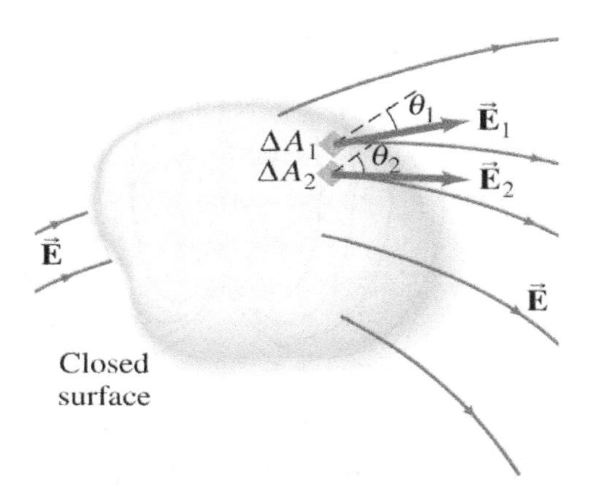

The net number of field lines through the surface is proportional to the charge enclosed and the flux. This gives *Gauss's Law* (or *Gauss's Flux Theorem*) that can be used to find the electric field when there is a high degree of symmetry in the system:

$$\sum E_{\perp} \Delta A = \frac{Q_{encl}}{\epsilon_0}$$

Magnetic Fields and Poles

Magnetism is a natural phenomenon that has essential applications in the modern world. For example, navigation with a compass, memory storage on a computer, and high-speed trains all require magnetism and an understanding of the properties of magnets.

The earliest assumption about magnetism was that it was associated with naturally occurring magnetic materials such as iron, cobalt, or nickel. It is known that there are ways in which magnetism can be produced.

On the atomic level, the electrons of material produce magnetism due to their motion (i.e., electrons spin and revolve). Electric currents can create magnets, the basis of electromagnetism.

All magnetic objects emit magnetic fields, which are visualized by magnetic field lines. These magnetic lines are similar to electric field lines and always start from the north pole (i.e., positive charge) and go towards the south pole (i.e., negative charge).

Unlike the electric field, magnetic fields always form closed loops because all magnets have a north pole and south pole.

The figure shows a bar magnet with the closed loop that magnetic fields created from the north pole to the south pole.

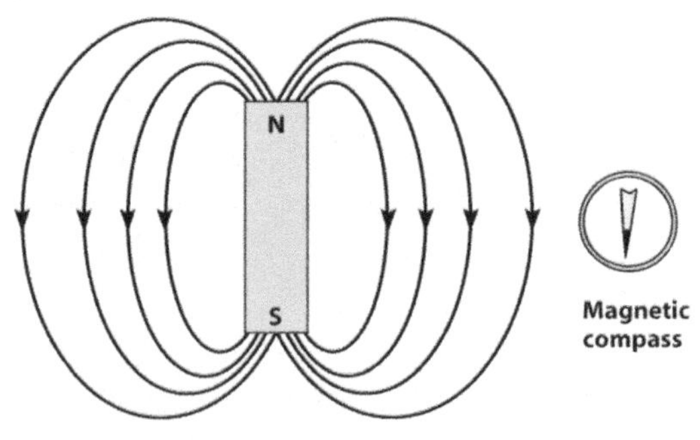

The Earth's magnetic field originates deep beneath the surface layer and is caused by currents of conductive elements in the Earth's molten core.

The magnetic field lines originate at the geographic south pole of Antarctica (magnetic north pole) and terminate at the geographic north pole (magnetic south pole).

A compass operates by aligning itself with the magnetic field of Earth (magnetic fields always form complete loops), allowing navigation with respect to a known axis.

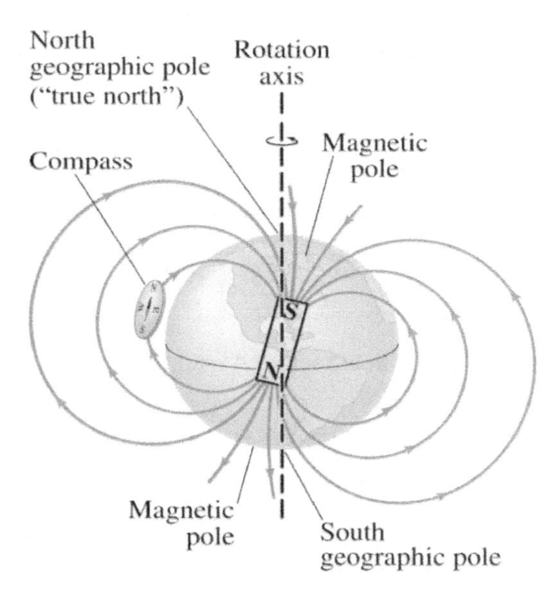

Another critical aspect of magnets is that they can never have only one pole; all magnets must have a north and south pole. If a magnet is broken into two halves, it will not produce two single polarity magnets. Instead, the two fragments each have a north and south pole. If these pieces are broken again, the process repeats.

Magnets also have some other similar behaviors to electrically charged particles. The poles of two different magnets exert a force on each according to their polarity. Opposite poles attract, while like poles repel.

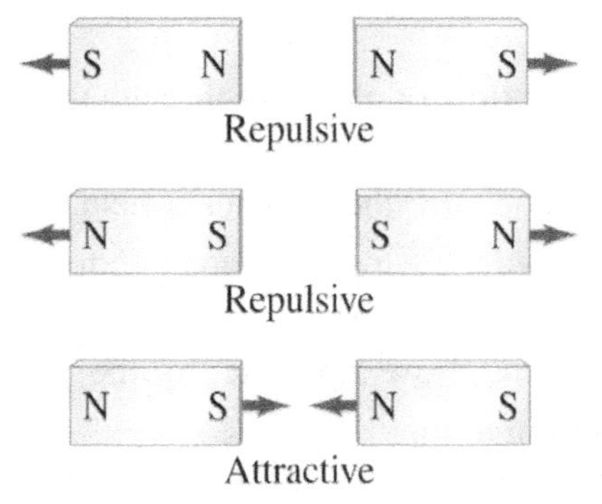

Magnet polarity determines attractive (opposite) or repulsive (similar)forces

Magnetic fields around a current-carrying wire

As discussed previously, the charge can be made to travel across a conductor if an electric potential difference exists. When this occurs, the traveling charge is an *electric current*, and is the rate of flow of charge through a conductor over time:

$$I = \frac{\Delta q}{\Delta t}$$

where I is the electric current given in amperes (A), and Δt is time (s).

Current is conventionally considered to be the movement of positive charge through a conductor. Although this is technically incorrect (protons are immobile, only electrons can "flow"), all diagrams and calculations of current follow this convention. The movement of electrons is opposite to that shown in illustrations.

When current flows through a wire, a magnetic field is formed that circulates the wire. The direction of the magnetic field can be found via the *right-hand rule*.

Observe the diagram below. According to the right-hand rule, the thumb goes in the direction of a positive charge (i.e., the direction of current by convention). The direction the fingers curl around in the direction of the magnetic field.

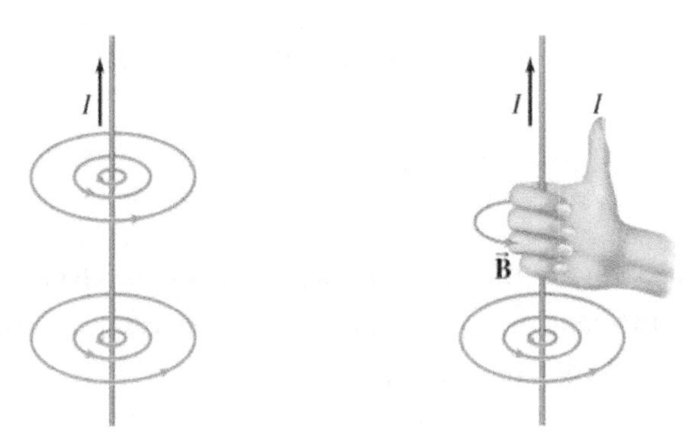

To understand the flow of electrons through the conductor, see the figure below.

Notice the direction of the magnetic field. The magnetic field is clockwise because current flows down the wire from top to bottom.

However, electrons flow opposite to the conventional current flow. Thus, electrons flow from bottom to top.

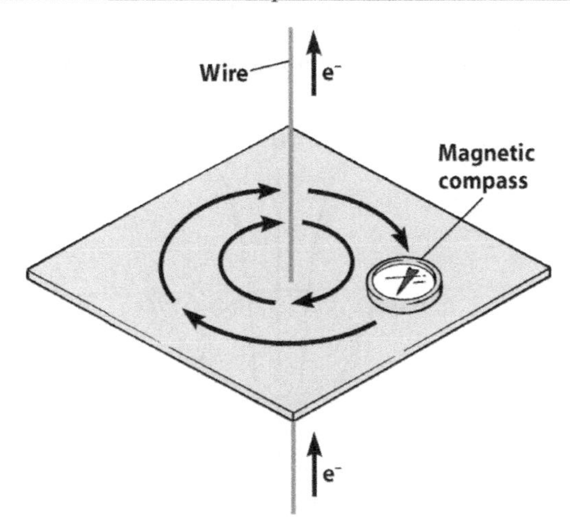

When a current (I) induces a magnetic field in a straight wire, the magnetic field (B) strength is calculated by *Ampère's Law*:

$$B = \frac{\mu_0 I}{2\pi r}$$

where μ_0 is the constant of permittivity of free space ($4\pi \times 10^{-7}$ T·m/A), I is the current (A), and r is the distance away from the wire (m).

Magnetic fields in solenoids

When a current-carrying wire is wrapped around, such that it forms a loop, it is a *solenoid.* Solenoids have many applications and are the basis for common electromagnets.

The primary importance of a solenoid is that the magnetic field can be manipulated, as seen in the diagrams:

Magnetic field shown from current flowing through solenoid

In the configuration below, the net magnetic field created by the loop is:

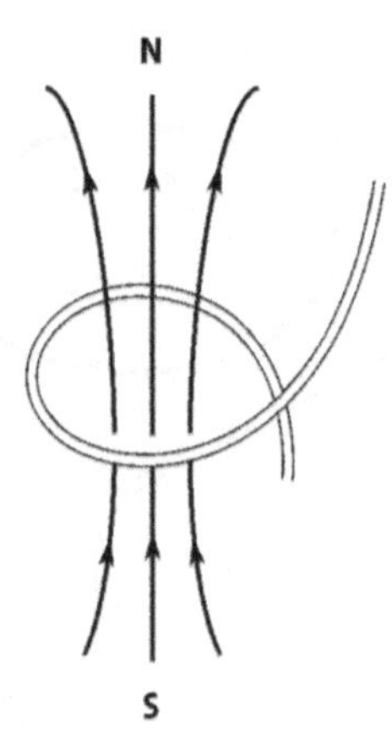

If the number of loops in the solenoid increases, the magnetic field increases proportionally with the number of loops.

If a piece of iron is inserted in the solenoid, the magnetic field dramatically increases.

From the equation for the magnetic field strength of a wire, the magnetic field strength through a solenoid (but not outside of it) can be expressed as:

$$B = \frac{\mu_0 \mu_r n I}{l}$$

where n is the number of loops, μ_r is the relative permittivity of core material ($\mu_r = 1$ in air), and l is the length of the solenoid (m).

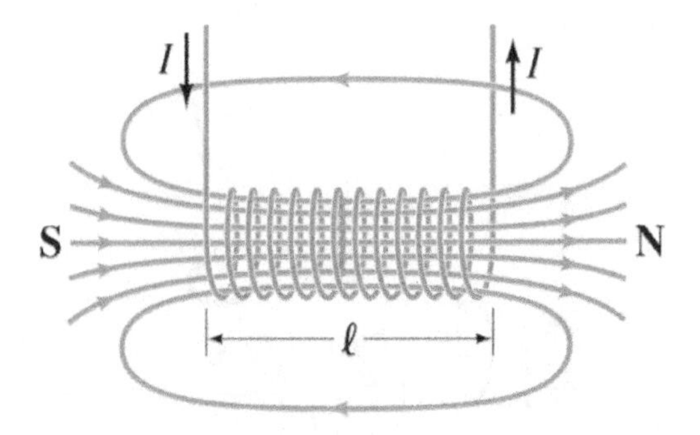

Magnetic Field Force

Particles

When a charged particle passes through a magnetic field with a specific velocity, which must be perpendicular to the magnetic field) it will experience a force from the magnetic field. This only occurs if the charge has a velocity perpendicular to the magnetic field. If its speed is zero or its direction is parallel to the field, then no force will be exerted:

$$F = qvB \sin(\theta)$$

where q is the electric charge (C), v is the velocity of the charge (m/s), B is the strength of the magnetic field given in the Tesla (T), and θ is the angle between the charge velocity and the magnetic field (degrees). $Sin(\theta)$ may be omitted, as θ is assumed to be 90°.

The magnetic force is always perpendicular to both the magnetic field and the velocity of the charge:

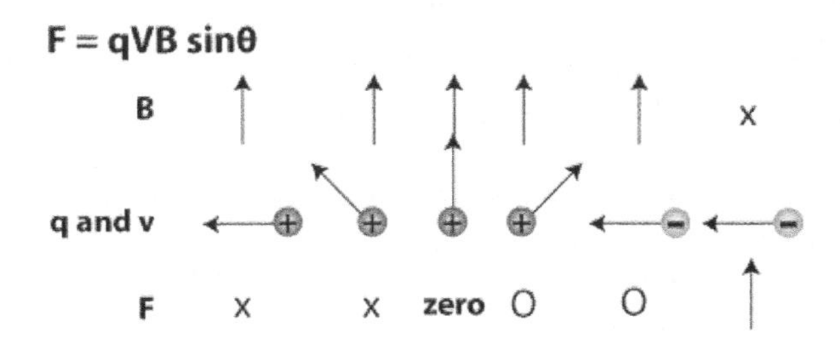

X represents into the page and O represents out of the page

From the right-hand rule (RHR), the thumb of the right-hand goes in the direction of a positive charge, the middle finger is the direction of the magnetic field, and the palm faces the direction of the force.

From the right-hand rule, if the charge is negative, then the direction of the force is in the opposite direction.

Physical Situation	Example	How to Orient Right Hand	Result
Magnetic field produced by current (RHR-1)	I I $\vec{B}$	Wrap fingers around wire with thumb pointing in direction of current I	Fingers curl in direction of $\vec{B}$
Force on electric current I due to magnetic field (RHR-2)	$\vec{F}$ $\vec{I}$ $\vec{B}$	Fingers first point straight along current I, then bend along magnetic field $\vec{B}$	Thumb points in direction of the force $\vec{F}$
Force on electric charge $+q$ due to magnetic field (RHR-3)	$\vec{F}$ $\vec{v}$ $\vec{B}$	Fingers point along particle's velocity $\vec{v}$, then along $\vec{B}$	Thumb points in direction of the force $\vec{F}$

Motion of charged particles in magnetic fields

When a charged particle travels through a magnetic field, it is deflected according to the magnitude and direction of the field. The deflection of the particle has important applications, such as mass spectrometers, because it can be related to the centripetal force to solve for the mass of the particle.

The equations for centripetal force and magnetic field force on a charged particle:

$$F_C = \frac{mv^2}{r} \qquad F_M = qvB \sin(\theta)$$

If the particle is assumed to be traveling perpendicular to the magnetic field, then the magnetic field force becomes:

$$F_M = qvB$$

If a particle with some velocity perpendicular to a magnetic field travels in a circular orbit, the mass can be solved for by:

$$\frac{mv^2}{r} = qvB$$

$$m = \frac{qBr}{v}$$

When $qvB < mv^2 / r$, there is not enough centripetal force, and the charged particle flies out of orbit.

When $qvB > mv^2 / r$, there is too much centripetal force, and the charged particle spirals inward.

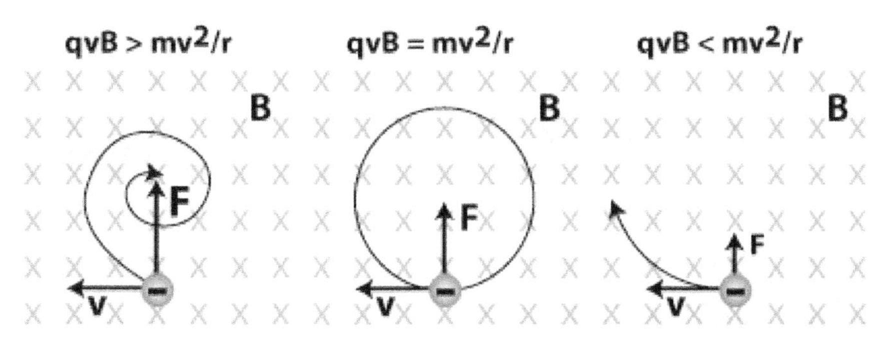

Charged particle as curved line trajectories based on magnetic field

Current-carrying wires

Wires carrying current have charged particles in motion through them.

A magnetic field exerts a calculable force on a current-carrying wire:

$$F = I\ell B \sin(\theta)$$

where I is the current through the wire (A), and ℓ is the length of wire exposed to the magnetic field (m).

Again, the direction of the force is given by the right-hand rule (RHR).

In the diagrams below, a length of current-carrying wire is exposed to a magnetic field.

In the diagram on the left, the current flow has with the magnetic field force exerting a downward force.

In the diagram on the right, the current is reversed, and the magnetic field force is opposite its original direction and point upwards.

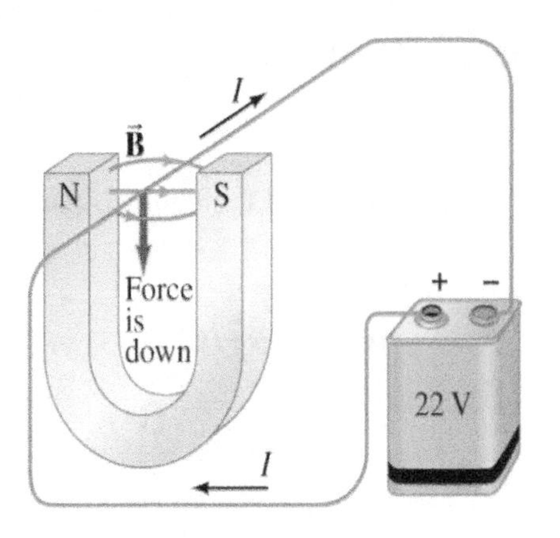

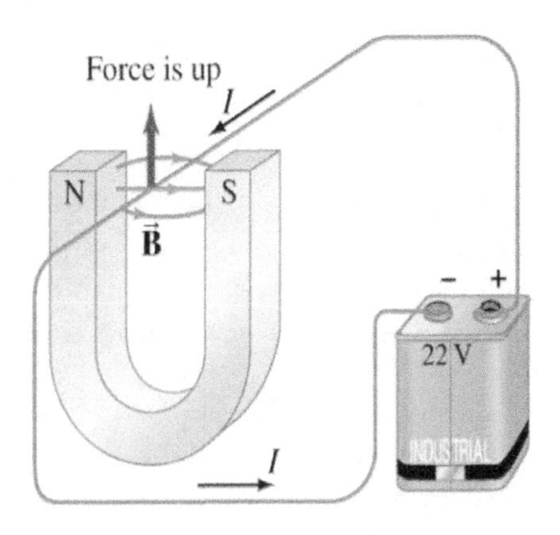

Parallel current-carrying wires

When two current-carrying wires, shown below, are parallel and close enough such that their magnetic fields interact, they exert a force on each other according to the magnitude and direction of their respective currents.

The magnitude of the magnetic field force per unit length of the parallel wire section is given by:

$$\frac{F}{l} = \frac{\mu_0}{2\pi}\frac{I_1 I_2}{d}$$

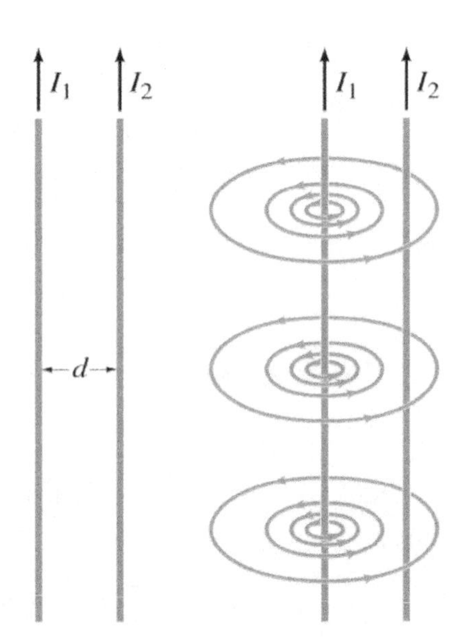

The direction of the force between the wires depends upon the direction of the currents in the wires with respect to the other.

Parallel currents (moving in the same direction) create an attractive force.

Anti-parallel currents (moving in opposite directions) create repulsive forces.

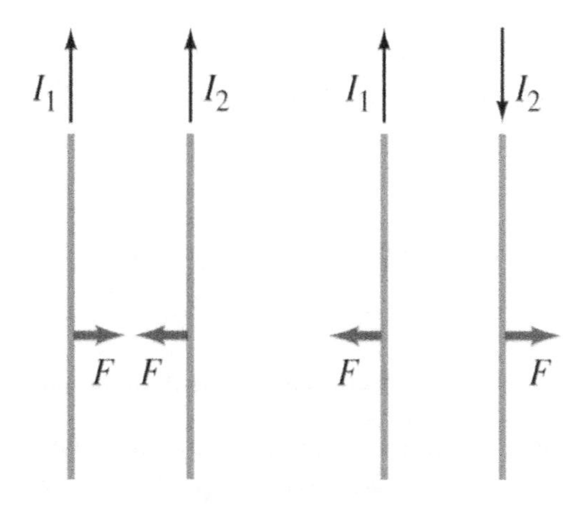

Force F exerted by current I flow in parallel (left) or antiparallel (right)

Faraday's Law

Michael Faraday (1791-1867) expressed the relationship that links the induced electromotive force (EMF) in a wire loop proportional to the rate of change of the magnetic flux through the loop. Faraday derived the *law of induction*. An understanding of magnetic flux is required to discuss the law of induction.

The magnetic flux (Φ_B) is similar to the electric flux and is defined as the amount of magnetic field passing through an area of a surface. The Weber (Wb) is the SI unit of magnetic flux, which is equal to volt-seconds. The Weber may also be given in tesla-meters-squared ($T \cdot m^2$).

The magnetic flux (Φ_B) equation is:

$$\Phi_B = B_{\perp}A = BA \cos(\theta)$$

where $B_{\perp}$ is the magnitude of the magnetic field that is perpendicular to the area (A).

If the field lines are not perpendicular to the area, then the right-hand equation must be used where θ is the angle between the field lines and the *vector normal to the surface* (i.e., *area vector*). The magnetic flux indicates the magnitude of the magnetic field that pierces through the coil.

The diagrams below demonstrate the varying amounts of magnetic flux through the coiled wire as it is rotated with respect to the magnetic field.

In the diagram on the left below, the magnetic flux is equal to zero because the coil *area vector* (i.e., the *coil area* vector points normal to the coil's cross-sectional area) is perpendicular, $\theta = 90°$, to the applied field.

The middle diagram has some magnetic flux because the *area vector* of the coil is at $\theta = 45°$ angle to the field.

The diagram on the right has the maximum flux because the *area vector* is entirely parallel, $\theta = 0°$, to the magnetic field.

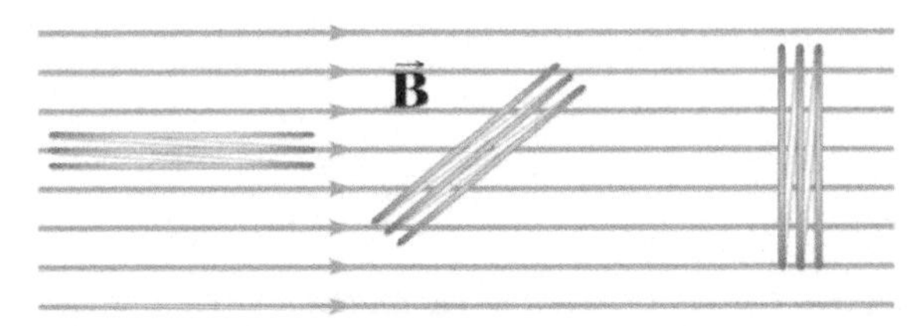

In the left panel above, $\theta = 90°$, $\Phi_B = 0$; in the center panel $\theta = 45°$, $\Phi_B = BA \cos 45°$; in the right panel $\theta = 0°$, $\Phi_B = BA$

Using an equation for magnetic flux, *Faraday's Law* can be applied to the voltage induced in a coiled wire under certain situations. Faraday's Law states that the induced voltage in a coil is proportional to the number of loops it contains, multiplied by the rate at which the magnetic flux changes within those loops.

The magnitude of the current is related to the resistance of the coil, its circuit, and the voltage induced. The more coils a wire has, the greater the induced voltage. However, the higher the voltage, the greater the current that pushes back against the magnet. The rapid motion of the magnet also induces a higher voltage.

This relationship is explained in the equation of Faraday's Law of Induction (i.e., Lenz's Law, named after the scientist who added the negative sign in 1834):

$$EMF = -N\frac{\Delta\Phi_B}{\Delta t}$$

The negative sign in Lenz's Law indicates that the induced EMF is oriented so that the current it creates makes a magnetic current that *opposes* the change in flux from the original ("external") field.

The diagrams below show a loop of wire exposed to a magnetic field. In the diagram on the right, the loop is stretched such that the area exposed to the flux changes. From Faraday's Law, an EMF is induced and current results through the loop.

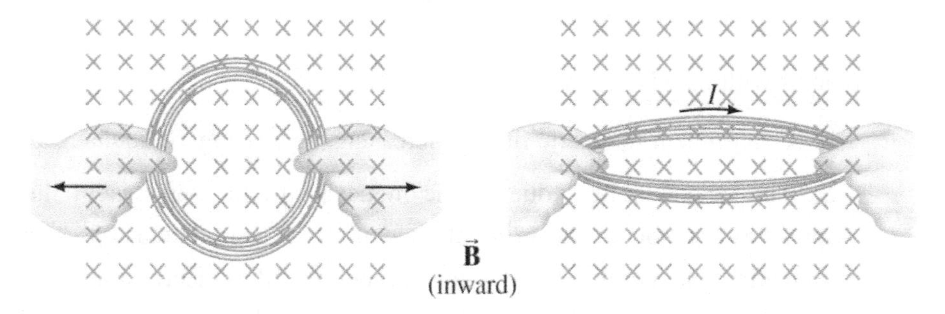

On the right, the flux through the coil decreases because the area decreased

A moving conductor can also form an EMF.

In the diagram below, the rectangle represents a conductor with a movable slide. If the slide is moved outward, then the magnetic flux changes.

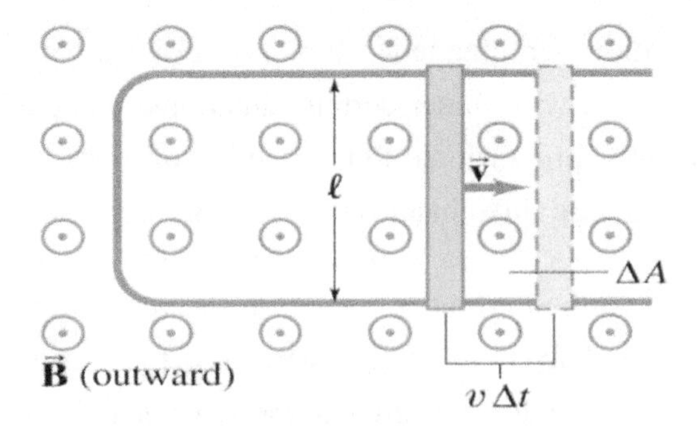

The induced current from the movement of the sliding bar causes a current to form, as seen in the following diagram.

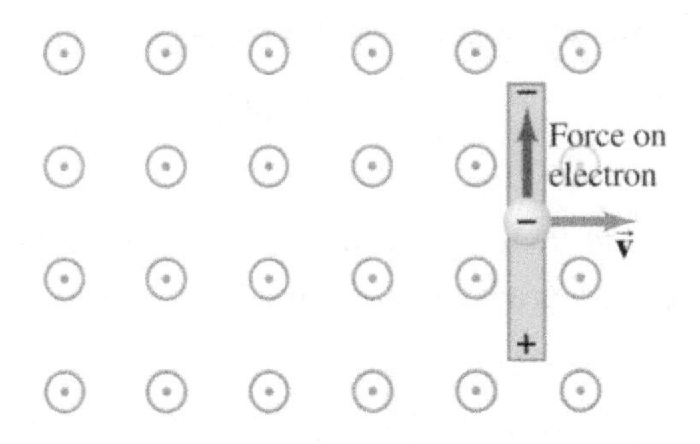

The induced EMF in such a situation has a magnitude of the form:

$$EMF = \frac{\Delta\Phi_B}{\Delta t} = \frac{B\Delta A}{\Delta t} = \frac{B\ell v\Delta t}{\Delta t} = B\ell v$$

Solving problems with Faraday's Law:

1. Determine if magnetic flux is increasing, decreasing, or unchanged.

2. The magnetic field due to the induced current points is in the opposite direction to the original field if the flux is increasing. The magnetic field points in the same direction if it is decreasing. The magnetic field is zero if the flux is not changing. This is the meaning of the negative sign in Lenz's Law.

3. Use the right-hand rule (RHR) to determine the direction of the resulting current. Point the thumb in the direction of a vector normal to the area vector on the side that the flux exits. The direction curled by the fingers as they close toward the palm in the direction of the current.

4. The external field and the field due to the induced current are different.

Torque on Current-Carrying Wire

When a wire carrying a current form a loop, the current reverses direction (x- and y-axes) as it travels in the loop and out of the loop.

When the loop is exposed to a magnetic field, a torque is generated about the axis of the loop due to the opposing forces on each side of the loop. This concept has many applications and is used in equipment such as galvanometers, motors, and loudspeakers.

A galvanometer uses the torque on a current loop to measure current.

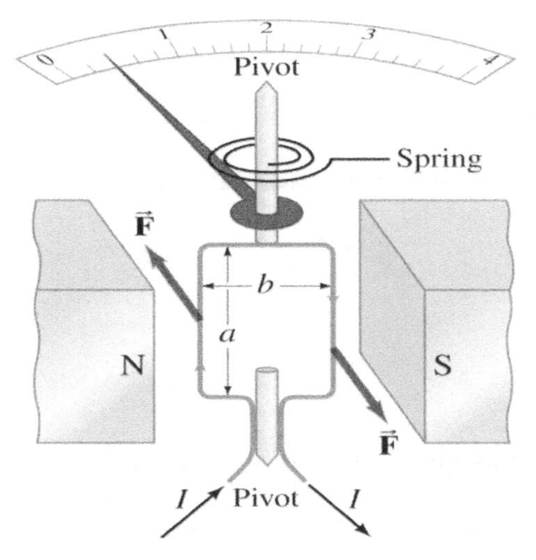

Galvanometer to measure current

If the current increases in the loop, the force from the magnetic field increases in proportion. The dial twists more because of increased torque.

The magnitude of the produced torque is calculated by:

$$\tau = nIAB \sin(\theta)$$

where the quantity nIA is the magnetic dipole moment, n is the number of loops (this equation also applies for a solenoid), I is the current (A), A is the area of the enclosed space (m^2) and θ is the angle of the loop area with respect to the magnetic field (degrees).

Properties of Electromagnetic Radiation

What is electromagnetic radiation?

Electromagnetic radiation is a form of energy that propagates through space as an oscillating electric and magnetic field. Classically, electromagnetic radiation is considered to be a wave; however, in quantum mechanics, electromagnetic radiation can be considered as packets of energy known as *photons*. Below is a figure of the wave nature of electromagnetic radiation.

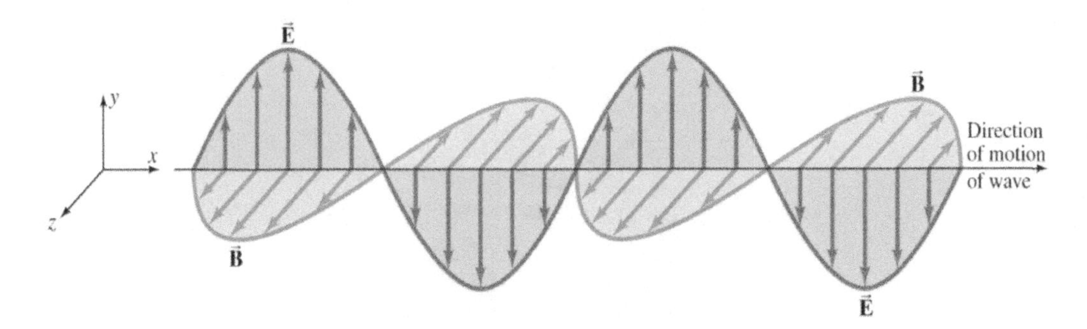

Radiation velocity equals constant *c* in a vacuum

Unlike sound and mechanical waves, electromagnetic radiation does not need a medium in which to propagate.

In a vacuum, electromagnetic radiation has a constant speed of:

$$c = 2.99792458 \times 10^8 \text{ m/s} \approx 3.0 \times 10^8 \text{ m/s}$$

This is the standard value for the *speed of light*. Nothing in the known universe travels faster than light in a vacuum. Electromagnetic radiation travels slower than this speed in other mediums, according to the value of the mediums' refractive index.

The speed of light in these mediums can be calculated by:

$$v = \frac{c}{n}$$

where v is the speed of light in the medium, and n is the refractive index of the medium.

Production, energy, and momentum of electromagnetic waves

Electromagnetic waves are produced by accelerating charges. The acceleration of charges produces oscillations of electric and magnetic fields (electromagnetic radiation), which propagate through space indefinitely until absorbed.

Like all waves, electromagnetic waves can be related by their speed, frequency, and wavelength by:

$$c = \lambda f$$

where c is the speed of light, λ is the wavelength (m), and f is the frequency of the electromagnetic radiation (Hz).

When considering electromagnetic radiation from the quantum mechanical viewpoint, the photons have discrete, quantifiable energies related to their frequency by:

$$E = hf = \frac{hc}{\lambda}$$

where E is the energy per photon (J or eV), and h is Planck's constant (6.626×10^{-34} J·s or 4.135×10^{-15} eV).

Electromagnetic radiation also carries momentum. Although the wave has no mass, it exerts a force as *radiation pressure.*

When the radiation is fully absorbed, the radiation pressure is at a minimum:

$$P_{minimum} = \frac{\bar{I}}{c}$$

When the radiation is fully reflected, the radiation pressure is at a maximum:

$$P_{maximum} = \frac{2\bar{I}}{c}$$

where P is the radiation pressure (N/m^2), and I is the intensity of the electromagnetic radiation (W/m^2).

Classification of Electromagnetic Spectrum, Photon Energy

Electromagnetic radiation varies significantly in properties and energies, depending on the frequency of the radiation. The figure below shows the spectrum of electromagnetic radiation according to wavelength and frequency. As frequency increases, the wavelength decreases (inversely proportional). The higher frequency forms of electromagnetic radiation (e.g., ultraviolet, X-rays, gamma rays) are known carcinogens (cancer-causing agents). These high-frequency forms of radiation have high energy, which can cause damage to cells and DNA.

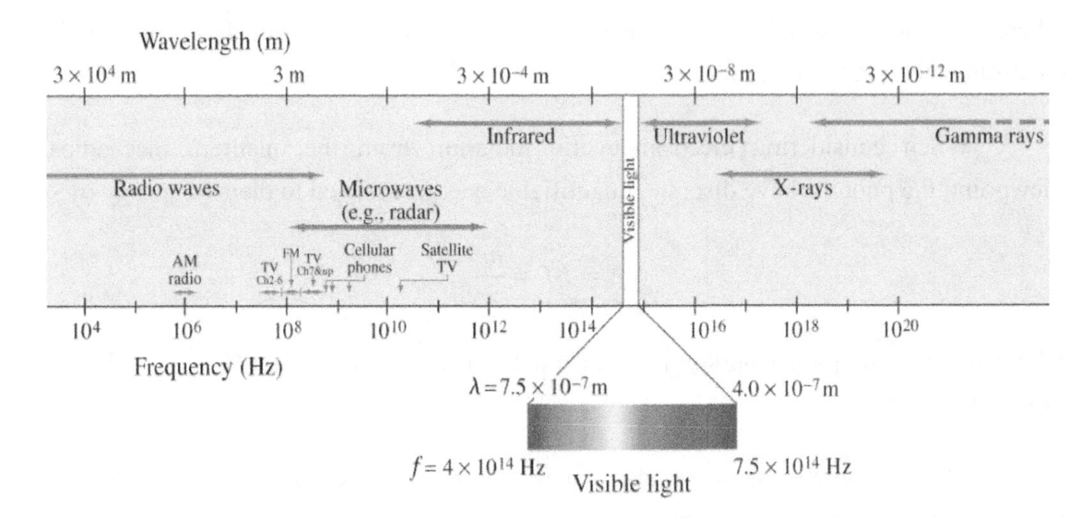

Lower frequency = longer wavelength = less energy	
Radio	Causes electronic oscillations in the antenna.
Microwave	Causes molecular rotation.
Infrared	Causes molecular vibration.
Visible	Excites electrons to higher energy orbitals. Visible light ranges from 400-700 nm. 400 being violet, 700 is red.
Ultraviolet	Breaks bonds and excites electrons to eject them, which is why UV is considered ionizing radiation.
X-rays	Ionizing radiation, the photoelectric effect.
Gamma rays	Even more energetic than X-rays.
Higher frequency = shorter wavelength = more energy	

Chapter Summary

Charge

- There are two kinds of electric charge—positive and negative.
- Charge is always conserved.
- The charge on an electron is: $e = 1.602 \times 10^{-19}$ C
- Charge is quantized in units of e (how many times greater than the charge on an electron).
- Conductors are materials in which electrons are free to move.
- Insulators are nonconductors, and thus do not allow electrons to move freely.
- Objects can be charged by conduction or induction.

Electric Fields

- Coulomb's Law gives the magnitude of the electrostatic force:

$$F = k \times \frac{Q_1 Q_2}{r^2}$$

- An electric field is a force per unit charge: $\vec{E} = \frac{\vec{F}}{q}$
- An electric field is given by a single point charge:

$$E = \frac{F}{q} = \frac{kqQ/r^2}{q} = k \times \frac{Q}{r^2}$$

- Electric field lines can represent electric fields.
- The static electric field inside a conductor is zero; the surface field is perpendicular to the surface.

- Electric flux (flow of a field through a closed surface):

$$\Phi_E = E_{\perp}A = EA_{\perp} = EA\ cos\ (\theta)$$

- Gauss's Law (electric flux through a closed surface):

$$\sum_{\substack{closed \\ surface}} E_{\perp}\Delta A = \frac{Q_{encl}}{\epsilon_0}$$

Electromagnetic Waves

- Maxwell's equations are the fundamental equations of electromagnetism.
- Electromagnetic waves are produced by accelerating charges.
- The propagation speed of electromagnetic waves is given by:

$$c = \frac{1}{\sqrt{\epsilon_0 \mu_0}}$$

- The fields are perpendicular to the direction of propagation.
- The wavelength and frequency of EM waves are related:

$$c = \lambda f$$

- The electromagnetic spectrum includes all wavelengths, from radio waves through visible light to gamma rays.

Magnets

- Magnets have north and south poles.
- Like poles repel, unlike attract.
- Electric currents produce magnetic fields.

Magnetic Fields

- Magnetic fields are given in units of Tesla (T).

- Parallel currents attract, anti-parallel currents repel.

- A magnetic field exerts a force on electric current in the form:

$$F = I\ell B \sin(\theta)$$

- A magnetic field exerts a force on a moving charge in the form:

$$F = qvB \sin(\theta)$$

- The magnitude of the field of a long, straight current-carrying wire:

$$B = \frac{\mu_0}{2\pi} \times \frac{I}{r}$$

- Ampère's Law (magnitude of a field):

$$\sum B_{||}\Delta\ell = \mu_0 I_{encl}$$

- The magnetic field inside a solenoid:

$$B = \frac{\mu_0 NI}{\ell}$$

Torque

- Torque on a current loop:

$$\tau = NIAB \sin(\theta)$$

Practice Questions

1. In the figure, $Q = 5.1$ nC. What is the magnitude of the electrical force on the charge Q? (Use Coulomb's constant $k = 9 \times 10^9$ N·m²/C²)

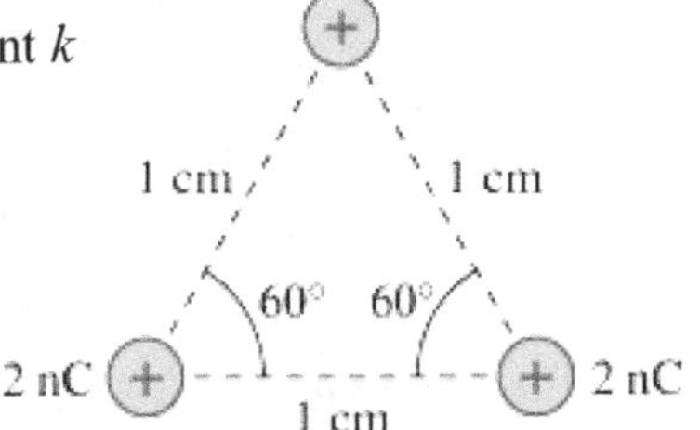

A. 4.2×10^{-3} N
B. 0.4×10^{-3} N
C. 1.6×10^{-3} N
D. 3.2×10^{-3} N
E. 7.1×10^{-3} N

2. Which one of the statements is correct?

A. The north pole of a magnet points towards Earth's geographic North Pole
B. The north pole of a magnet points towards Earth's geographic South Pole
C. Earth's geographic North Pole is the north pole of Earth's magnetic field
D. Earth's geographic South Pole is the south pole of Earth's magnetic field
E. None of the above

3. Two uncharged metal spheres, A and B, are mounted on insulating support rods. A third metal sphere, C, carrying a positive charge, is then placed near B. A copper wire is momentarily connected between A and B, and then removed. Finally, sphere C is removed. In this final state:

A. spheres A and B both carry equal positive charges
B. sphere A carries a negative charge and B carries a positive charge
C. sphere A carries a positive charge and B carries a negative charge
D. spheres A and B both carry positive charges, but B's charge is greater
E. sphere A remains uncharged, and B carries a positive charge

4. Two charges separated by 1 m exert a 1 N force on each other. What is the force on each charge when they are pulled to a separation distance of 3 m?

A. 3 N **B.** 0 N **C.** 9 N **D.** 0.33 N **E.** 0.11 N

5. A balloon after being rubbed on a wool rug can stick to a wall. This illustrates that the balloon has:

I. magnetism II. net charge III. capacitance

A. I only **B.** II only **C.** III only **D.** I, II and III **E.** I and III only

6. The diagram shows two unequal charges $+q$ and $-Q$, of opposite sign. Charge Q has a greater magnitude than charge q. Point X is midway between the charges.

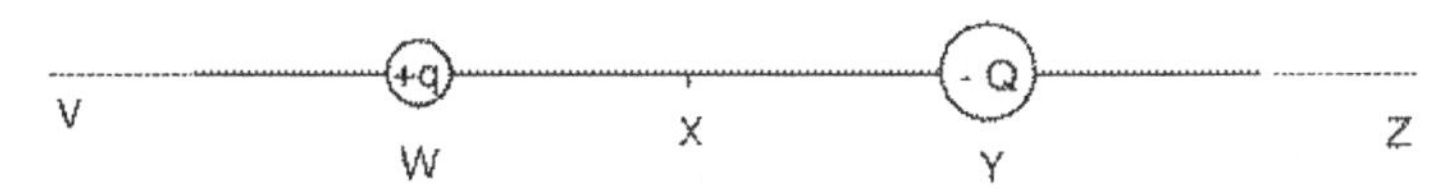

In what section of the line is the point where the resultant electric field could equal zero?

A. VW **B.** WX **C.** XY **D.** YZ **E.** VX

7. One coulomb of charge passes through a 6 V battery. Which is the correct value for the increase of some property of the battery?

A. 6 watts **B.** 6 ohms **C.** 6 amps **D.** 6 J **E.** 6 N

8. What travels through a conductor at near the speed of light when a current is established?

A. Protons **C.** An electric field
B. Photons **D.** Electrons **E.** Neutrons

9. Which statement is accurate for a proton that moves in a direction perpendicular to the electric field lines?

A. it is moving from high potential to low potential and gaining electric potential energy
B. it is moving from high potential to low potential and losing electric potential energy
C. it is moving from low potential to high potential and gaining electric potential energy
D. it is moving from low potential to high potential and losing electric potential energy
E. both its electric potential and electric potential energy remain constant

10. Light having a frequency in a vacuum of 3×10^{14} Hz enters a liquid with a refractive index of 2. What is the frequency of the light in this liquid?

A. 6×10^{14} Hz **C.** 12×10^{14} Hz
B. 1.5×10^{14} Hz **D.** 3×10^{14} Hz **E.** 9×10^{14} Hz

11. With all other factors remaining constant, how does doubling the number of loops of wire in a coil affect the induced emf?

A. The induced emf increases by a factor of $\sqrt{2}$
B. The induced emf doubles
C. There is no change in the induced emf
D. The induced emf quadruples
E. The induced emf increases by a factor of 3

12. A proton travels to the right and encounters a region S, which contains an electric field or a magnetic field or both. The proton is observed to bend up the page. Which of the statements is correct regarding region S?

I. There is a magnetic field pointing into the page
II. There is a magnetic field pointing out of the page
III. There is an electric field pointing up the page
IV. There is an electric field pointing down the page.

A. I only **B.** II only **C.** I and III **D.** II and IV **E.** II and III

13. An object with a 6 μC charge is accelerating at 0.006 m/s^2 due to an electric field. If the object has a mass of 2 μg, what is the magnitude of the electric field?

A. 0.002 N/C **B.** –0.005 N/C **C.** 2 N/C **D.** –2 N/C **E.** –0.07 N/C

14. Two Gaussian surfaces, A and B, enclose the same positive charge $+Q$. The Gaussian surface A has an area two times greater than surface B. Compared to the flux of the electric field through Gaussian surface B, the flux of the electric field through surface A is:

A. two times smaller
B. equal
C. two times larger
D. four times larger
E. four times smaller

15. Which of these electromagnetic waves has the shortest wavelength?

A. γ rays
B. Visible light
C. Radio waves
D. Infrared
E. Microwaves

Solutions

1. C is correct.

$F_e = kQ_1Q_2 / r^2$

$F_e = [(9 \times 10^9 \text{ N·m}^2/\text{C}^2)·(5.1 \times 10^{-9} \text{ C})(2 \times 10^{-9} \text{ C})] / (0.1 \text{ m})^2$

$F_e = 9.18 \times 10^{-4} \text{ N}$

$F_e \sin(60°)$ represents the force from one of the positive 2 nC charges.

Double to find the total force:

$F_{total} = 2F_e \sin(60°)$

$F_{total} = 2(9.18 \times 10^{-4} \text{ N}) \sin(60°)$

$F_{total} = 1.6 \times 10^{-3} \text{ N}$

The sine of the angle is used since only the vertical forces are added because the horizontal forces are equal and opposite, and therefore they cancel.

2. E is correct.

None of the above statements are true.

3. C is correct.

When the positively charged sphere C is near sphere B, it polarizes the sphere causing its negative charge to migrate towards C and a positive charge to build on the other side of sphere B.

The wire between sphere A and sphere B allows negative charge to flow to B and create a net positive charge on sphere A. Once the wire is removed and sphere C is removed, sphere A will have a net positive charge, and B has a net negative charge.

4. E is correct.

Coulomb's Law:

$F_e = kQ_1Q_2 / r^2$

If r is increased by a factor of 3:

$$F_{\text{new}} = kQ_1Q_2 / (3r)^2$$

$$F_{\text{new}} = kQ_1Q_2 / (9r^2)$$

$$F_{\text{new}} = (1/9)kQ_1Q_2 / r^2$$

$$F_{\text{new}} = F_{\text{original}}\,(1/9)$$

$$F_{\text{new}} = (1\text{ N})\cdot(1/9)$$

$$F_{\text{new}} = 0.11\text{ N}$$

5. B is correct.

The balloon sticks to the wall because the rubbing on the wool transferred charges to the balloon, leading to an electrostatic force.

6. A is correct.

If the charge Q is of a greater magnitude than charge q, then the electric field points toward Q (because it is negative) in section W to Z.

In section VW, the electric field is the difference in magnitude between q and Q.

If Q has a large enough charge, then the difference could equal zero, and there is no electric field.

7. D is correct.

A volt is defined as the potential difference that causes 1 C of charge to increase potential energy by 1 J. Therefore, moving 1 C through 6 V causes the potential energy of the battery to increase by 6 J.

8. C is correct.

The current is caused by a voltage (potential difference) across a conductor.

Whenever a voltage exists an electric field exists, and this travels at near the speed of light. Electrons do not travel quickly when a current is established and only travel at their drift speed, which is proportional to voltage.

9. E is correct.

A proton moving perpendicular to electric field lines does not get close to the charges creating the electric field. Thus, its electric potential and potential energy remain constant because these values are related to distance from other charges.

Electric Potential Energy:

$$U = kQq / r$$

Electric Potential:

$$V = kQ / r$$

10. D is correct.

Only the wavelength of light changes in different mediums.

The frequency does not change in different mediums.

11. B is correct.

Faraday's Law states that the electromotive force (emf) in a coil is:

$$\text{emf} = N\Delta B\text{A} \cos \theta / \Delta t$$

where N = number of loops of wire.

If N doubles, the emf also doubles.

12. C is correct.

There is a force on the proton up to the page. The electric field points in the direction of the force on positive particles, therefore it is pointed upwards.

The right-hand rule determines the direction of the magnetic field.

$F = qvB$, where q is a charge, v is velocity and B is a magnetic field

When F is oriented upwards, the curling fingers from the direction of velocity gives B into the page.

13. A is correct.

Convert all units to their correct form:

$$F = ma$$

$$F = (2 \times 10^{-6}\ \text{kg})\cdot(0.006\ \text{m/s}^2)$$

$$F = 1.2 \times 10^{-8}\ \text{N}$$

Substituting into the equation for electric field:

$$E = F / q$$

$$E = (1.2 \times 10^{-8}\ \text{N}) / (6 \times 10^{-6}\ \text{C})$$

$$E = 0.002\ \text{N/C}$$

Note: $1\ \text{N} = 1\ \text{kg}\cdot\text{m/s}^2$, not $1\ \text{g}\cdot\text{m/s}^2$

14. B is correct.

$$\Phi = Q / E_0$$

For an enclosed charge, the area of the surface does not affect the flux.

15. A is correct.

γ rays are the electromagnetic waves listed that have the shortest wavelength.

Chapter 10

Fluids and Solids

FLUIDS

- **Density, Specific Gravity**
- **Archimedes' Principle: Buoyancy**
- **Pressure**
- **Continuity Equation (Av = constant)**
- **Bernoulli's Equation**
- **Viscosity Poiseuille flow**
- **Surface Tension**

SOLIDS

- **Stress and Types of Stress**
- **Elasticity: Elementary Properties**
- **Thermal Expansion Coefficient**

FLUIDS

Density, Specific Gravity

The three standard states of matter are solid, liquid, and gas.

A *liquid* has a fixed volume but can be any shape.

A *gas* can be any shape and can be compressed (variable volume).

Since liquids and gases are both able to "flow," they are fluids.

The *density* (ρ) of an object is its mass per unit volume. The equation for density:

$$\rho = \frac{m}{V}$$

where ρ is density, m is mass, and V is volume.

Density is expressed in SI units of kg/m^3; it may also be given in g/cm^3.

To convert g/cm^3 to kg/m^3, use dimensional analysis.

However, it is easier to multiply by 1,000 (cm^3kg/m^3g):

$$\left(\frac{g}{cm^3}\right) \cdot \left(\frac{100\ cm}{1\ m}\right)^3 \cdot \left(\frac{1\ kg}{1{,}000\ g}\right) = \left(\frac{g}{cm^3}\right) \cdot \left(\frac{1{,}000\ cm^3 kg}{m^3 g}\right) = \frac{kg}{m^3}$$

The *specific gravity* of a substance is the ratio of its density to the density of the reference substance.

Mostly, specific gravity is referenced against water.

Thus, the specific gravity of water is equal to 1.

The equation for specific gravity is written as follows:

$$\text{specific gravity} = \frac{\rho_{\text{object}}}{\rho_{\text{water}}}$$

Archimedes' Principle: Buoyancy

Archimedes' principle of buoyancy (250 B.C.) states that *an object wholly or partially submerged in a fluid will be buoyed up by force equal to the weight of the fluid that it displaces.*

The equation for *buoyancy* is:

$$F_{\text{buoyant}} = \rho_{\text{fluid}} V_{\text{object}} g$$

where F_{buoyant} is the buoyant force (N), V is the volume of the fluid displaced (m^3), ρ is the density of the fluid it displaces (kg/m^3), and g is the value of gravity (9.8 m/s^2).

For floating objects, the fraction of the object that is submerged is the ratio of the object's density to that of the fluid.

$$\frac{V_{\text{sub}}}{V_{\text{object}}} = \frac{\rho_{\text{object}}}{\rho_{\text{fluid}}}$$

The buoyancy force counteracts the weight of the object in the fluid, making objects in water feel lighter than when out of the water. Objects float when their densities are less than that of the fluid in which they are submerged.

If an object's density is less than that of water, there is an upward net force on it, and the object rises until it is partially out of the water.

The figure below depicts a log submerged in water $\rho = 1{,}000$ kg/m^2.

When fully submerged, the net force on the log is greater than zero. Thus, the log floats until the buoyant force is equal to the weight of the log (shown below):

$$F_{net} = F_B - m_o g$$

$$F_{\text{net}} = \rho_F V_{\text{displaced}} g - m_o g$$

$$F_{\text{net}} = \left(1{,}000 \frac{\text{kg}}{\text{m}^3}\right) \cdot (2 \text{ m}^3) g - (1{,}200 \text{ kg}) g$$

$$F_{\text{net}} = (2{,}000 \text{ kg}) g - (1{,}200 \text{ kg}) g$$

$$F_{\text{net}} = (800 \text{ kg}) g$$

$$F_{\text{net}} > 0$$

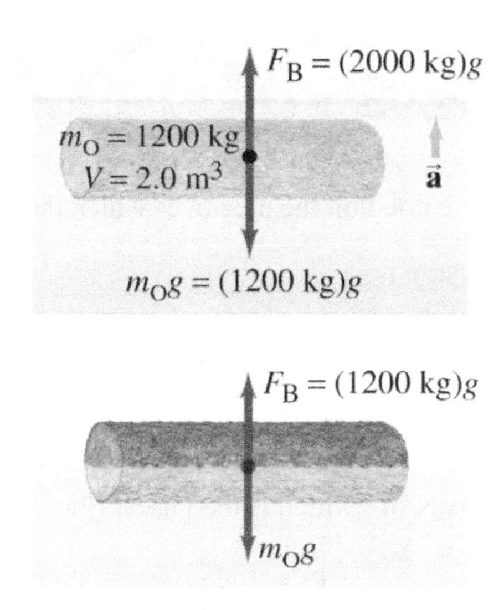

This principle of buoyancy works in the air. Hot-air and helium balloons rise. Both hot air and helium have densities less than air at room temperature. The difference in densities causes the hot air or helium to exert a buoyant force and the balloon.

The balloon rises because the buoyant force is greater than the weight of the object it is carrying. The graphic below shows that the net force on the object is the difference between the buoyant force and the gravitational force.

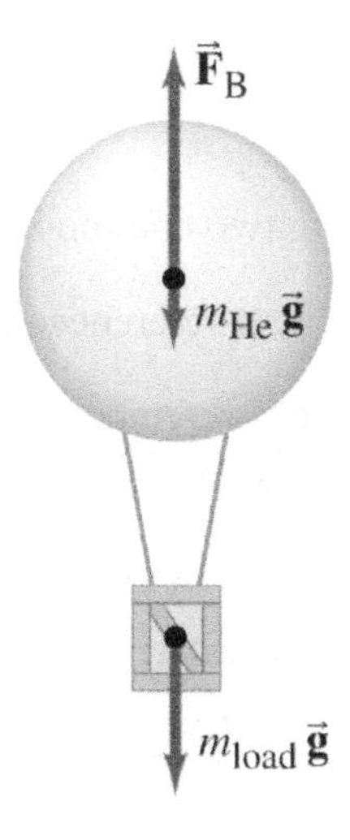

An object floats perfectly (i.e., the top is level with the surface of the fluid) when the buoyant force is equal to the weight of the object.

An object will rise upward out of the fluid when the buoyant force is greater than the weight of the object, and an object will sink when the buoyant force is less than the weight of the object.

Pressure

Pressure is the force exerted on the area over which the force acts.

The equation for pressure is:

$$P = \frac{F}{A}$$

where P is pressure (N/m^2), F is a force (N), and A is area (m^2).

The unit of pressure is N/m^2, which is the Pascal (Pa).

$$1 \text{ Pa} = 1 \text{ N/m}^2$$

Anything that exerts a force can exert pressure; therefore, the concept of pressure applies to solid objects, liquids, and gasses. When discussing fluids, the pressure exerted by the atmosphere, and measurements of pressure relative to it, are of concern.

At sea level, the atmospheric pressure is about 1.013×10^5 N/m^2; this is one atmosphere (1 atm). Another unit of pressure is the bar (1 bar = 1.00×10^5 N/m^2). Standard atmospheric pressure is slightly more than 1 bar.

Atmospheric pressure does not crush organisms because their cells maintain an internal pressure that balances it.

The pressure is measured using two common types.

Absolute pressure is pressure zero-referenced against a perfect vacuum.

Gauge pressure is pressure zero-referenced against atmospheric pressure. Most gauges measure pressure above the atmospheric pressure. Hence, most systems measure gauge pressure.

For example, if there are two different pressure dials (with one displaying absolute pressure and the other displaying gauge pressure) at sea level, what would the reading be?

The absolute pressure dial would read 1.013×10^5 Pa, and the gauge pressure dial would read 0 Pa. This difference is because the atmosphere exerts 1.013×10^5 Pa of pressure over a perfect vacuum; thus, the absolute pressure gives this value. The gauge pressure reads 0 Pa because it is referenced to atmospheric pressure and requires pressures above this value to give a reading.

The equation for gauge pressure is the difference between absolute pressure and atmospheric pressure.

$$P_{\text{gauge}} = P_{\text{abs}} - P_{\text{atm}}$$

A reading for absolute pressure used the equation:

$$P_{\text{abs}} = P_{\text{atm}} + P_{\text{gauge}}$$

Pressure Versus Depth ($P = pgh$)

Below the surface of a liquid, pressure increases linearly with depth. Liquid pressure on an object is found only using the density of the liquid and the distance of the object below the surface. The pressure on an object is not determined by using the density of the object in the liquid nor shape of the container.

The equation for liquid pressure is expressed as:

$$P = \frac{F}{A} = \frac{\rho A h g}{A}$$

This simplifies to:

$$P_{\text{liquid}} = \rho g h$$

where P_{liquid} is the pressure exerted by the liquid (N/m^2 or Pa), g is the acceleration of gravity, ρ is the density of the liquid, and h is the distance below the surface of the liquid (m).

Liquid pressure is useful in the operation of some types of pressure gauges.

For example, the gauge pictured below is an open-tube manometer. The pressure in the open end is open to atmospheric pressure, and the pressure that is being measured causes the fluid to rise until the pressures on both sides of the tube are equal.

From the liquid pressure equation, the pressure being measured is calculated as:

$$P_0 = P_{atm}$$

$$P = P_{atm} + P_{liquid}$$

$$P = P_{atm} + \rho g \Delta h$$

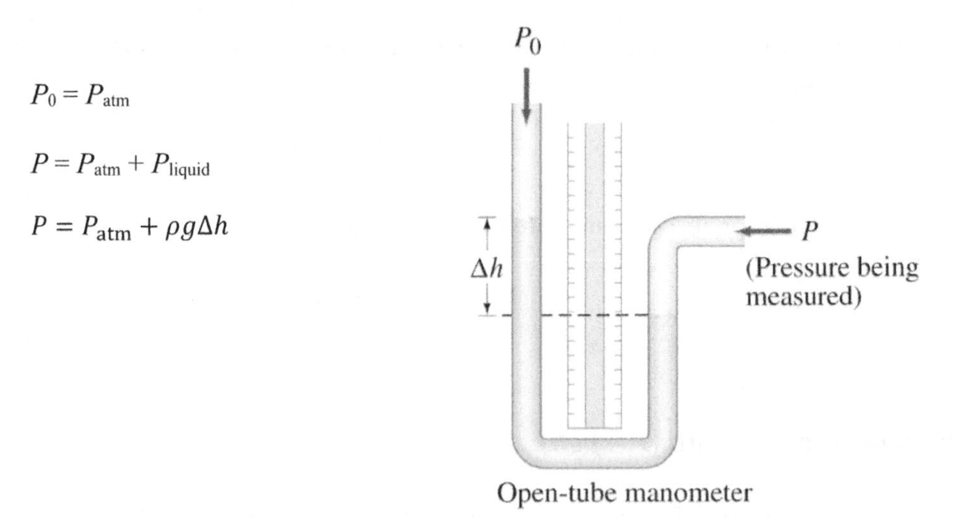

Open-tube manometer

In this example, the pressure measured is in absolute pressure because the equation includes atmospheric pressure. This is an important distinction because if evaluating the pressure on an object at depth, the gauge pressure gives only the liquid pressure on the object.

In contrast, the absolute pressure gives the gauge pressure plus atmospheric pressure so that it would be more accurate.

The figure below has two different shaped containers holding water at equal levels. If points 1 and 2 are at equal height, what are the gauge and absolute pressure at points 1 and 2?

Notice that although the containers are different shapes, the two points are at equal depths within the water. Therefore, they have the same gauge and absolute pressures. The shape of the container does not influence the liquid pressure; only the depth below the liquid contributes to liquid pressure.

Thus, the gauge and absolute pressure can be calculated as:

$$P_1 = P_2$$

$$P_{gauge} = \rho g \Delta h$$

Pascal's Law

Pascal's Law (1647-1648) states that *a pressure exerted on an incompressible liquid transmits equally to all parts of the liquid.* This principle has important implications because it explains why pressure differentials are not created within a closed volume of liquid.

A small cube is within a beaker of water. For this example, consider the sides small enough that liquid pressure due to depth is equal on the top and bottom of the cube.

The pressure on all sides of the object is equal because the pressure is the same in every direction in a fluid at a given depth. If this were not the case, the fluid would flow because of the natural tendency of particle movement from high pressure to low pressure.

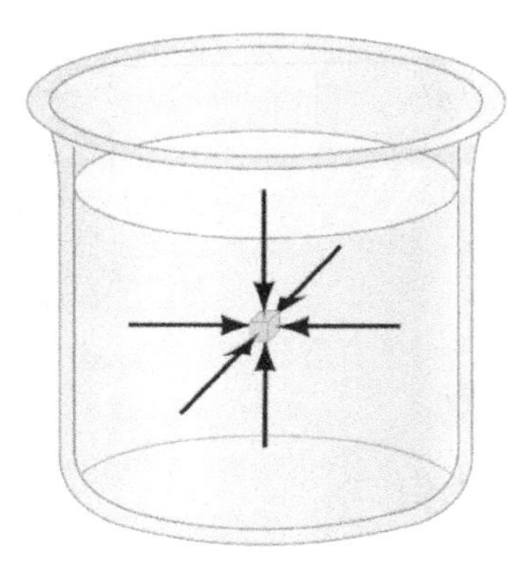

Although the pressures across an incompressible liquid are the same, the forces vary depending on the area they are applied over (because of $P = F / A$). This observation is useful because it allows the development of systems that impart mechanical advantage.

If the external pressure is applied to a confined fluid, the pressure at every point within the fluid increases by that amount. This principle is used in hydraulic lifts and hydraulic brakes. The fluid is confined, so when pressure is applied, the fluid is compressed, which raises the pressure throughout the system.

In the system below, a small force is applied over a small area (F_1 over A_1). The output force (F_2) at the location with the larger area (A_2) will be higher.

Energy must be conserved; thus, the work done on one end is the same as the work done at the other end.

The volume of liquid displaced must be conserved so that the volume displaced is equal at both the input and output.

Work Equivalence

$$W_1 = W_2$$

$$F_1 d_1 = F_2 d_2$$

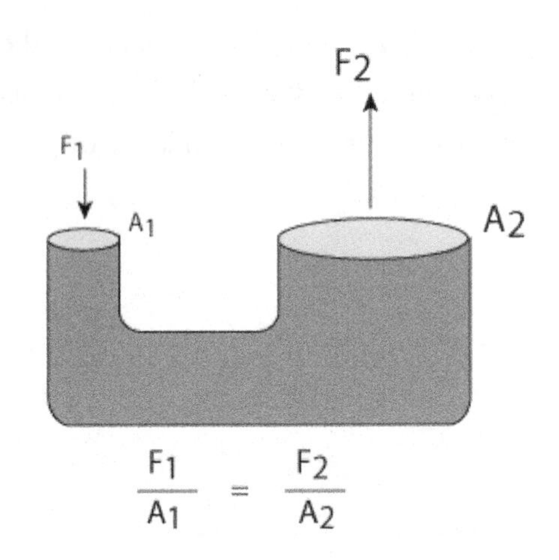

Volume Equivalence

$$V_1 = V_2$$

$$A_1 d_1 = A_2 d_2$$

Continuity Equation (*Av = constant*)

When liquids flow through pipes, the flow is considered incompressible, and the volume of the liquid remains constant regardless of pressure. As such, the volume of liquid flowing through a pipe in a given amount of time must always be equal.

This principle is the conservation of volume and can be expressed by the *continuity equation*, which describes the volume flow rate of an incompressible liquid.

Often, the continuity equation is used to calculate the speed of the fluid flowing through pipes, and is given as:

$$A_1 v_1 = A_2 v_2$$

where A is an area of the pipe's cross-section (m^2), and v is the velocity of the fluid through that point (m/s).

Imagine an incompressible liquid flowing through a pipe that has a diameter reduction. The volume of fluid passing through the constriction over time cannot change.

The *continuity equation* can solve for the resulting velocity of the liquid.

$$A_1 v_1 = A_2 v_2$$

$$v_2 = \frac{A_1 v_1}{A_2}$$

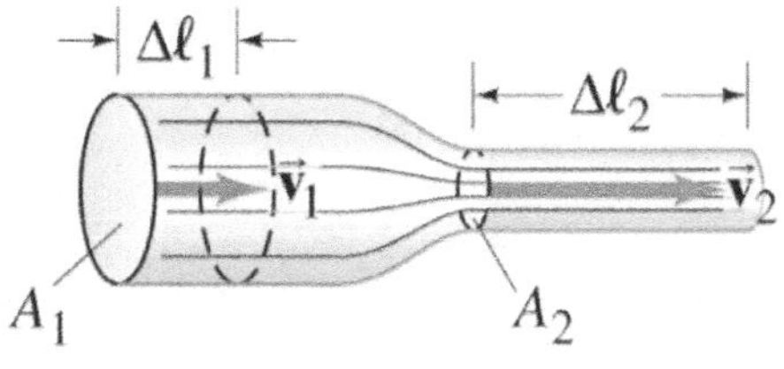

$\Delta \ell_1$ $\Delta \ell_2$ $\vec{v}_1$ $\vec{v}_2$ A_1 A_2

Bernoulli's Equation

The *Bernoulli equation* (1738) is a statement of conservation of energy within flowing fluids. More accurately, the Bernoulli equation states that *the pressure energy, kinetic energy, and potential energy of a flowing fluid is conserved.*

The *Bernoulli equation* is:

$$P_1 + \rho g h_1 + \frac{1}{2}\rho v_1^2 = P_2 + \rho g h_2 + \frac{1}{2}\rho v_2^2$$

where P is pressure (Pa), ρ is the density of the liquid, g is the acceleration due to gravity, h is the height of the fluid (m), and v is the velocity of the fluid through that point (m/s).

The Bernoulli equation is particularly helpful in analyzing the fluid flow across pressure or height differentials, and the energy needed to transport volumes of fluid.

In the diagram below, water is flowing into the left side at a pressure of P_1, a velocity of v_1, through an area of A_1, and at the height of y_1. As the fluid moves to the right, the pipe rises in height and decreases in the area (i.e., circumference).

Bernoulli's Principle, given a few starting or ending conditions, can be used to find the other values.

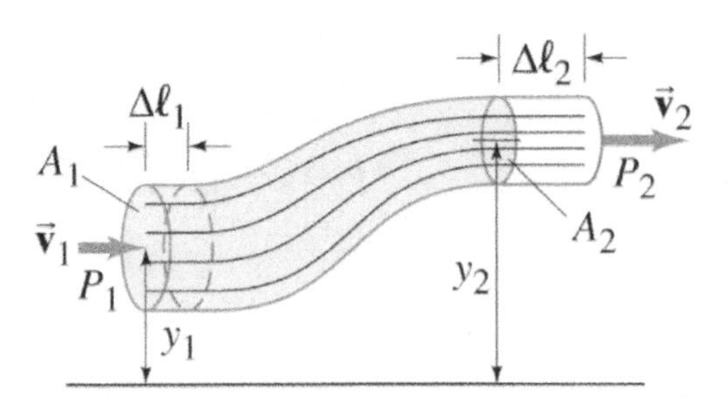

Bernoulli's principle correlates the speed of a fluid through an opening to the height of the fluid above the opening.

As shown below, suppose there is a spigot at the bottom of an open tank of water. The speed of this fluid can be found by using Bernoulli's principle and making a few assumptions.

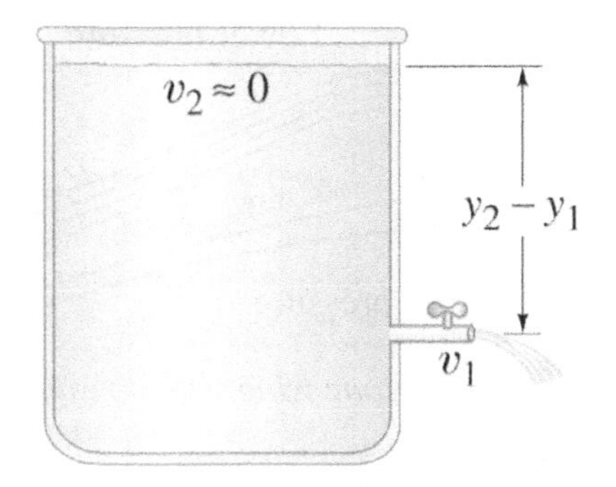

$$P_1 + \rho g y_1 + \frac{1}{2}\rho v_1^2 = P_2 + \rho g y_2 + \frac{1}{2}\rho v_2^2$$

Assume that the top of the container is much larger than the opening of the spigot:

$$A_{\text{top}} \gg A_{\text{spigot}}$$

By the continuity equation, the velocity of fluid flow through the top is much smaller than through the spigot. The flow through the spigot can be set to zero:

$$v_{\text{top}} \ll v_{\text{spigot}}$$

$$v_{\text{top}} \approx 0$$

Atmospheric pressure acts on both the top of the container and at the end of the spigot. This pressure is equal on both sides of the equation. It cancels, leaving:

$$g y_1 + \frac{1}{2} v_1^2 = g y_2$$

Assume the height of the spigot as the reference height and set it equal to zero. Then solve for the velocity of fluid flow out of the spigot:

$$\frac{1}{2} v_1^2 = g y_2 \quad \text{or} \quad v_1 = \sqrt{2 g y_2}$$

This expression is the same as for the conservation of mechanical energy into kinetic energy. The Bernoulli equation can be used to explain the process of lift for airplane wings. When air strikes an airplane wing, the air must separate to flow over both the top and bottom of the wing.

Due to the wing curvature, air flowing over the top surface has a slightly higher speed than air flowing over the bottom surface. The speed difference causes a pressure imbalance. The top of the wing experiences less pressure than the bottom of the wing.

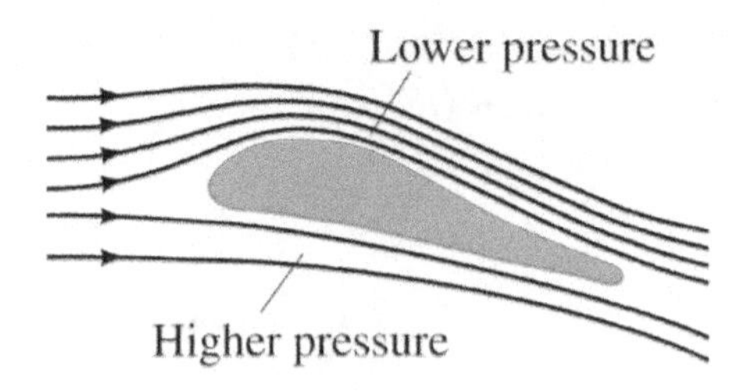

Schematic of air flowing over an airplane wing with a resulting pressure differential

If the height of the wing is negligible, then the Bernoulli equation reduces to:

$$P_T + \frac{1}{2}\rho v_T^2 = P_B + \frac{1}{2}\rho v_B^2$$

The flow speed of air over the top of the wing is higher than the flow speed under the bottom. Thus, by conservation of energy:

$$v_T > v_B \text{ and } P_T < P_B$$

Venturi effect, pivot tube

The *Venturi effect* (1797) describes the drop in fluid pressure that occurs when a fluid flows through a constriction in a pipe. From the continuity principle, flow through a constriction in a tube increases the velocity of the flow to conserve volume.

From the Bernoulli principle, the higher velocity gas must also experience a drop in pressure. This phenomenon is the Venturi effect and can be used to measure fluid flow by measuring pressure differences.

Measuring the velocity of fluid flow through a pipe uses a constriction placed in the pipe. Two pressure meters are attached with one before the constriction and the other at the constriction. If the areas of the pipe and the constriction are known, the velocity of the fluid flow through the pipe is solved by both the continuity and Bernoulli equation.

$$A_1 v_1 = A_2 v_2$$

$$P_1 + \frac{1}{2}\rho v_1^2 = P_2 + \frac{1}{2}\rho v_2^2$$

Viscosity Poiseuille Flow

When water poured out of a cup is compared to honey poured out of a jar, the pattern of flow is substantially different. Although water and honey are both liquids, they behave differently when flowing, due to their *viscosity*.

The *viscosity* of a fluid is a measure of the internal frictional force. All fluids have some measure of viscosity. Viscosity may be high (e.g., honey), or low (e.g., water). The characteristic viscosity of a fluid is a known value and is represented by the symbol η with units of (Pa·s).

When high viscous fluids flow through a pipe, the behavior is markedly different than when a fluid with lower viscosity flows through the pipe. Specifically, the flow from high viscous fluids forms a front that is shaped like a parabola bulging outward. A parabola shape is observed in the fluid because the frictional forces on the fluid from the sides of the pipe are much higher than the frictional forces in the center of the fluid.

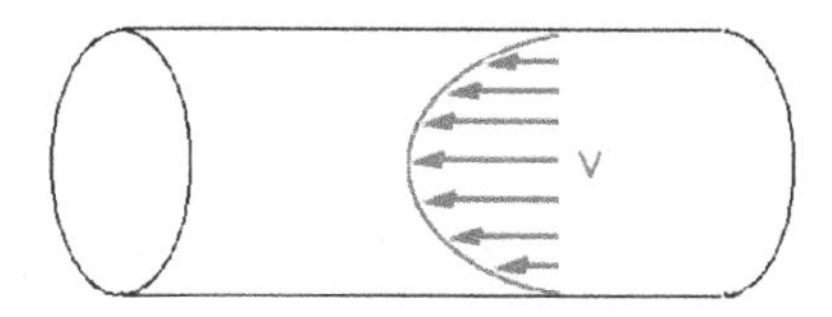

Parabola shape observed for high viscous fluids

Along with the parabola flow shape, the flow rate of a fluid through a pipe must account for the viscosity of the liquid.

Poiseuille's Law calculates this flow rate:

$$\dot{V} = \frac{\Delta P \pi r^4}{8 \eta L}$$

where $\dot{V}$ is the volumetric flow rate (m^3/s), ΔP is the pressure differential across the pipe (Pa), r is the radius of the pipe (m), η is the viscosity of the pipe (Pa·s), and L is the length of the pipe (m).

Poiseuille's Law (1838) has significant consequences for blood flow. If the radius of an artery is half what it should be (i.e., occlusion), the pressure must increase by a factor of 16 to maintain the same blood flow. Usually, the human heart cannot work hard enough to reach 16 times the output. Blood pressure increases (hypertension) as the heart pumps (cardiac stress) to satisfy this demand for typical blood flow with blockage.

Surface Tension

Sometimes small objects, such as paperclips, can be placed lightly on the surface of the water, and they float. However, if pushed down, they readily sink. These objects have higher densities than water, but how do they float on the surface? This phenomenon is due to the attraction between the liquid water molecules known as *surface tension*.

Surface tension is due to the cohesive forces within the water molecules. This cohesion occurs because the water molecules are often electrostatically attracted and stick to neighboring water molecules.

When a small object is placed gently on top of the water, it may not weigh enough to break the surface tension. Therefore, it floats despite its higher density than the water. Some insects, for example, can walk on water, not because they are less dense than the water, but because of the resistive force of surface tension.

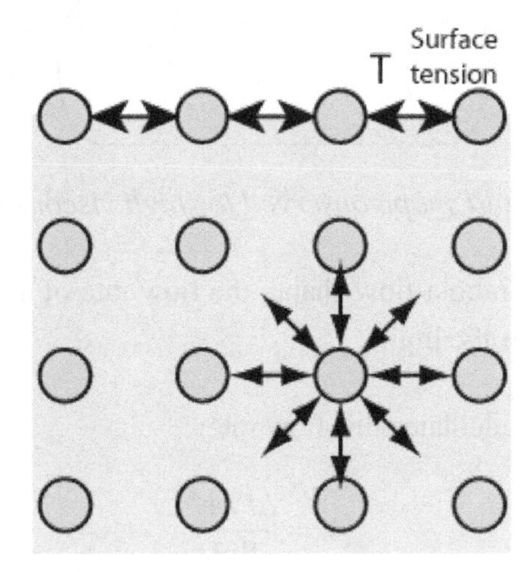

Often, the surface of a liquid at rest in a container is not perfectly flat. This phenomenon is observed in a small container, such as a graduated cylinder, in which the liquid curves up or down as it contacts the walls of the container. This curved surface for the liquid is the *meniscus* and results from surface tension and adhesion.

Adhesion is the property of a substance to be attracted to a different substance. Unlike cohesion, adhesion is caused by several different mechanisms and may or may not be due to electrostatic forces, as discussed for water molecules. The behavior of a liquid in a container is determined by the cohesive and adhesive forces the liquid.

If the adhesive force between the liquid and the container is greater than the cohesive force between the liquid molecules, the liquid curves up to meet the edges of the container, creating a concave meniscus.

If the cohesive force is greater than the adhesive force, the liquid curves downward, creating a convex meniscus. For example, water molecules are more attracted to glass than they are to each other (adhesion > cohesion), so the surface of the water curves upward toward the walls of its container. Mercury has adhesion < cohesion, and so its surface curves downward against the walls of the container.

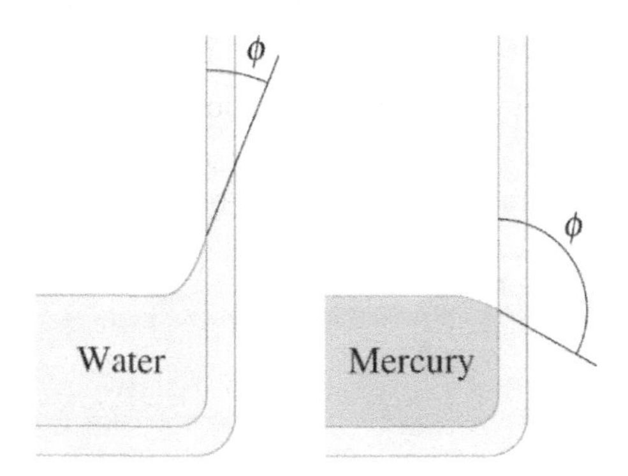

If a narrow tube is placed in a fluid, the fluid exhibits *capillarity*. This phenomenon is also due to the cohesive and adhesive forces. If a narrow tube is placed in water (adhesion > cohesion), then the water travels up the tube some distance, as shown in the figure below on the left. This movement is because the adhesive forces pull the water along the edge, and the cohesive forces pull water up the tube as well.

If the same tube is placed in mercury (adhesion < cohesion), then the cohesive forces of the mercury attempt to keep the mercury together, rather than climb the wall of the tube. Mercury becomes lower in the capillary tube with regards to the height of the mercury level around it, as shown in the figure below on the right.

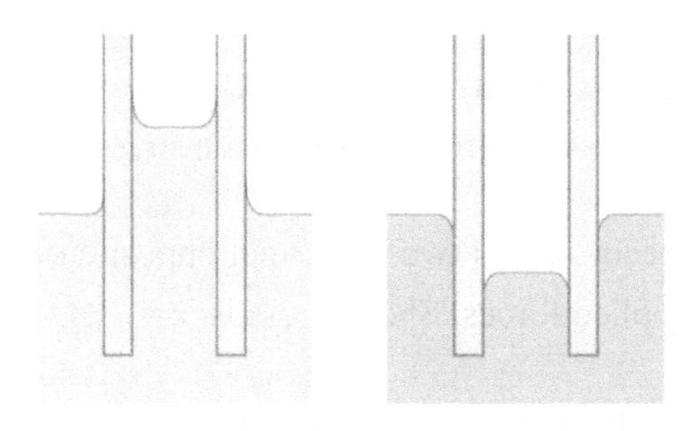

Capillary tube in water (on left) compared to in mercury (on right)

Stress and Types of Stress

When a solid material has a force applied to it, it resists deformation (up to a maximum applied force) and retains its shape. The resistance to deformation is due to stress within the material, which applies a restoring force per unit area.

Stress has the form: stress = force / area

All solid materials can undergo three basic types of stress: tension, compression, and shear stress. These three types of stress are important because they dictate how structures are built, the loads they can handle, and how the failure of materials may occur.

The figure below illustrates the three types of applied stress:

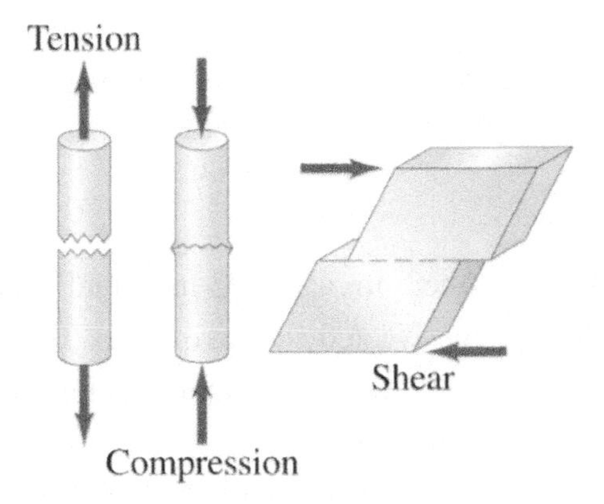

Tensile stress results when a force acts on a material in a manner such that it pulls it apart. Tensile stress is expressed as: $\sigma = F / A$

where σ is the stress an object experiences (N/m^2), F is the force exerted on an object (N), and A is the area over which it is applied (m^2).

Compressive stress results when a force acts to compress the material. The same formula expresses compressive stress as for tensile stress: $\sigma = F / A$

Shear stress occurs when two forces act opposite such that the material slides on a plane between the applied forces. Shear stress is: $\tau = F/A$

where τ is the stress an object experiences (N/m^2), F is the force exerted on an object (N), and A is the area over which it is applied (m^2)

Elasticity: Elementary Properties

Elasticity is an essential property in tensile and compressive stress. When a material undergoes tensile or compressive deformation, elasticity is the measure of a material's ability to return to its original shape after distortion.

Rubber bands have high elasticity because, after moderate stretching, a rubber band returns to its original shape.

Conversely, if a low-elasticity material is deformed, such as a piece of ceramic pottery, it experiences little deformation before snapping.

Their elastic properties determine the behaviors of these two materials.

Tension and compression

When materials undergo *tension*, they stretch according to the force acting upon them. As mentioned earlier, all materials have some degree of elasticity. After this inherent value is exceeded, they stretch according to the tensile stress applied.

When a solid material undergoes tension, its elasticity is described as either being ductile or brittle.

If the materials tend to deform (stretch) under tensile stress, then the material is *ductile*.

However, if the material undergoes little deformation before breaking, the material is *brittle*.

Compression is the inverse of tension and occurs when force points inward on a material, compacting it. Similarly, if the applied force is high enough, the material deforms.

Malleability is the propensity for a material to deform under compression.

If the material is highly malleable, it can be compressed and deformed easily. Non-malleable materials break rather than deform. Regardless, two properties indicate the elasticity of material in either tension or compression: strain and Young's Modulus.

Strain, given by the lowercase Greek letter epsilon (ε), is the change in length of an object in the direction of the applied force, divided by the original length.

$$\varepsilon = \frac{\Delta l}{l_0}$$

where ε is the strain, Δl is the change in length (m), and l_0 is the original length (m)

Young's modulus (elastic modulus) is a measure of the force per unit needed to stretch material. Young's modulus is expressed as the ratio of stress divided by strain:

$$E = \frac{\sigma}{\varepsilon} = \frac{F l_0}{A \Delta l}$$

Most structural materials are considered somewhat elastic, and the behavior of these materials under tensile or compressive stress is important.

The image below displays a general force vs. elongation graph of an elastic material undergoing tension or compression:

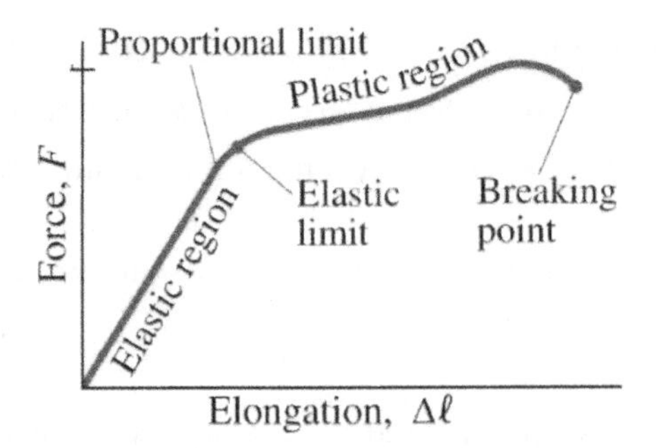

The force vs. elongation graph shows an elastic and plastic region

Elastic materials under stress display a pattern of force vs. elongation (or compression), as seen above. When the maximum amount of stress is applied to the material, it no longer deforms elastically but deforms plastically.

The *elastic limit* (yield strength) is the point is for the maximum stress a material can sustain before all subsequent deformation becomes permanent (i.e., plastic deformation).

After passing the elastic limit, the material continues to deform plastically until the ultimate strength is reached, after which the material breaks apart in two pieces.

Shear

Like tension and compression, all materials resist shearing to a certain degree. This resistance is governed by the shear strain and the *shear modulus*, which, like Young's modulus, is used to measure the stiffness of materials.

The figure below shows a block undergoing shearing forces and deforming according to shear forces:

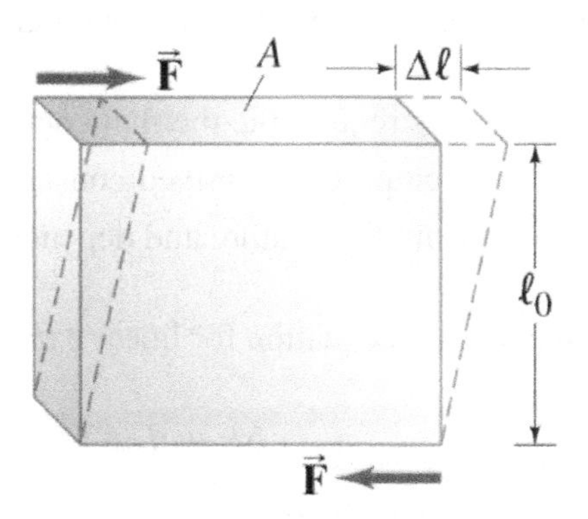

The shear strain is given by:

$$\varepsilon_{shear} = \frac{\Delta l}{l_0}$$

Therefore, the Shear Modulus can be written as:

$$S = \frac{\tau}{\varepsilon_{shear}} = \frac{Fl_0}{A\Delta l}$$

When an object undergoes shear forces, the shear strain, and shear modulus are calculated with Δl and l_0, which are perpendicular.

Thermal Expansion Coefficient

Most materials expand when the temperature rises, and contract when the temperature falls; this phenomenon is *thermal expansion*.

In solid materials, the magnitude of thermal expansion can be calculated to account for linear expansion, area expansion, and volumetric expansion.

All three calculations require the thermal coefficient of the material. The thermal expansion coefficient is an empirically derived constant that describes the expansion of a given material during a rise in temperature and depends on the geometry of the material.

The thermal expansion equation for linear expansion is:

$$\Delta L = \alpha L_0 \Delta T$$

where ΔL is the change in length (m), L_0 is the initial length (m), ΔT is the change in temperature (K), and α is the coefficient of linear expansion(K^{-1})

The equations for the area and volume expansions are below:

Area Expansion: $\Delta A = \gamma A_0 \Delta T$

where ΔA is the change in the area (m^2), and γ is the coefficient of area expansion (K^{-1}).

Volume Expansion: $\Delta V = \beta V_0 \Delta T$

where ΔV is the change in volume (m^3), and β is the coefficient of volume expansion (K^{-1}).

Chapter Summary

- States of matter: solid, liquid, and gas phases. Liquids and gases are fluids.
- Density is mass per unit volume and is given by $\rho = \frac{m}{v}$
- The pressure is force per unit area and is given by $P = \frac{F}{A}$
- External pressure applied to a confined fluid is transmitted throughout the fluid.
- Gauge pressure is the total pressure referenced at the atmospheric pressure, expressed as $P_{gauge} = \rho g \Delta h$
- An object submerged partly or wholly in a fluid is buoyed by force equal to the weight of the fluid that the object displaces.
- The buoyant force is an upward force experienced by an object submerged in a fluid, due to displacement.

  The buoyant force is given by $F_{buoyant} = \rho V g$, where V is the volume of the fluid displaced.
- The Continuity Equation states that the flow rate through a pipe (area times velocity) is constant so that $A_1 \upsilon_1 = A_2 \upsilon_2$. This expresses the idea that a larger cross-sectional area of pipe experiences fluids traveling at a lower velocity.
- Bernoulli's Equation is a statement of conservation of energy. It states:

$$P + \rho g y + \frac{1}{2} \rho \upsilon^2 = \text{constant}$$

- Viscosity is an internal frictional force within fluids.
- The resistance to the deformation of solid material is due to stress. The three basic types of stress are tension, compression, and shear. Stress is given by:

$$\text{stress} = \frac{\text{force}}{\text{area}}$$

- *Tensile stress* results when a force acts on the material to pull it apart.

  Compressive stress results when a force acts to compress the material.

  Shear stress occurs when two forces act opposite such that the material slides on a plane between the applied forces.

- *Elasticity* is the measure of a material's ability to return to its original shape after distortion after undergoing tensile or compressive deformation.

- *Strain*, given by the lowercase Greek letter epsilon (ε), is the change in length of an object in the direction of the applied force, divided by the original length.

$$\varepsilon = \frac{\Delta l}{l_0}$$

  where ε is the strain, Δl is the change in length (m), and l_0 is the original length (m)

- *Young's modulus* (elastic modulus) is a measure of the force per unit needed to stretch material. It is expressed as the ratio of stress divided by strain, given as:

$$E = \frac{\sigma}{\varepsilon} = \frac{F l_0}{A \Delta l}$$

- Most materials expand when the temperature rises, and contract when the temperature falls; this phenomenon is *thermal expansion*.

  The thermal expansion equation for linear expansion is given as:

$$\Delta L = \alpha L_0 \Delta T$$

  where ΔL is the change in length (m), L_0 is the initial length (m), ΔT is the change in temperature (K), and α is the coefficient of linear expansion (K^{-1})

Practice Questions

1. The Bernoulli effect describes the lift force on an airplane wing. Wings must be designed to ensure those air molecules:

A. move more rapidly past the lower surface of the wing than past the upper surface
B. flow around wings that are smooth enough for an easy flow of the air
C. are deflected upward when they hit the wing
D. are deflected downward when they hit the wing
E. move more rapidly past the upper surface of the wing than past the lower surface

2. Two horizontal pipes (A and B) are the same length, but pipe B has twice the diameter of pipe A. Water undergoes viscous flow in both pipes, subject to the same pressure difference across the lengths of the pipes. If the flow rate in pipe A is Q, what is the flow rate in pipe B?

A. 2Q **B.** 4Q **C.** 8Q **D.** 16Q **E.** $\sqrt{2}$Q

Questions **3-5** are based on the following:

A pressurized cylindrical tank is 5 m in diameter. Water exits from the pipe at point C with a velocity of 13 m/s. Point A is 10 m above point B, and point C is 3 m above point B. The cross-sectional area of the pipe at point B is 0.08 m^2, and the pipe narrows to a cross-sectional area of 0.04 m^2 at point C. Assume an ideal fluid in laminar flow. The density of water is 1,000 kg/m^3. (Use the acceleration due to gravity g = 9.8 m/s^2)

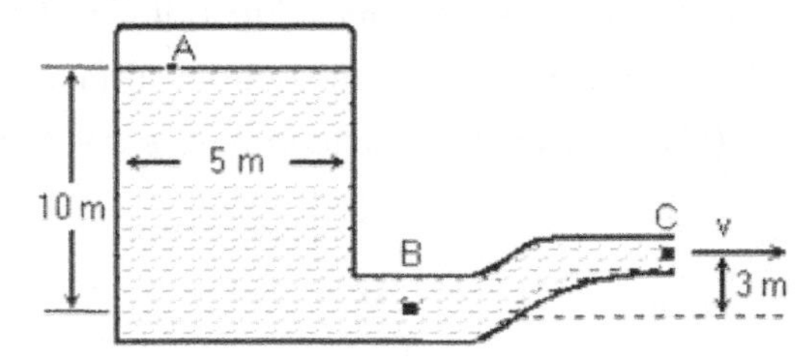

3. What is the mass flow rate in the pipe at point C?

A. 320 kg/s
B. 440 kg/s
C. 570 kg/s
D. 610 kg/s
E. 520 kg/s

4. What is the rate at which the water is falling into the tank?

A. 12 mm/s **B.** 15 mm/s **C.** 86 mm/s **D.** 44 mm/s **E.** 26 mm/s

5. What is the gauge pressure in the pipe at point B?

A. 82 kPa **B.** 167 kPa **C.** 71 kPa **D.** 98 kPa **E.** 45 kPa

6. A closed cubical chamber resting on the floor contains oil and a piston. If the piston is pushed down hard enough to increase the pressure just below the piston by an amount ΔP, which of the statements is correct?

A. The pressure at the top of the oil increases by less than ΔP
B. The pressure on the sides of the chamber remains the same.
C. The pressure in the oil increases by less than ΔP
D. The pressure at the bottom of the oil increases by more than ΔP
E. The increase in the force on the top of the chamber equals the increase in the force on the bottom of the chamber

7. What is the gauge pressure in the water at the deepest point of the Pacific Ocean, which is 11,030 m? (Use the density of seawater $\rho = 1{,}025$ kg/m^3 and the acceleration due to gravity $g = 9.8$ m/s^2)

A. 1.1×10^8 Pa **C.** 4.2×10^7 Pa
B. 3.1×10^8 Pa **D.** 7.6×10^7 Pa **E.** 5.4×10^8 Pa

8. A cubical box with 25 cm sides is immersed in a fluid. The pressure at the top surface of the box is 108 kPa, and the pressure on the bottom surface is 114 kPa. What is the density of the fluid? (Use acceleration due to gravity $g = 9.8$ m/s^2)

A. 980 kg/m^3 **C.** 2,452 kg/m^3
B. 1,736 kg/m^3 **D.** 2,794 kg/m^3 **E.** 1,360 kg/m^3

9. A polar bear of mass 240 kg stands on a floating ice 100 cm thick. What is the minimum area of the ice that will support the bear? (Use the specific gravity of ice = 0.98 and the specific gravity of saltwater = 1.03)

A. 2.6 m^2 **B.** 4.9 m^2 **C.** 4.8 m^2 **D.** 11.2 m^2 **E.** 9.1 m^2

10. A 14,000 N car is raised using a hydraulic lift. The lift consists of a U-tube with arms of unequal areas, initially at the same level. The lift is filled with oil with a density of 750 kg/m^3 with tight-fitting pistons at each end. The narrower arm has a radius of 6 cm, while the wider arm of the U-tube has a radius of 16 cm. The car rests on the piston on the wider arm of the U-tube. What is the force that must be applied to the smaller piston to lift the car after it has been raised 1.5 m? (Ignore the weight of the pistons and use the acceleration due to gravity $g = 9.8$ m/s^2)

A. 4,568 N **B.** 3,832 N **C.** 2,094 N **D.** 1,379 N **E.** 1,748 N

11. An object has a volume of 4.2 m^3 and weighs 41,800 N. What is its apparent weight in water? (Use acceleration due to gravity $g = 9.8$ m/s^2 and density of water $\rho = 1{,}000$ kg/m^3)

A. 1,140 N **B.** 230 N **C.** 800 N **D.** 640 N **E.** 2,360 N

12. A solid sphere of mass 9.2 kg, made of metal whose density is 3,650 kg/m^3, hangs by a cord. When the sphere is immersed in a liquid of unknown density, the tension in the cord is 42 N. What is the density of the liquid? (Use acceleration due to gravity $g = 9.8$ m/s^2)

A. 1,612 kg/m^3
B. 1,468 kg/m^3
C. 1,950 kg/m^3
D. 1,742 kg/m^3
E. 1,144 kg/m^3

13. The tensile strength for a steel wire is $2{,}860 \times 10^6$ N/m^2. What is the maximum load that can be applied to a wire made of this steel if its diameter is 3.5 mm?

A. 46.4 kN **B.** 19.7 kN **C.** 33.2 kN **D.** 58.5 kN **E.** 27.5 kN

14. A grade of motor oil, which has a viscosity of 0.3 N·s/m^2, is flowing through a 1 m long tube with a radius of 3.2 mm. What is the average speed of the oil, if the drop in pressure over the length of the tube is 225 kPa?

A. 0.82 m/s **B.** 0.96 m/s **C.** 1.2 m/s **D.** 1.4 m/s **E.** 1.7 m/s

15. Ice has a lower density than water because ice:

A. molecules vibrate at lower rates than water molecules
B. is made of open-structured, hexagonal crystals
C. is denser and therefore sinks when in liquid water
D. molecules are more compact than when in the solid state
E. density decreases with decreasing temperature

Solutions

1. E is correct. According to the Bernoulli effect, air moving with a higher velocity exerts less pressure against a surface than air with a lower velocity.

Thus, to produce lift, the pressure on the underside of a wing should be higher, and the air should be slower on the bottom surface of the wing compared to air on the top surface.

2. D is correct. Poiseuille's Law:

$Q = \pi \Delta P r^4 / 8\eta L$

$D_B = 2D_A$

$r_B = 2r_A$

$Q_A = \pi \Delta P r_A{}^4 / 8\eta L$

$Q_B = \pi \Delta P(2r_A)^4 / 8\eta L$

$Q_B = 16(\pi \Delta P r_A{}^4 / 8\eta L)$

$Q_B = 16Q_A$

3. E is correct. Mass flow rate $\dot{m}$ is:

$\dot{m} = \rho v A_C$

where ρ = density of the fluid, v = velocity of flow, A_C = cross-sectional area

The dimensions of the tank are irrelevant to this answer, so $\dot{m}$ is calculated as:

$\dot{m} = (1{,}000 \text{ kg/m}^3)\cdot(13 \text{ m/s})\cdot(0.04 \text{ m}^2)$

$\dot{m} = 520 \text{ kg/s}$

4. E is correct. Relationship between area and velocity of both exists as:

$A_1 v_1 = A_2 v_2$

$v_1 = A_2 v_2 / A_1$

$v_1 = [(0.04 \text{ m}^2)\cdot(13 \text{ m/s})] / [(\pi / 4)\cdot(5 \text{ m/s})]$

$v_1 = 0.26 \text{ m/s} = 26 \text{ mm/s}$

5. D is correct. Gauge pressure is the pressure reference against the surrounding air pressure. Because of this, it is valid to ignore both the pressure in the tank and atmospheric pressure.

$$P = \rho g h$$

$$P = (10^3 \text{ kg/m})\cdot(9.8 \text{ m/s}^2)\cdot(10 \text{ m})$$

$$P = 98{,}000 \text{ Pa} = 98 \text{ kPa}$$

6. E is correct. According to Pascal's Law, the pressure is transmitted undiminished in an enclosed static fluid. Thus the pressure increases by ΔP everywhere in the oil. If the chamber is cubic, the top and bottom sides have the same area and experience the same increase in force:

$$A_{\text{top}} = A_{\text{bottom}}$$

$$\Delta \text{P}_{\text{top}} = \Delta \text{P}_{\text{bottom}}$$

$$P = F / A$$

Thus, force is directly proportional to pressure, and if the area of the top is equal to the area of the bottom:

$$\Delta F_{\text{top}} = \Delta F_{\text{bottom}}$$

7. A is correct. Gauge pressure is referenced relative to atmospheric pressure and is calculated by:

$$P_{\text{gauge}} = \rho g h$$

$$P_{\text{gauge}} = (1{,}025 \text{ kg/m}^3)\cdot(9.8 \text{ m/s}^2)\cdot(11{,}030 \text{ m})$$

$$P_{\text{gauge}} = 1.1 \times 10^8 \text{ Pa}$$

8. C is correct.

$$P_{\text{top}} = \rho g h$$

$$P_{\text{top}} = 108 \times 10^3 \text{ Pa}$$

$$h = (1 / \rho)\cdot(108 \times 10^3 \text{ Pa} / 9.8 \text{ m/s}^2)$$

$$h = (1 / \rho)\cdot(11{,}020 \text{ kg/m}^2)$$

$P_{\text{bottom}} = \rho g(h + 25\text{ cm})$

$\rho g(h + 25\text{ cm}) = 114 \times 10^3\text{ Pa}$

$h = (1 / \rho)\cdot(114 \times 10^3\text{ Pa} / 9.8\text{ m/s}^2) - 0.25\text{ m}$

$h = (1 / \rho)\cdot(11{,}633) - 0.25\text{ m}$

Set equal and solve for ρ:

$(1 / \rho)\cdot(11{,}020\text{ kg/m}^2) = (1 / \rho)\cdot(11{,}633\text{ kg/m}^2) - 0.25\text{ m}$

$(11{,}020\text{ kg/m}^2) = (11{,}633\text{ kg/m}^2) - 0.25\text{ m}(\rho)$

$0.25\text{ m}(\rho) = 613\text{ kg/m}^2$

$\rho = (613\text{ kg/m}^2) / (0.25\text{ m})$

$\rho = 2{,}452\text{ kg/m}^3$

9. C is correct. Solve for the density of ice and saltwater:

$SG = \rho_{\text{substance}} / \rho_{\text{water}}$

$SG_{\text{ice}} = 0.98 = \rho_{\text{ice}} / 10^3\text{ kg/m}^3$

$\rho_{\text{ice}} = 980\text{ kg/m}^3$

$SG_{\text{saltwater}} = 1.03 = \rho_{\text{saltwater}} / 10^3\text{ kg/m}^3$

$\rho_{\text{saltwater}} = 1{,}030\text{ kg/m}^3$

Solve for the volume of ice:

$F_B = F_{\text{bear}} + F_{\text{ice}}$

$\rho_{\text{saltwater}} V_{\text{ice}} g = m_{\text{bear}} g + \rho_{\text{ice}} V_{\text{ice}} g$, cancel g from all terms

$V_{\text{ice}} = m_{\text{bear}} / (\rho_{\text{saltwater}} + \rho_{\text{ice}})$

$V_{\text{ice}} = (240\text{ kg}) / (1{,}030\text{ kg/m}^3 - 980\text{ kg/m}^3)$

$V_{\text{ice}} = 4.8\text{ m}^3$

Solve for the area of ice:

$A = V / h$

$A = (4.8\text{ m}^3) / (1\text{ m})$

$A = 4.8\text{ m}^2$

10. C is correct.

$P_1 = P_2 + \rho gh$

$F_1 / A_1 = F_2 / A_2 + \rho gh$

$F_1 = A_1(F_2 / A_2 + \rho gh)$

$F_1 = \pi(0.06 \text{ m})^2 \cdot [(14{,}000 \text{ N}) / \pi(0.16 \text{ m})^2 + (750 \text{ kg/m}^3) \cdot (9.8 \text{ m/s}^2) \cdot (1.5 \text{ m})]$

$F_1 = \pi(0.0036 \text{ m}^2) \cdot [(14{,}000 \text{ N}) / \pi(0.0256 \text{ m}^2) + (750 \text{ kg/m}^3) \cdot (9.8 \text{ m/s}^2) \cdot (1.5 \text{ m})]$

$F_1 = (0.0036 \text{ m}^2) \cdot [(14{,}000 \text{ N}) / (0.0256 \text{ m}^2) + (11{,}025 \text{ N})\pi]$

$F_1 = 1{,}969 \text{ N} + 125 \text{ N}$

$F_1 = 2{,}094 \text{ N}$

11. D is correct.

$F_B = \rho Vg$

$F_{net} = F_{object} - F_B$

$F_{net} = (41{,}800 \text{ N}) - (1{,}000 \text{ kg/m}^3)(4.2 \text{ m}^3)(9.8 \text{ m/s}^2)$

$F_{net} = (41{,}800 \text{ N}) - (41{,}160 \text{ N})$

$F_{net} = 640 \text{ N}$

12. C is correct.

$F_{net} = F_{object} - F_B$

$F_{net} = mg - \rho_{fluid} Vg$

$\rho_{fluid} = -(F_{net} - mg) / V_{sphere}g$

$\rho_{fluid} = -[42 \text{ N} - (9.2 \text{ kg}) \cdot (9.8 \text{ m/s}^2)] / [(9.2 \text{ kg} / 3{,}650 \text{ kg/m3}) \cdot (9.8 \text{ m/s}^2)]$

$\rho_{fluid} = 1{,}950 \text{ kg/m}^3$

13. E is correct. tensile strength = F_{max} / A

$F_{max} = \text{A(tensile strength)}$

$F_{max} = (\pi / 4) \cdot (0.0035 \text{ m})^2 \cdot (2{,}860 \times 10^6 \text{ N/m}^2)$

$F_{max} = 27{,}516 \text{ N} = 27.5 \text{ kN}$

14. B is correct. By Poiseuille's Law, the volumetric flow rate of a fluid is given by:

$V = \Delta PAr^2 / 8\eta L$

The volumetric flow rate is the volume of fluid that passes a point per unit time:

$V = Av$

where v is the speed of the fluid.

Therefore:

$Av = \Delta PAr^2 / 8\eta L$

$v = \Delta Pr^2 / 8\eta L$

$v = (225 \times 10^3\ \text{Pa})\cdot(0.0032\ \text{m})^2 / [8\ (0.3\ \text{Ns/m}^2)\cdot(1\ \text{m})]$

$v = 0.96\ \text{m/s}$

15. B is correct.

For most substances, the solid form is denser than the liquid phase. Therefore, a block of most solids sinks in the liquid. For pure water, a block of ice (solid phase) floats in liquid water because ice is less dense than liquid water.

Like other substances, when liquid water is cooled from room temperature, it becomes increasingly dense. However, at approximately 4 °C (39 °F), water reaches its maximum density, and as it is cooled further, it expands the distance between molecules and becomes less dense. This phenomenon is negative thermal expansion and is attributed to strong intermolecular interactions that are orientation dependent.

The density of water is about 1 g/cm^3 and depends on the temperature. When frozen, the density of water is decreased by about 9%. This decreased density of water is due to the decrease in intermolecular vibrations, which allows water molecules to form stable hydrogen bonds with other water molecules around.

As these hydrogen bonds form, molecules are locking into positions to form hexagonal structures. Even though hydrogen bonds are shorter in the crystal than in the liquid, this position locking decreases the average coordination number of water molecules as the liquid reaches the solid phase.

Chapter 11

Heat and Thermodynamics

- **The Basics of Thermal Physics**
- **Heat, Temperature and Thermal Energy**
- **Heat Capacity and Specific Heat**
- **Heat Transfer**
- **Thermodynamic Systems**
- **The Zeroth Law of Thermodynamics**
- **First Law of Thermodynamics: Conservation of Energy**
- **Second Law of Thermodynamics: Entropy**
- **The Third Law of Thermodynamics**
- **Enthalpy**
- **Gibbs Free Energy**
- **PV diagrams**
- **The Carnot Cycle**
- **The Kinetic Theory of Gases**
- **Coefficient of Linear Expansion**
- **Phase Diagram: Pressure and Temperature**
- **Latent Heat**

The Basics of Thermal Physics

Energy is defined as the ability to do work, and the energy can be classified as either kinetic energy (KE) or potential energy (PE). Kinetic energy is the energy of an object in motion; for example, a rolling ball, moving the car, and dropped coin all have kinetic energy. Potential energy is defined as the potential to do work.

As discussed earlier, there are many forms of potential energy, including gravitational potential energy, spring potential energy, chemical potential energy. The hammer in the figure below initially has gravitational potential energy. When the hand holding the hammer relaxes, the hammer begins to move downward, converting the potential energy (energy of position) into kinetic energy (energy of motion).

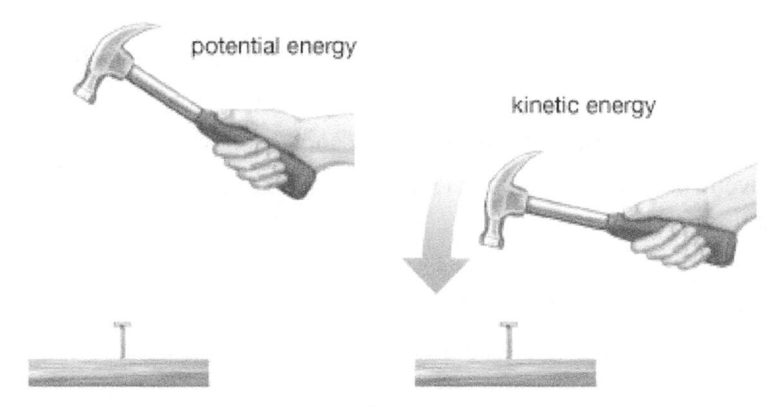

Thermal physics is the study of heat and the energy it retains. In all situations, the conservation of energy holds, meaning there is no net loss of energy. However, in most real-life examples, non-conservative forces play a role, and the energy is not always converted into useful work.

When forces act on an object, the energy they use by acting over a distance cannot be regained—the energy disappears into thin air. When a book is pushed along a table, the opposing force of friction is exerted on the book.

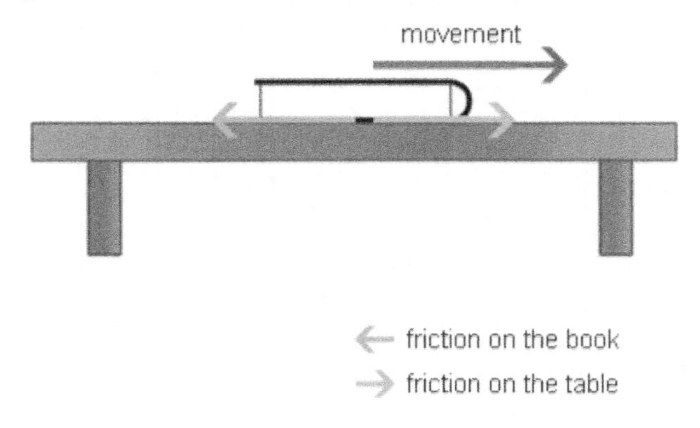

Depending on how much effort is exerted to push the book, the book travels a certain distance, and then come to rest as it steadily decreases in velocity. The book comes to rest because there is a force of friction opposing its motion.

As the book's velocity decreases, so do its kinetic energy ($\frac{1}{2}mv^2$). Conservation of energy must hold, so the loss in kinetic energy is energy being converted into heat. In general, neither the surface of the counter nor the bottom of the book feels warm because more energy transfer is needed to raise the temperature, so the difference is detected.

If the process of sliding the book is repeated many times, the bottom of the book and the surface of the table begins to feel warm to the touch. Enough energy had been deposited into the book to raise the temperature noticeably.

On a microscopic level, thermal energy is the energy related to the vibration of molecules and atoms. Every bit of matter is made of billions and trillions of small molecules, and even in solids, these molecules and atoms never are entirely still. The amount of movement an object has in its molecules determines how much thermal energy the object holds.

The figure below illustrates a closed container of gas, in which the gas molecules are in random motion throughout the container.

The thermal energy (or internal energy; the sum of all the kinetic energies of the particles) of the gas is directly proportional to the individual kinetic energy (KE) of the molecules.

The higher kinetic energy of the molecules, the more thermal energy the system possesses. With less kinetic energy, the gas has less thermal energy.

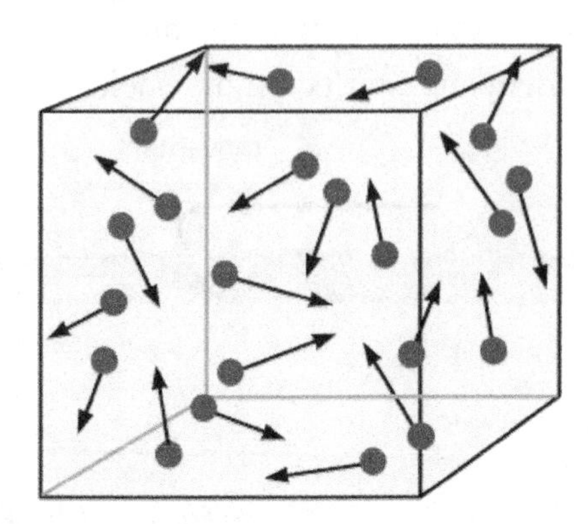

Gas molecules exhibit kinetic energy from random motion in the container

Heat, Temperature and Thermal Energy

Heat and temperature are often used interchangeably in everyday vernacular. However, this usage of interchanging heat and temperature is incorrect. *Heat* is a form of energy and exists independently from any medium; heat is a way to transfer energy. *Temperature* is a measure of the kinetic energy in a specific medium and is thus dependent on the material itself.

Temperature and heat are connected principles, but heat is the total energy due to the molecular motion. In contrast, the temperature is the average measure of the kinetic energy stored within a substance. Some substances respond drastically to a change in thermal energy, and some have a subtler reaction. This response depends on the material and heat capacity, which is discussed later.

The quantity of thermal energy that is transmitted from one body to another is a measure of heat. An object does not "contain" heat, but it can have a certain amount of thermal energy.

Heat can be measured in joules (J) but is frequently measured in calories (cal).

Conveniently, *calories* relate heat to change in temperature, which is often useful when solving problems in thermal physics. The definition of a calorie (lowercase *c*) is the amount of heat needed to raise the temperature of one gram of water by one degree Celsius (g/°C).

$$1\ cal = 1\ \frac{g}{°\mathrm{C}} = 4.19\ J$$

Food Calories (uppercase *C*) are not equivalent to heat calories. This is similar to the relationship between kilograms and grams (1 Calorie = 1,000 calories). Another difference to note is that while nutrition Calories measure stored energy just as calories do, they also measure the energy stored in the chemical bonds of food that are broken down and stored when digested.

Calories are a measure of energy in a substance and can be converted to their equivalents of conventional energy. Calories contained in 5 lbs. of spaghetti have enough energy to brew a pot of coffee; calories in one piece of cheesecake can power a 60 W incandescent light bulb for 1.5 hours; 217 Big Macs contain enough energy to drive a vehicle for 88 miles.

In the United States, the unit of temperature is the *Fahrenheit* (°F). However, the SI unit for temperature is *Celsius* (°C). This is a more straightforward system to remember, as water freezes at 0 °C and boils at 100 °C (instead of the 32 °F and 212 °F). The conversion between the two scales may be computed using the relationship:

$$°F = \frac{9}{5}°C + 32$$

The *Kelvin* is another unit of temperature often used in scientific calculations. This scale is often used when working with gases, as usually, low temperatures are in question. Kelvin (K) is the measure of the absolute temperature.

Absolute zero is the coldest theoretical temperature any substance can have, and it is equal to 0 K. A temperature of 0 K is equivalent to –273 °C!

The figure below shows three different thermometers displaying equal temperatures in the three standard units of temperature.

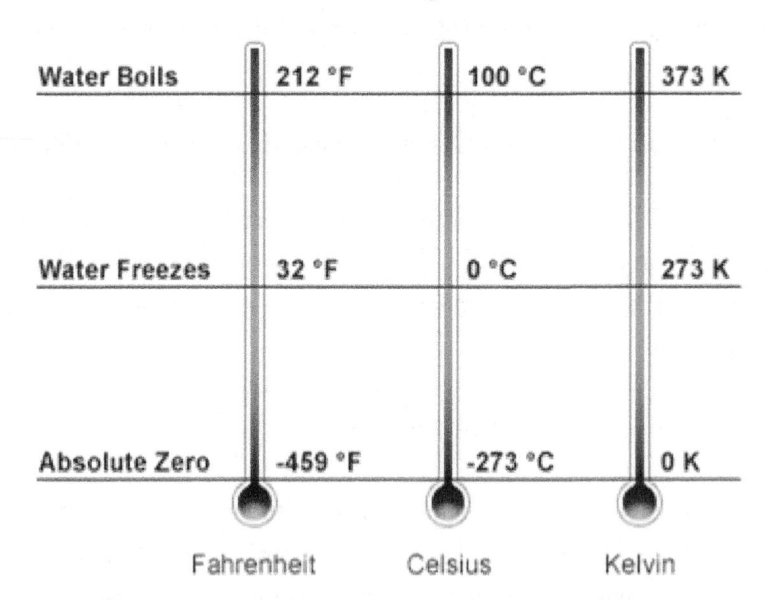

Comparison between the three temperature scales

Temperature itself is the average amount of kinetic energy of the particles (i.e., molecules and atoms) that make up a material. At absolute zero, these particles will theoretically have no kinetic energy, and therefore no movement.

Kelvin and Celsius use the same step in their degree measurements, which can make many calculations easier. For example, a temperature difference is the same in

Kelvin as it is in Celsius, so when asked to calculate a temperature difference in either of these units, the values do not need to be converted.

If required to convert between Kelvin and Celsius, the conversion can be performed by adding 273 to the Celsius temperature to get Kelvin:

$$K = °C + 273$$

The temperature of absolute zero may never be attained. Scientists have achieved temperatures close to absolute zero but have never fully reached it.

In 2003, the lowest temperature recorded was by a group of scientists from MIT (Massachusetts Institute of Technology). They successfully cooled sodium gas to a temperature of half-a-billionth of a degree above absolute zero, but not at absolute zero.

Heat Capacity and Specific Heat

The *heat capacity C* of a substance is the ratio of the absorbed heat energy to the resulting temperature change. The SI unit for heat capacity is the joule per Kelvin (J/K). When measuring the heat capacity of a substance, a variable must be held constant, since the heat capacity of a system depends on the temperature, the pressure, and the volume of the system.

For a constant reading, gases and liquids are typically measured at constant volume and subjected to a pressure. When pressure is held constant, the process to define the heat capacity is an *isobaric process*; when the volume is held constant, it is an *isochoric process*. These are noted with either a c_p or a c_v, respectively.

A similar concept is a *specific heat c* for a substance. Specific heat is the measurement of heat needed to raise the temperature of 1 gram of a substance by 1 °C. Notice that the heat capacity and the specific heat of a substance are similar.

However, specific heat is the ratio of energy absorbed to the temperature rise per unit of mass of the substance, and heat capacity is the ratio of energy absorbed for the temperature rise. The specific heat *c* is a characteristic of the substance and thus does not change. Below is a table of the specific heats of several substances:

Specific heats of selected materials	
Material	***c*(J/kg·K)**
Aluminum	879
Concrete	850
Diamond	509
Glass	840
Helium	5,193
Water	4,181

Substances such as metals (e.g., copper) have a low specific heat because it does not take much energy to transfer the heat and excite the molecules. Therefore, raising the value of their average kinetic energy (raising their temperature). Materials that are difficult to heat, such as rubber, have a much higher specific heat because more energy is required to raise the temperature by the same amount.

This is the reason for the *sea breeze effect*. The previous table shows that water has a much higher heat capacity than other common materials found on land. It takes more energy to raise an ocean by 1 °C than it does to raise city sidewalks by the same amount. Thus, during the day, land warms faster than water when subjected to the same amount of thermal energy from the sun.

The hot land heats the air above it, causing it to rise and create a low-pressure system above it. The cooler water creates a high-pressure system, due to air cooling down and descending. The difference in pressures creates a flow of air from the sea to the land during the day as a sea breeze.

At night, the land cools faster than the water (i.e., lower specific heat capacity), and the effect is reversed as air flows from land to the sea.

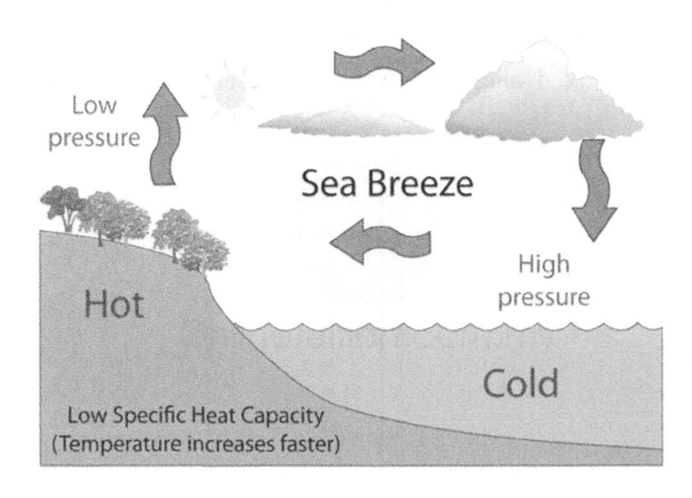

The specific heat of a substance (*c*), heat (*Q*), and temperature are be related by:

$$Q = mc\Delta T$$

where Q is the heat transferred to the material (J), m is the mass of the object being heated (kg), and ΔT is the change in temperature (K).

For example, 3,200 J of heat is added to 1.0 kg of water (c = 4,190 J/kg·°C) at an initial temperature of 10 °C. By rearranging the equation for specific heat, the increase in temperature is solved.

$$\Delta T = \frac{Q}{mc}$$

$$\Delta T = \frac{3{,}200\,J}{1.0\,kg \cdot 4{,}190\,\frac{J}{kg} \cdot {}^{\circ}\mathrm{C}}$$

$$\Delta T = 0.76\ {}^{\circ}C$$

The symbol ΔT represents the change in temperature. This value for the change (not the absolute value) must be either added to the initial temperature to determine the final temperature or subtracted from the final to find the initial.

A reaction must be contained within a system so that all the products can be observed. A *calorimeter* is considered an ideally insulated system. The heat liberated by the reaction process is transferred to either the other substances within the calorimeter or the calorimeter itself. No heat is lost to the surroundings.

Since no thermal energy is lost in a calorimeter, the heat of a reaction can be measured by inserting a thermometer into a calorimeter and recording the temperature readings before and after the reaction. A stirrer ensures that all contents of the calorimeter are well mixed and react uniformly. Below is a simple calorimeter to demonstrate the concept:

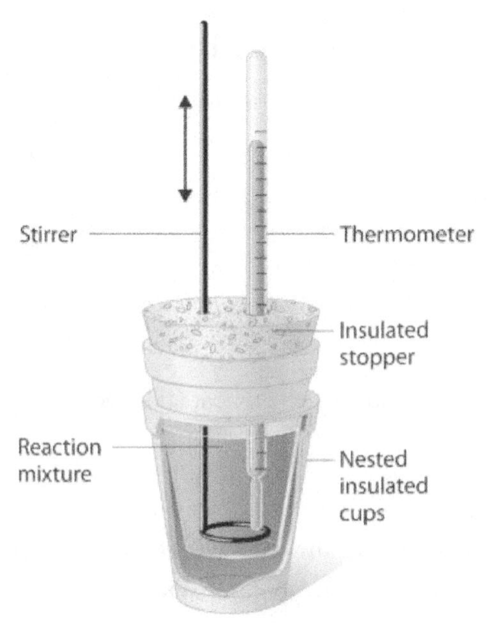

Components of a simple calorimeter

Heat Transfer

Heat can be transferred between two objects in three ways: conduction, convection, or radiation.

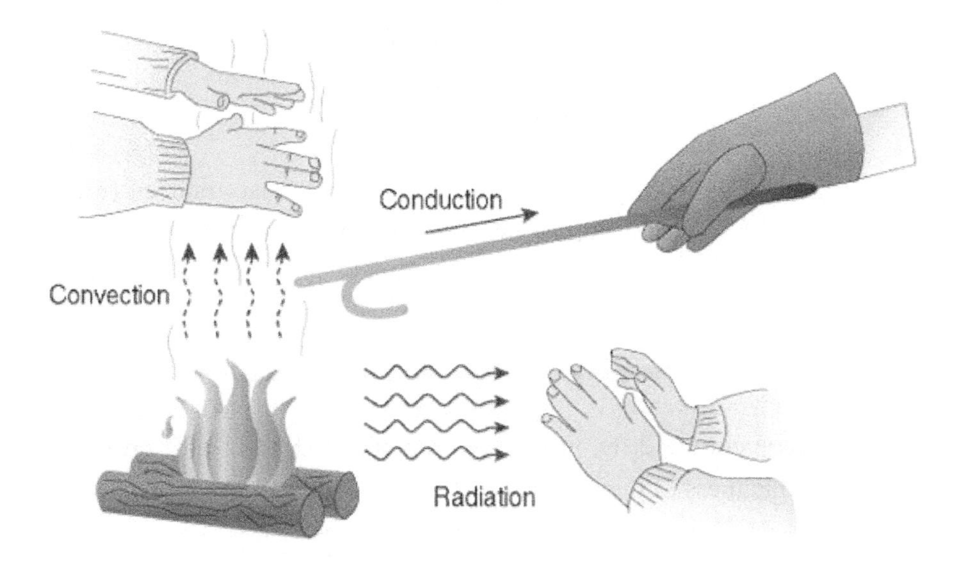

Conduction occurs when the heat is transferred through direct contact. On a cold winter night, when a person wraps their hands around a cup of hot cocoa, their fingers become warm because they are in direct contact with the hot mug, which is in direct contact with the hot liquid. The same process is happening at a molecular level.

The liquid in the mug contains heat energy, and the molecules of the hot liquid are in constant motion (high kinetic energy). These molecules continuously collide with the molecules of the mug, increasing their motion. The action of the collisions gradually transfers some of the energy from the hot liquid molecules to the mug molecules.

Through the same mechanism of conduction, the mug then transfers heat to a person's hands. The person perceives the transfer of energy as an increase in temperature; thus, their hands feel warm.

The figure below illustrates the process of conduction (heat transfer by direct contact) from the hot liquid to the walls of a mug:

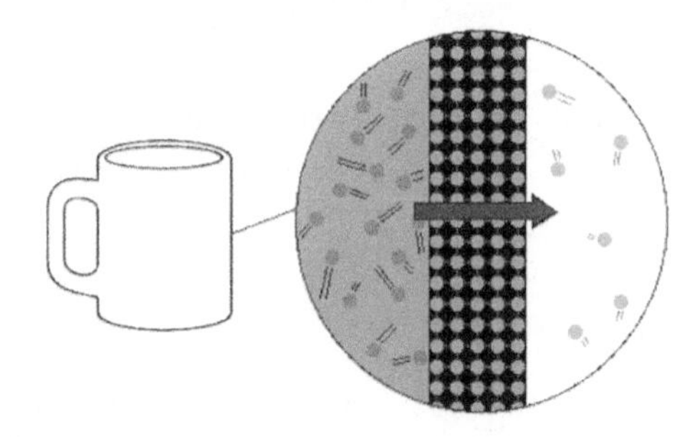

When heat transfers by conduction, the rate of heat transfer (H) can be calculated. The rate of heat transfer is measured in joules per second (J/s) or watts (W). This describes the amount of energy (heat) conducted during a set time interval:

$$H = \frac{kA\Delta T}{t}$$

where ΔT is the temperature difference across the object (K), A is the cross-sectional area (m^2), t is the thickness of the material (m), and k is the thermal conductivity of the material (J/s·m·k). This quantity for the rate of heat transfer (H) is a characteristic of the material and is usually is provided in the problem. Below is a table of select values for thermal conductivities.

Thermal Conductivities of Selected Materials	
Material	**k (J/s·m·K)**
Aluminum	237
Concrete	1
Copper	386
Glass	0.9
Stainless Steel	16.5
Water	0.6

When calculating heat conduction through an object, it is essential to note that the cross-sectional area refers to the area through which the heat is being conducted. For the example of the mug, the heat is being conducted through its outer curved surface.

Therefore, the cross-sectional area is the circumference of the mug times its height. The thickness, in this case, is the thickness of the mug itself. When the mug is placed on a table, and the heat conduction to the table is calculated, the cross-sectional area is the area of the bottom of the mug, instead of the area of the sidewalls.

Convection is the process of transferring heat through the flow of energized molecules from one place to another. This is done through the movement of fluids. Gas is also a fluid, as it can flow and change shape depending on its container.

A pot of water is placed on a hot stove. The water at the bottom of the pot warms first. As the water warms, it rises and is replaced by cooler water. This circulation of the heated water transfers the thermal energy from the bottom of the pot to the top. This movement of molecules results in a closed pattern of fluid flow known as *convection currents*.

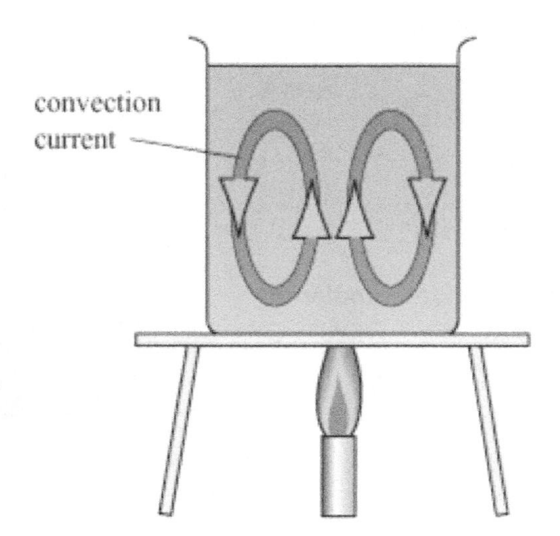

Convention current represented by the circulating water

A convection oven works similarly, but with gas molecules instead of liquid molecules. Air is blown past a heating element to the oven chamber. When passing through the heating element, the air molecules are energized by the hot element. The energized air molecules are then circulated throughout the oven by airflow provided by the fan. This process continues until all the molecules are energized to the same amount.

Although convection is similar to conduction, notice the difference. Conduction is the flow of heat between two materials in direct contact, while convection is the flow of energized (e.g., air) molecules. When the energized molecules settle, they then deposit their energy by direct contact using conduction.

Radiation is the transfer of energy through electromagnetic waves. Radiation is emitted from all objects or surfaces with heat energy. Objects with more energy emit more radiation energy.

For example, a hot piece of metal radiates heat in the form of infrared electromagnetic waves. As the metal gets hot, it also begins to radiate heat in the visual spectrum, thus making the metal appear "red hot."

The Sun is another well-known source of radiation. The sun releases electromagnetic waves that travel through space, then through the atmosphere to heat the Earth. Similarly, holding hands next to a fire heats the hands through radiation.

Notice that radiation is the transfer of heat through electromagnetic waves, and therefore does not require a medium; thus, radiation can occur in a vacuum.

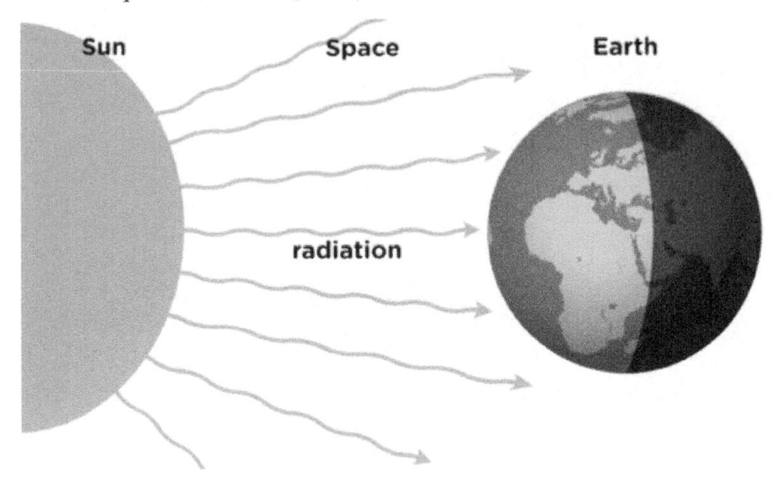

Radiation emitted from the sun travels to the Earth's surface

The power radiated by an object is related to its temperature:

$$P = e\sigma A(T^4 - T_C^4)$$

where P is the power radiated (W), e is the emissivity of the radiator, σ is Stefan's constant equal to 5.67×10^{-8} W/m^2·K, A is the area of the radiating surface (m^2), T is the temperature of the radiator (K), and T_C is the temperature of the surroundings (K).

Thermodynamic Systems

A *thermodynamic system* describes thermodynamic processes and is a quantity of matter within a boundary. Thermodynamic systems can be isolated, closed, or open. An *isolated system* has no transfer of heat, work, or matter with its surroundings. A *closed system* may transfer heat and work to its surroundings, but never transfers matter. An *open system* can transfer heat, work, or matter to its surroundings.

State functions, (i.e., state quantities), are a group of equations representative of the properties of a system, which depend *only* on the current state of the system, not the path used to achieve the current state.

Thermodynamic state functions include enthalpy, entropy, and Gibbs free energy.

System	Exchanges with surrounding	Total amt. of Energy	Example	Illustration
Open	Energy & Matter	Does not remain constant	Solution kept in an open flask	Matter (water vapor); Energy; Open system; Energy
Closed	Only Energy	Does not remain constant	Solution kept in a sealed flask	Energy; Closed system; Energy
Isolated	Neither energy nor matter	Remains constant	Sealed flask kept in a thermos flask	Isolated system

The Zeroth Law of Thermodynamics

The Zeroth Law of Thermodynamics is named because it is the logical predecessor of the First and Second Laws of Thermodynamics. The Zeroth Law states that *if two systems are in thermal equilibrium with a third system, the two initial systems are also in thermal equilibrium.*

All three systems in thermal equilibrium contain the same amount of heat energy (same temperature) and do not exchange heat. This observation is the fundamental principle underlying thermodynamics. The figure below demonstrates the Zeroth Law of Thermodynamics.

If system A is in equilibrium with B, and system C is in equilibrium with B, then system A and C must be in equilibrium, and all three systems must be at the same temperature.

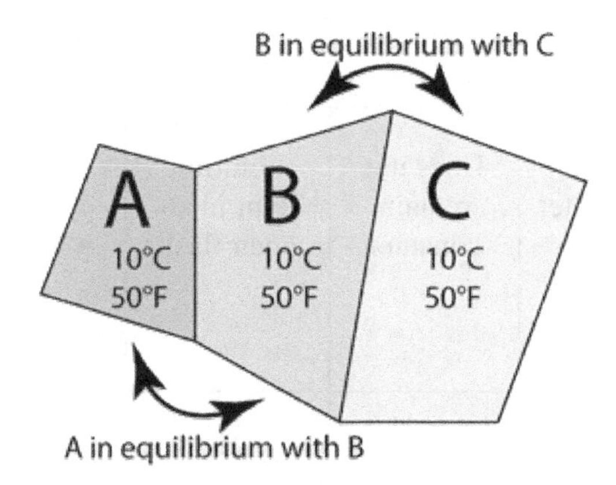

It is important to notice, however, that although no net transfer of heat occurs between systems in equilibrium, thermal energy is technically transferring between the related systems.

The exchange of thermal energy requires that every unit of energy that is passed from any system has the same value of energy passed back into the system. This holds even if the two systems contain materials with different specific heats. This observation of thermal energy transfer means that there must be a property that can be considered the same upon which heat transfer depends, and that property is temperature.

First Law of Thermodynamics: Conservation of Energy

Energy, like matter, is always conserved. Energy can neither be created nor destroyed. The *First Law of Thermodynamics* relates closely to the *internal energy* of a substance.

The internal energy refers to the microscopic energy of the disordered motion of particles. This internal energy cannot be seen and has no apparent effect on the motion of the object, but it does play a part in the thermodynamic properties of the substance. Internal energy is represented by the symbol (U).

A glass of water does not look like it contains kinetic or potential energy. On the atomic scale, the water molecules are whizzing around, moving at hundreds of meters per second. If the water is dumped out of the glass, its internal energy has no apparent effect on its motion.

The First Law of thermodynamics states that *the internal energy of a system increases if heat is added to the system, or if work is done on the system.*

The difference between the amount of heat added to the system (Q) and the work done by the system (W) is equal to the total change in internal energy (ΔU):

$$\Delta U = \Delta Q - \Delta W$$

As with other energy, the unit of internal energy is the joule (J), although it is sometimes expressed using calories (cal). By convention, work done on a system is positive, and work done by a system is negative.

Occasionally, the above equation is written as: $\Delta U = \Delta Q + \Delta W$, and ΔW is defined as the work done on the system. Be sure to note which form is being used.

Using the quantities in the $\Delta U = \Delta Q + \Delta W$ equation, it can be determined what a thermodynamic system is experiencing.

For example, in a chemical reaction, if ΔQ is a quantity less than zero ($\Delta Q < 0$), the reaction is losing heat. This reaction is *exothermic,* and heat is a product.

If ΔQ is a quantity greater than zero ($\Delta Q > 0$), the reaction is absorbing heat from its surroundings. This is an *endothermic* reaction, and heat is a reactant.

The figure below depicts both an endothermic reaction (absorbing heat) and an exothermic reaction (producing heat).

On the left, the reaction is exothermic because heat is produced and released to the surroundings of the system.

On the right, the reaction is endothermic because heat is absorbed from the surroundings.

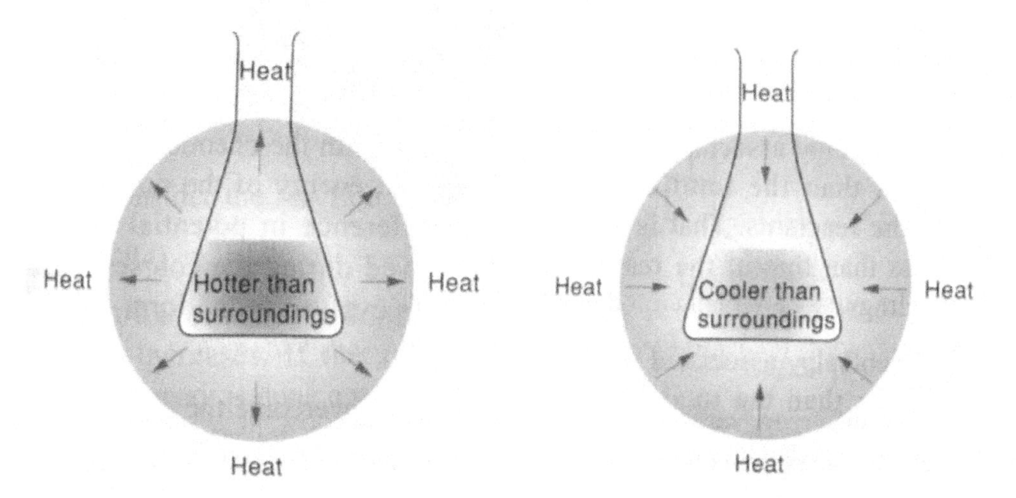

An exothermic process (releasing heat) is shown on the left and an endothermic process (absorbing heat) is shown on the right.

Second Law of Thermodynamics: Entropy

The *Second Law of Thermodynamics* states *that in an isolated system (no transfer of heat, work, or matter), a process either increases in entropy or stays constant, but entropy never decreases*. *Entropy* is a measure of the system's thermal energy per unit temperature that is unavailable to perform useful work. Work requires ordered molecular motion. Entropy is a measure of the randomness or disorder of a system.

The only exception to the Second Law of Thermodynamics (i.e., entropy cannot decrease) is an open system where heat, work, or matter can be transferred to surroundings. This exception is because the system's entropy decrease is related to an increase in the entropy of the surroundings.

Regarding heat flow, the Second Law states that heat flows spontaneously from a hot object to a cold object, but it never spontaneously flows in the opposite direction. A bowl of ice cream never gets colder in a warm room; the ice cream always gets warmer.

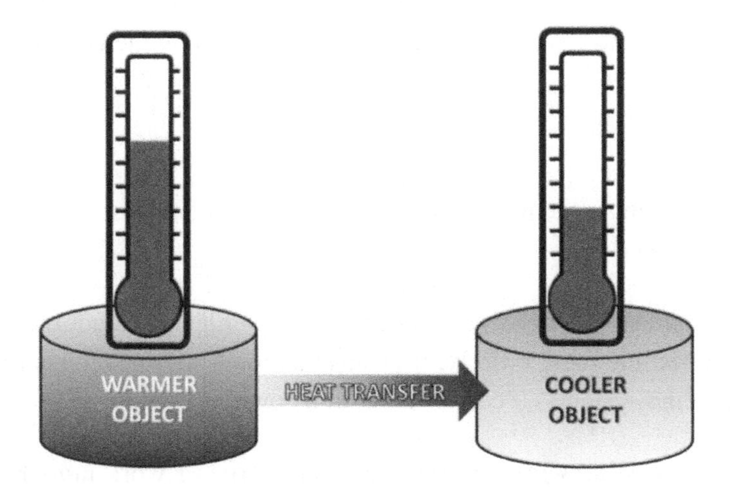

Entropy is a measure of disorder within a system. In reality, it is more a measure of multiplicity and likelihood—how likely a system resembles the same state twice, or how likely a reaction is to happen.

The entropy of a natural system never decreases, as it is the natural tendency of all things to move towards maximum disorder—maximum entropy.

For example, a gas expands to fill a space, because there are more ways it can arrange its given particles—more multiplicities.

A solid has less entropy than a liquid, which has much less entropy than a gas.

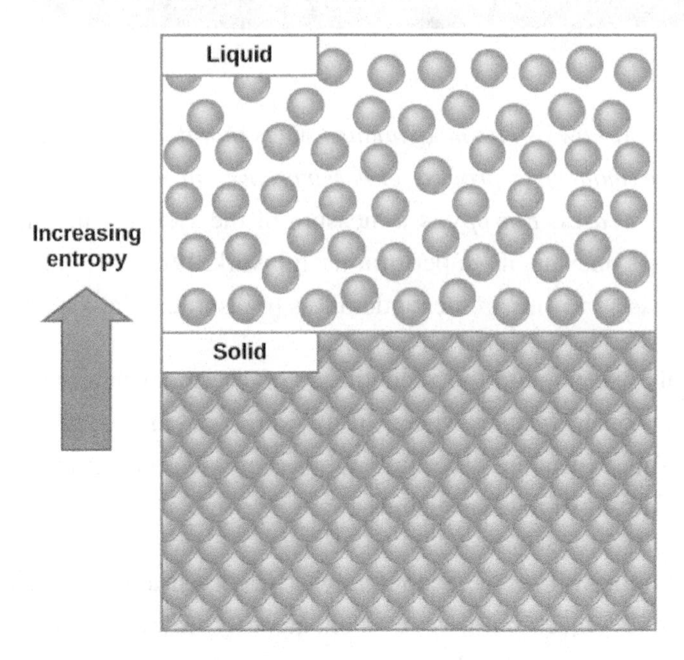

According to the Second Law of Thermodynamics (i.e., entropy can stay constant or increase but never decrease), entropy is given by the equation:

$$\Delta S \geq \frac{Q}{T}$$

where ΔS is the change in entropy of a process (J/K), Q is the heat transferred (J), and T is the temperature (K).

If a system strives to increase in entropy (disorder) does not mean that the system cannot become more orderly. It means that to do so, outside energy must be transferred into the system—it cannot do it on its own.

A *nonspontaneous process* requires that energy or work must be added to the system for a decrease in entropy to occur.

A *spontaneous process* moves towards randomness and requires no energy input for the increase in entropy.

For example, a valve connects a container with air and another container in a vacuum. When the valve is opened, the air flows from the container with air to the container in a vacuum. No work or energy is needed for this to happen. Therefore the process is spontaneous; at the end of the reaction, the entropy has increased.

For the reverse to occur, energy must be added to the system to re-pressurize the original container with air. This process is an example of a nonspontaneous reaction.

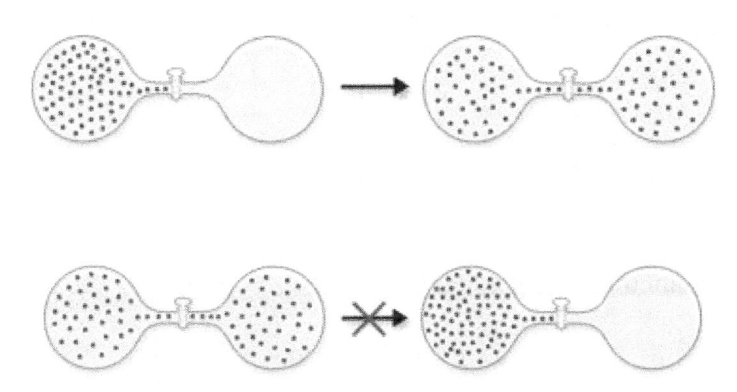

Another essential idea is that the entropy of the universe always increases when a spontaneous process occurs.

The entropy of the universe is expressed as:

$$\Delta S_{universe} > 0$$

The Third Law of Thermodynamics

The Third Law of Thermodynamics states that *as the temperature of a system approaches absolute zero, the entropy approaches a constant value*.

Typically, this constant value is zero. In this definition, the entropy of a system is related to the number of microstates possible. Microstates refer to all the atomic configurations possible. The equation represents this relationship:

$$S = k \cdot ln\ (\Omega)$$

where S is the entropy (J/K), k is Boltzmann's constant (1.381×10^{-23} J/K), and Ω is the number of microstates.

Only one microstate is possible at absolute zero, and so Ω equals 1:

$$S = k \cdot ln(1) = k \cdot 0 = 0$$

Therefore, the entropy of a system at absolute zero is zero. However, some systems have more than one minimum energy state. That is, as their temperature approaches absolute zero, the entropy levels off at a value other than zero.

It is impossible to cool any process to absolute zero in a finite number of steps.

In the image on the left, $T = 0$ can be reached following an infinite number of steps (step lines between X_1 and X_2).

In the image on the right, $T = 0$ is reached by a finite number of steps, each getting closer and closer to zero, but never to zero.

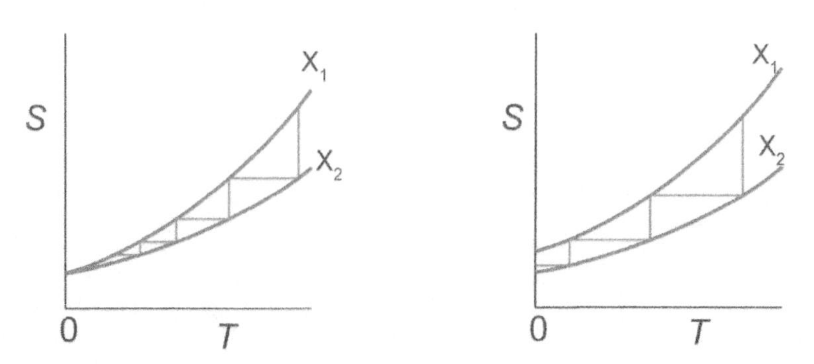

Entropy vs. temperature graphs illustrating the step to approach absolute zero

Enthalpy

The *enthalpy* (ΔH) of a system is the amount of heat used at constant pressure in a system or reaction and is given in SI units of Joules. It describes the energy changes of the system, and can be written as:

$$\Delta U = \Delta H - P\Delta V$$

where ΔH is the change in enthalpy of a system (J), P is the pressure (Pa), and ΔV is the change in volume (m^3).

During an isobaric process (pressure is held constant), such as that conducted in an open container, the change in enthalpy equals the amount of heat transferred during the process.

Most reactions do not generate many gaseous products, and as a result, there is little work associated with the reaction. The change in energy of the system is equal to the change in enthalpy:

$$\Delta E \approx \Delta H$$

The enthalpy is equal to the transfer of thermal energy (heat).

However, this is only valid at constant pressure.

An *exothermic reaction* has a change in enthalpy less than zero ($\Delta H < 0$).

An *endothermic reaction* has a change in enthalpy greater than zero ($\Delta H > 0$).

Since enthalpy (H) is a state function, it represents a property of a system. It can be used to describe a reaction or an equation for a reaction.

Hess' Law states that the change of enthalpy in a reaction is equal to the sum of the products' enthalpy, minus the sum of the reactants' enthalpy:

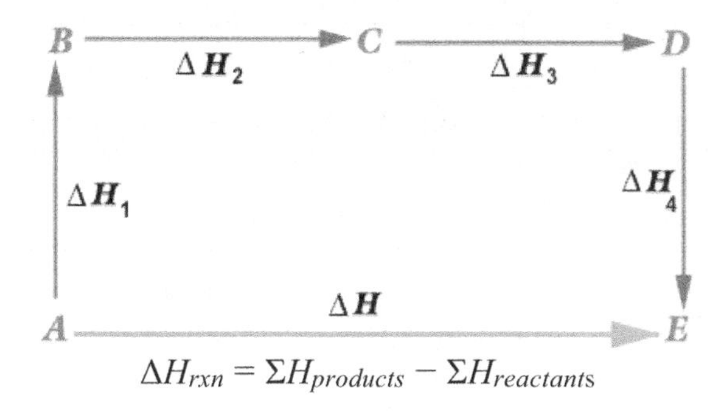

$$\Delta H_{rxn} = \Sigma H_{products} - \Sigma H_{reactants}$$

Gibbs Free Energy

Josiah Willard Gibbs (1839-1903) was an American physicist who developed the concept of free energy for thermodynamics.

The equation for Gibbs free energy ΔG is:

$$\Delta G = \Delta H - T\Delta S$$

where G is Gibbs free energy, measured in Joules (J).

If at a given temperature and pressure, the change in G is negative, then the reaction is spontaneous.

If G is equal to zero, the reaction is at equilibrium.

If G is positive, the reaction is nonspontaneous.

Because Gibbs free energy ΔG depends on signs, its value can be calculated using estimates of values.

The table gives an approximation for the value of the free energy and if the reaction is spontaneous or nonspontaneous.

ΔH	ΔS	Result
-	+	Spontaneous at all temperatures
+	+	Spontaneous at high temperatures
-	-	Spontaneous at low temperatures
+	-	Not spontaneous at any temperatures

PV Diagrams

A *PV diagram* shows the thermodynamic process by graphing pressure against volume. The work done by the system is equal to the area under the curve.

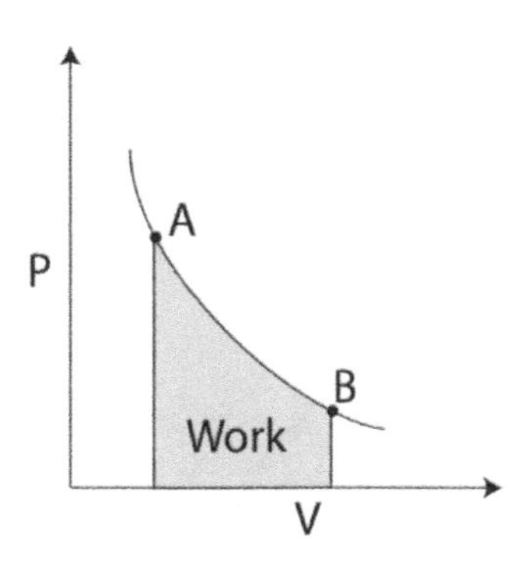

A process is labeled: adiabatic, isothermal, isobaric, and isochoric.

An *isothermal reaction* keeps temperature constant ($\Delta T = 0$). The change in internal energy must stay the same, as an increase in thermal energy (heat) increases temperature ($\Delta U = 0$).

The pressure vs. volume (PV) diagram of an isothermal process is below:

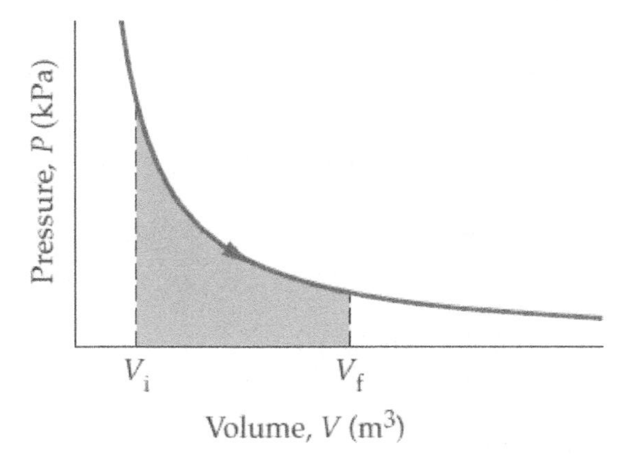

An *adiabatic process* has no transfer of heat ($Q = 0$). The change in energy is equal to the work. This process is often confused with the isothermal process because it is sometimes assumed that no transfer of heat means no change in temperature. However, this is not true—in isothermal instances, heat energy must be allowed into or out of the reaction to maintain a constant temperature.

An adiabatic process, unlike an isothermal reaction, occurs quickly and has a PV graph that has the same shape but is much steeper than the isothermal process. Adiabatic processes include a tire pump or a sharp expulsion of breath. Adiabatic PV diagrams are often shown in reference to isothermal lines called *isotherms*.

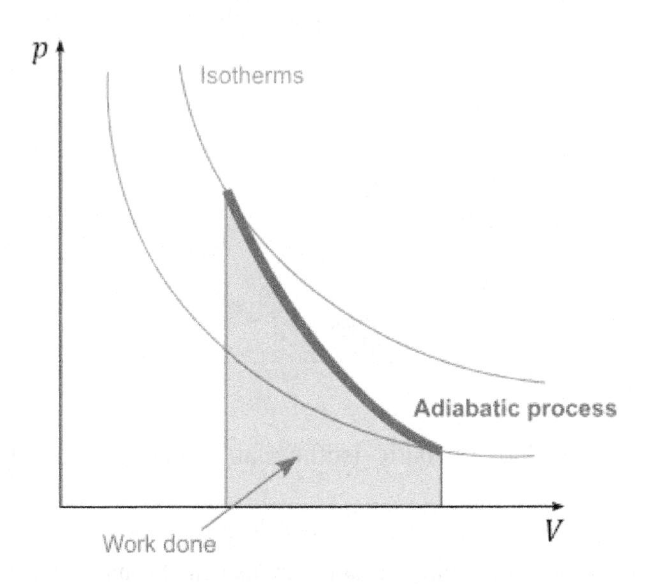

Isobaric reactions keep the pressure constant. When this occurs, the work is equal to the pressure times the change in volume ($W = P\Delta V$). The graph of an isobaric reaction is a horizontal line.

Some examples of isobaric reactions include a piston in an engine, or a flexible container open to Earth's atmosphere. When pressure is constant, the PV graph of an isobaric process is a horizontal straight line.

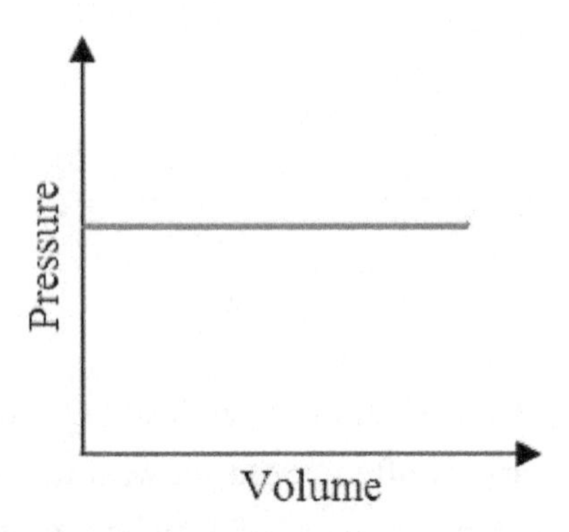

An *isochoric process* is an isovolumetric process because the volume is constant. Since there is no movement, the work is zero, and all the change in energy is due to the amount of heat added to the system ($W = 0$, $\Delta E = Q$).

Isochoric processes include those inside a closed and rigid container or a constant volume thermometer. The graph of an isochoric reaction is a vertical line.

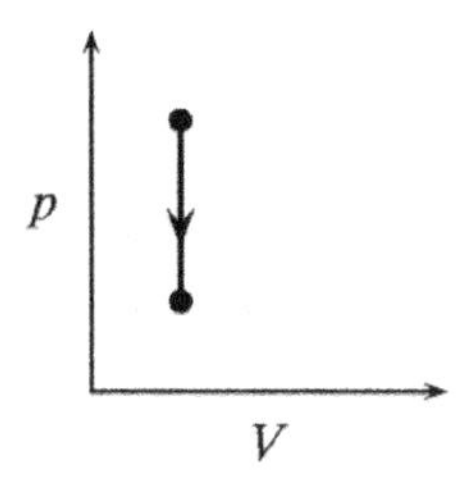

The processes, with relationship to the Ideal Gas Law (PV = nRT), is below.

Process	Constant	PV Diagram	Ideal Gas Law	First Law of Thermodynamics
Isobaric	Pressure	Horizontal line	V α T	$\Delta U = Q + W$
Isochoric	Volume	Vertical line	P α T	$\Delta U = Q$
Isothermal	Temperature	Curved line	PV α T	$\Delta U = 0$
Adiabatic	No heat exchanged	Curved line (jumps to different isotherm)	PV = nRT (only "nR" are constant)	$\Delta U = W$

Types of Work Cycles

In cyclic reactions, the work done is the area enclosed in the graph. These reactions are often comprised of different processes, as shown below. From the diagram, work is done by the gas when the volume increases, and pressure decreases and is done on the gas when volume decreases, and pressure increases.

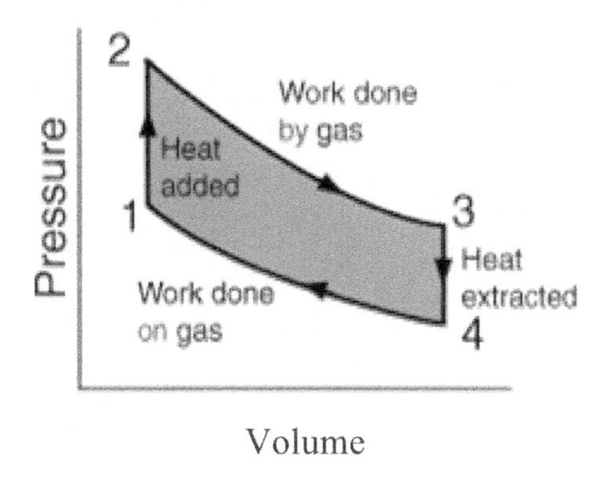

The Carnot Cycle

The *Kelvin-Planck statement* of the Second Law of Thermodynamics references heat engines. It states that *it is not possible to extract an amount of heat from a "hot reservoir" (Q_2) and use all of it to do work.*

Some of this heat must be exhausted to a "cold reservoir" (Q_1).

A perfect heat engine would use all the energy taken from the hot reservoir to do work. This is not possible because all real heat engines lose some heat to the outside environment.

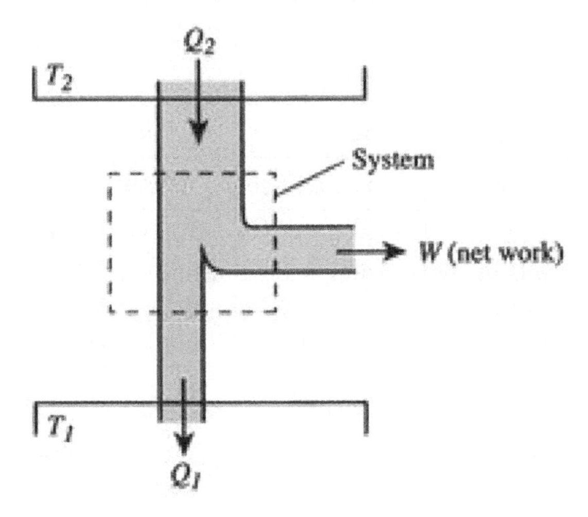

The *Carnot Cycle* represents the most efficient heat engine cycle. It is made of two isothermal processes and two adiabatic processes, as shown below.

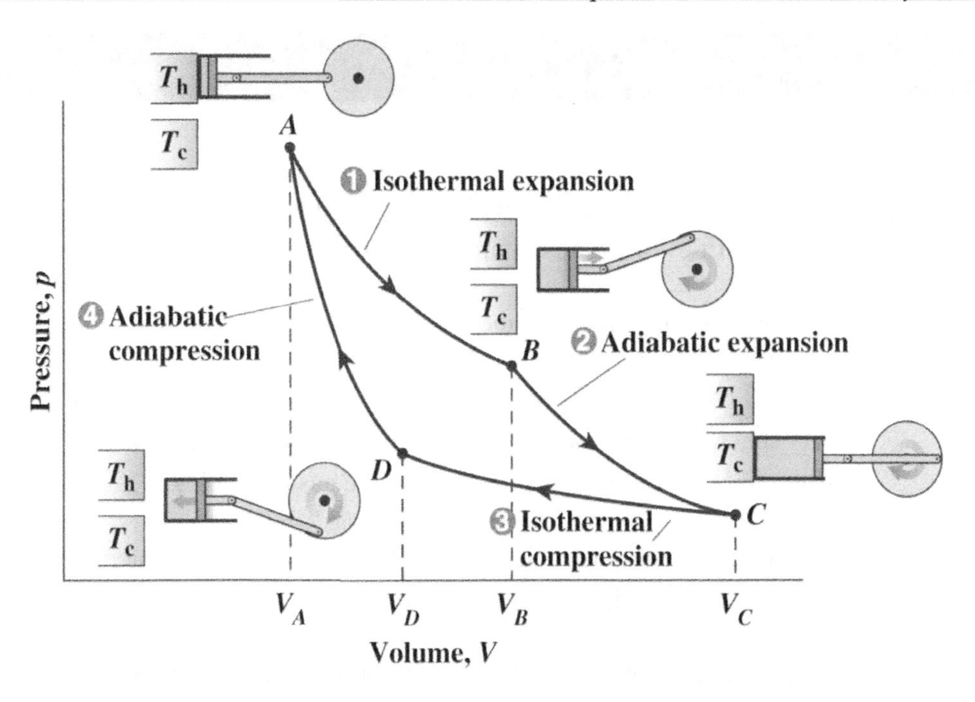

The Carnot Cycle is an idealized situation and can never be replicated in real life. From the Second Law, an equation of the *Carnot efficiency* sets a limiting value on the fraction of heat taken from the hot reservoir, which can be used for work:

$$n = \frac{T_H T_C}{T_H} \times 100\%$$

where n is the Carnot efficiency, T_H is the temperature of the hot reservoir (K), and T_C is the temperature of the cold reservoir (K).

For the Carnot efficiency, the entire process must be reversible, and the change in entropy must be zero. This is impossible, as no process of a real engine is reversible, and every process involves an increase in entropy because maximum entropy is the natural state of the universe.

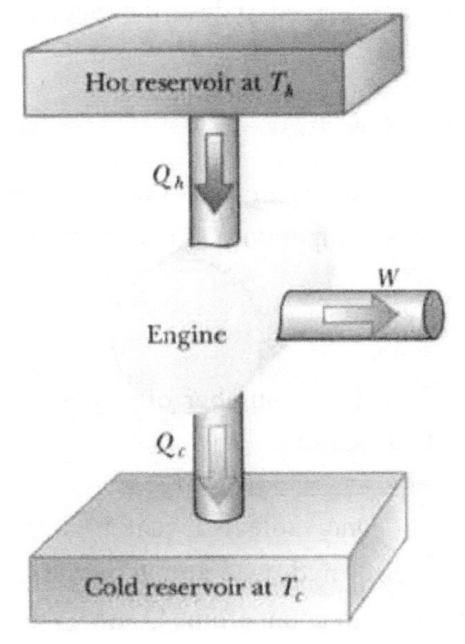

The Kinetic Theory of Gases

The *Kinetic Theory of Gases* relates the microscopic values of atomic kinetic energy to palpable quantities like temperature and pressure.

There are four basic premises for the kinetic theory of gases:

1. Gases are comprised of molecules. Molecules may be treated as perfect spheres of mass, and the space between each of these masses is many times greater than the diameter of the gas molecules.

2. The motion of each gas molecule is random. There is no order or pattern to either the direction or magnitude of the velocity.

3. Molecules follow Newton's Laws of Motion. Each gas molecule moves in a straight line with a constant velocity. If gas molecules collide, the molecules exert an equal but opposite force on the other.

4. The collisions of molecules are perfectly elastic—they lose no kinetic energy.

These rules display gases as ideal substances and only approximate their actual behavior. However, their description is remarkably accurate. From these assumptions, laws are derived to describe the behavior of gases.

Ideal Gas Law

The *Ideal Gas Law* explains the relationship between pressure (P), volume (V), and temperature (T). The gas law is expressed as:

$$PV = nRT$$

where n is the number of moles, and R is the universal gas constant with a value of 8.314 J/ mol·K.

One mole is equal to 6.023×10^{23} molecules. Technically, a mole is the number of hydrogen atoms in one gram of hydrogen. Because atoms are so small, it is practical to count atoms in moles than to count them individually.

The Ideal Gas Law applies to the pressure exerted by a gas on a cylinder with a moving wall.

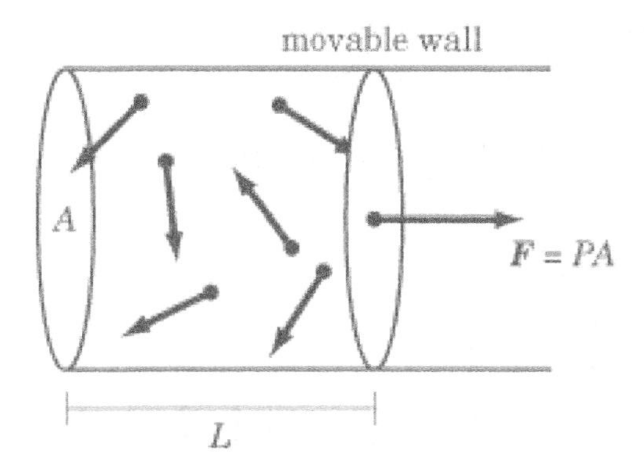

Since pressure is equal to $P = F/A$, the force that the gas exerts on the wall is equal to $F = PA$. If this force moves the wall back a length of L, then the volume of the cylinder increases by $\Delta V = LA$. By solving for A and substituting this into the equation for force, the result is $F = P\Delta VL$. This is equal to $P\Delta V = FL$.

Previously, work was defined as the force multiplied by the distance traveled. By pushing the wall, a distance of L with a force of F, the gas has done work equal to FL. When gas does work, it symbolizes a change in energy. If a change in PV is equal to a change in energy, then PV is the total energy of the gas. For the ideal gas law, this means that nRT is the expression for the total kinetic energy of the gas molecules.

The ideal gas law (PV = nRT) can be related to the number of molecules (N) and the Boltzmann's constant (k), which has a value of 1.381×10^{-23} J/K:

$$PV = NkT$$

Other gas laws derive from the Ideal Gas Law by holding specific variables constant. The number of moles (n) and the gas constant (R) are constant, so four other gas laws result from the Ideal Gas Law:

Boyle's Law,

Charles' Law,

Combined Gas Law, and

Closed Container Law.

Boyle's Law (isothermal process)

For *Boyle's Law*, the temperature of a gas is held constant. It states that an increase in pressure causes a decrease in volume, or that a decrease in pressure causes an increase in volume. Boyle's Law is expressed as:

$$P_1V_1 = P_2V_2$$

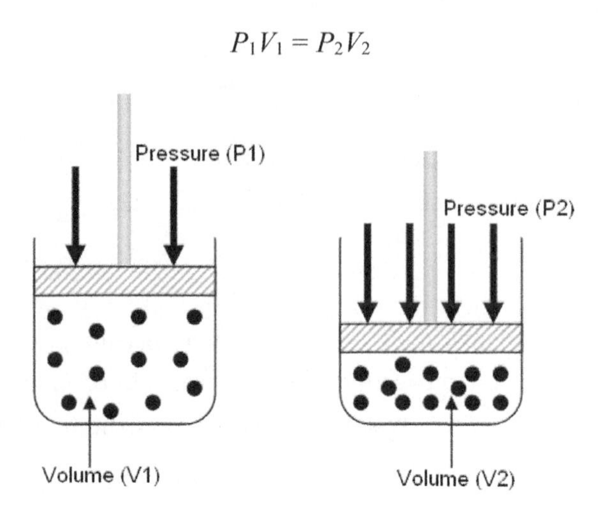

Charles' Law (isobaric process)

Charles' Law describes the behavior of a gas under constant pressure. In this relationship, volume and temperature are directly proportional; when the temperature increases, the volume increases; when the temperature decreases, the volume decreases. Charles' Law is expressed as:

$$\frac{V_1}{T_2} = \frac{V_2}{T_2}$$

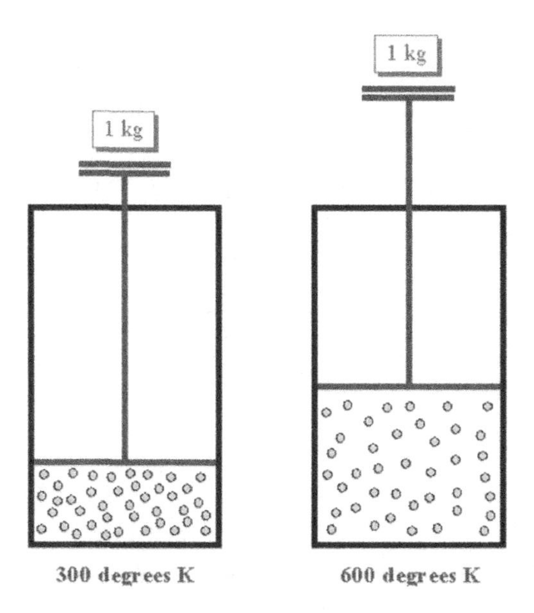

Closed Container Law (constant volume)

The *Closed Container Law* describes the behavior of a gas under constant volume. In such cases, pressure and temperature are directly proportional:

$$\frac{P_i}{T_i} = \frac{P_f}{T_f}$$

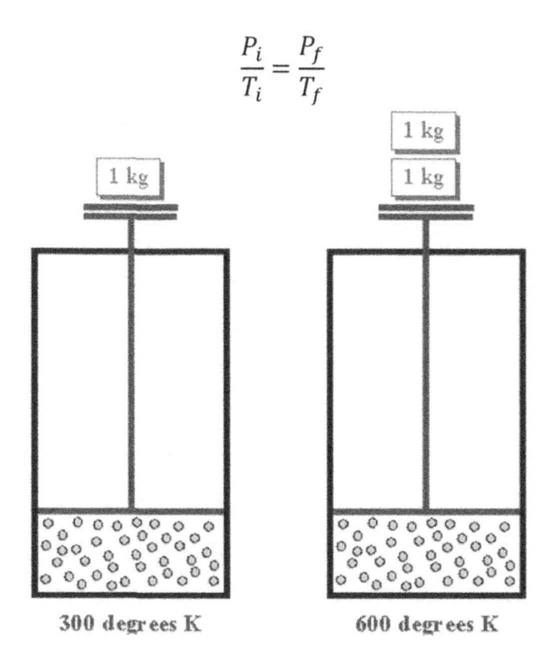

Combined Gas Law

The *Combined Gas Law* is a combination of Boyle's and Charles' Laws.

The *Combined Gas Law* relates the pressure, volume, and temperature of a gas, and is expressed as:

$$\frac{P_i V_i}{T_i} = \frac{P_f V_f}{T_f}$$

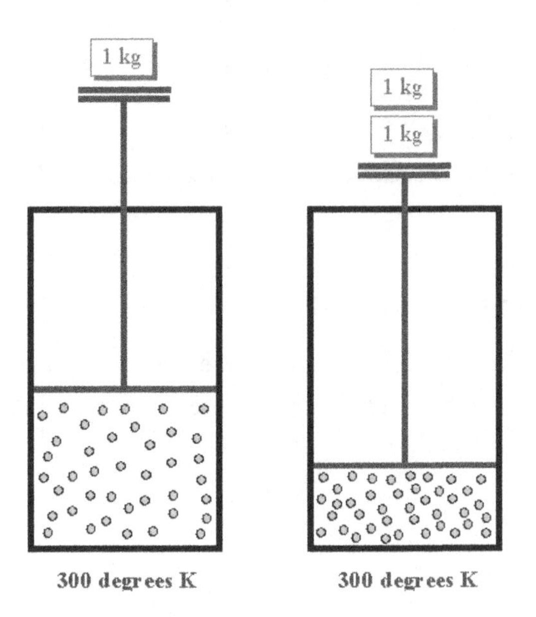

Coefficient of Linear Expansion

When material changes temperature, it shrinks or expands, depending upon the temperature difference. As the temperature increases, the molecules within the material gain energy and vibrate at higher rates. This vibration increases the distance between the molecules and causes the material to expand. If the material does not experience a phase change (liquid to gas), this expansion of material can be related to the change in temperature. The *coefficient of expansion* is a measurement of the expansion or contraction per unit of length of a material that occurs when the temperature is increased or decreased by 1 °C. This term is also known as *expansivity*.

The *coefficient of thermal expansion* (α) is given as:

$$\alpha_{linear} = \frac{\Delta l}{l_i \cdot \Delta T}$$

where α_{linear} is the coefficient of thermal expansion (K^{-1}), Δl is the change in length (m), l_i is the initial length (m), and ΔT is the change in temperature (K).

In the figure below, a bar of material is at an initial temperature and has a length denoted as *L*. If the bar is heated such that its temperature rises, it expands linearly by some length ΔL.

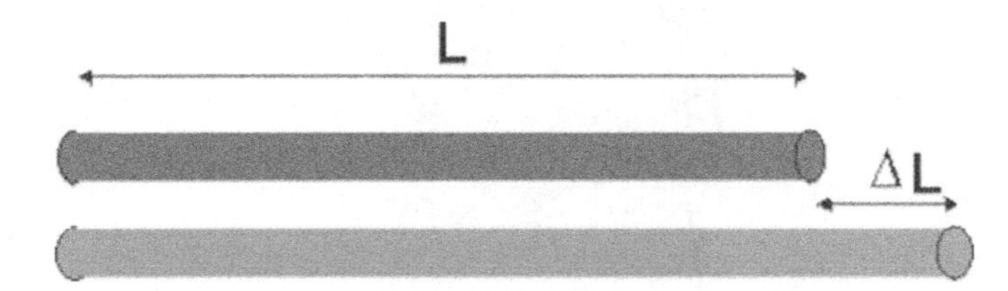

Objects can also undergo area expansion and volume expansion due to the same principles. The *coefficient of area expansion* is expressed as:

$$\alpha_{area} = \frac{\Delta A}{A_i \cdot \Delta T}$$

where α_{area} is the coefficient of thermal expansion (K^{-1}), ΔA is the change in the area (m^2), A_i is the initial area (m^2), and ΔT is the change in temperature (K).

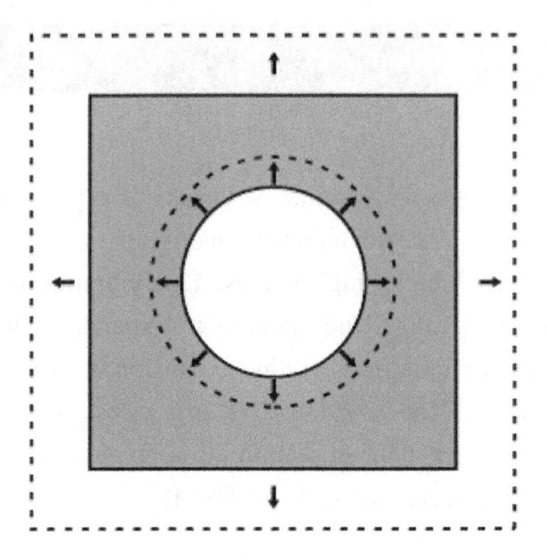

The *coefficient of volume expansion* is expressed as:

$$\alpha_{volume} = \frac{\Delta V}{V_i \cdot \Delta T}$$

where α_{volume} is the coefficient of thermal expansion (K^{-1}), ΔV is the change in volume (m^3), V_i is the initial volume (m^3), and ΔT is the change in temperature (K).

When designing products or structures that experience changes in temperature, thermal expansion is extremely important.

If the product design uses different materials with different thermal expansion coefficients, the design must also allow for a varying amount of component expansion and contraction. If not, the components experience a considerable amount of stress when they attempt to expand and cannot.

This makes the design and construction of bridges and aircraft difficult.

Bridges in regions that get relatively cold have metallic joints within the surface where the bridge starts and ends.

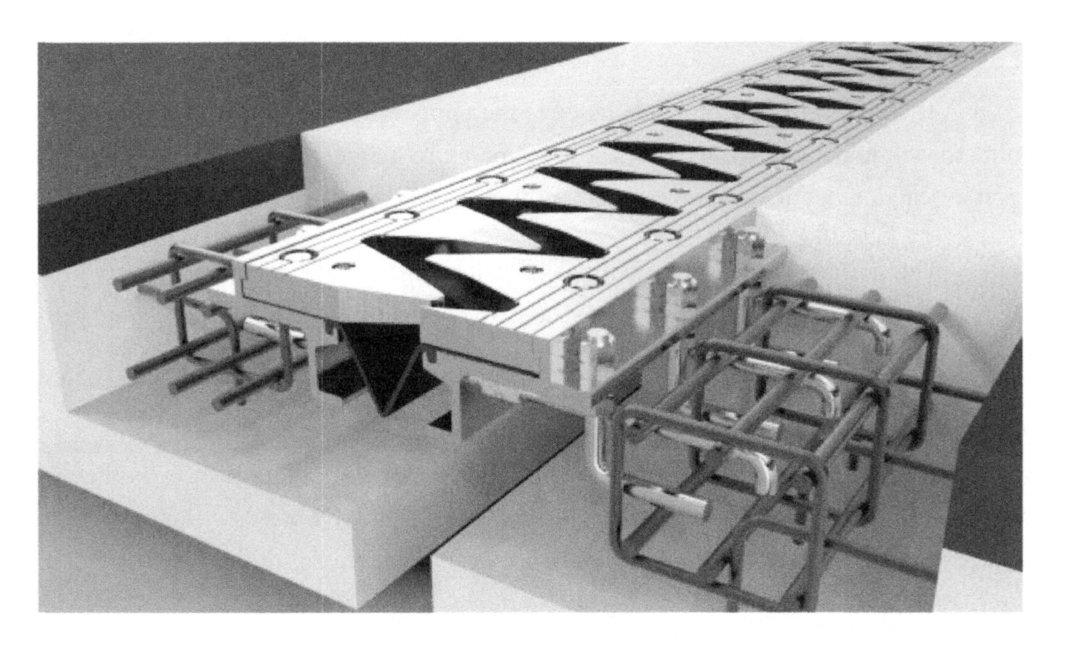

Expansion joints permit for variation in material size without failure

In the summer, these expansion joints might be almost completely closed. In the cold of winter, the thermal contraction of the bridge materials changes. When subjected to the cold air above and below causes the bridge to "shrink" and open a space up between these metal joints. This prevents the bridge from cracking or breaking because of inherent material strain during fluctuations in temperature.

Phase Diagram: Pressure and Temperature

When an ice block is placed in the sun, it slowly absorbs the thermal energy through radiation and undergoes a series of phase changes. That is, the ice cube changes from a solid to a liquid to a gas.

The *melting point* of a substance is the temperature at which it changes from a solid to a liquid, and it has the same temperature range as the *freezing point*; the reverse reaction occurs at the same temperature.

The *boiling point* (vaporization) of a substance is the temperature at which it changes phase from a liquid to a gas, and it is the same temperature range as the *condensation point*.

From the graph below, once the ice cube absorbs enough heat to melt (i.e., reaches 0 °C), its temperature remains constant. Melting is points B to C on the graph where all molecules have reached a temperature of 0 °C. Only when this occurs does the ice melt as illustrated by the horizontal plateau from B to C in the graph.

When the liquid water changes to a gas, the liquid absorbs enough heat to vaporize (i.e., reaches 100 °C). The temperature of the liquid remains constant during point D to E until all molecules reach a temperature of 100 °C. Only when this occurs does the liquid vaporize as illustrated by the horizontal plateau from D to E.

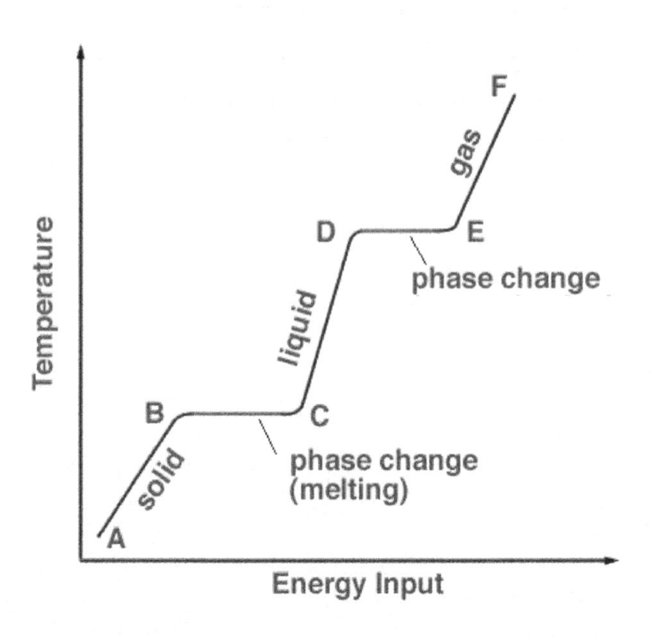

Although the energy (heat) input increases with temperature, it does not remain constant as the temperature does during a phase change. As energy is continuously added to the process, the substance uses the added heat to transition each molecule before increasing the temperature. Technically, heat converts the potential energy stored within each atom to kinetic energy before it changes phase.

A *phase diagram* displays the phases of a substance. A phase diagram is a graph plotted as pressure vs. temperature. The graph is split into three disproportionate parts, which represent the three phases of the substance. Each point on the graph represents a possible combination of pressure and temperature for the system.

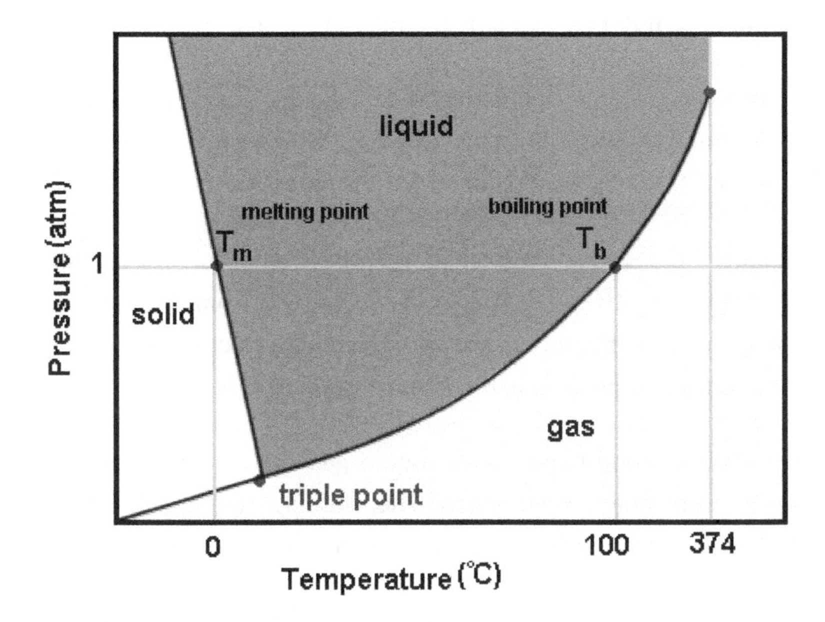

The *melting point* and *boiling point* of a substance are found by drawing a horizontal line at 1 atm of pressure, which is atmospheric pressure. Where the horizontal line intersects with a line on the phase diagram, the substance changes phases at that temperature.

The *triple point* is the point on the graph at which all three lines intersect and at which all three phases exist simultaneously.

The *critical point* on a phase diagram is the temperature, above which the substance is always a gas, regardless of pressure.

Latent Heat

The amount of heat required for a phase change is the *latent heat*. The latent heat has been reported for most substances. In general, only computing the total heat is required for the reaction.

The heat required to convert a substance from one phase to another is:

$$Q = m \cdot l$$

where l is the latent heat (J/kg), and m is the mass of the substance (kg).

If the particles become excited (solid to a liquid, liquid to a gas), the change in entropy is always positive, because heat is added to the system. For this process, the latent heat of fusion is used. The total heat required for this process is the heat of fusion.

If the particles become less excited (gas to a liquid, liquid to solid), the change in entropy is always negative, because heat is removed from the system. A negative change in entropy results in a more ordered molecular structure of the substance. The total heat required for this process is the *heat of vaporization*.

Occasionally, a solid bypasses the liquid phase and transitions directly into the gas phase; this phenomenon is *sublimation*. The heat required for sublimation is the sum of the heat of fusion and the heat of vaporization. The reverse reaction of sublimation is *deposition*.

$$Q_{sub} = Q_{fus} + Q_{vap}$$

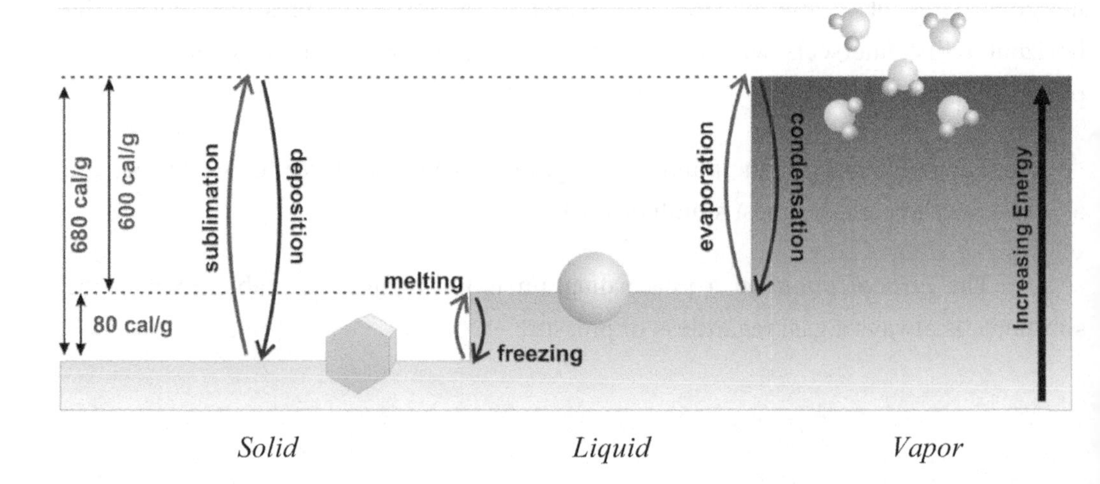

Latent heats are sometimes given in molar heat, or joules per mol (J/mol). Divide by the molecular weight of the substance if given the mass in grams. If any of these processes are reversed, the latent heat is assigned a negative sign (reverse process).

For a process with multiple phase changes, the total heat of the reaction is the sum of the separate heats needed for each phase change and the addition of heat. For these types of problems, it is essential to consider the First Law of Thermodynamics, which states that heat must be conserved (heat gained = heat loss), and the equation for heat ($Q = mc\Delta T$) where c is the specific heat capacity, and m is the mass.

To calculate the change in heat (change in entropy) when 2 kg of ice at 0 °C is warmed to 120 °C, a few equations are needed, using the known values.

- melting point = 0 °C
- boiling point = 100 °C
- latent heat of fusion = 3.33×10^5 J/kg
- latent heat of vaporization = 2.26×10^6 J/kg
- specific heat capacity = 4,186 J/kg °C

Heat required to phase change the ice into liquid water:

$$Q = m\, l_{\text{fusion}} = (2\text{kg})\cdot(3.33 \times 10^5\ \text{J/kg}) = 6.66 \times 10^5\ \text{J}$$

Heat required to heat the water to its boiling point:

$$Q = \text{mc}\Delta\text{T} = (2\ \text{kg})\cdot(4186\ \text{J/kg °C})\cdot(100\ \text{°C} - 0\ \text{°C}) = 837{,}200\ \text{J}$$

Heat required to phase change the liquid into gas:

$$Q = m\, l_{\text{vap}} = (2\text{kg})\cdot(2.26 \times 10^6\ \text{J/kg}) = 4.52 \times 10^6\ \text{J}$$

Heat required to heat the water to 120 °C:

$$Q = mc\Delta\text{T} = (2\ \text{kg})\cdot(4{,}186\ \text{J/kg °C})\cdot(120\ \text{°C} - 100\ \text{°C}) = 167{,}440\ \text{J}$$

The total heat required for the process to occur is the sum of the separate heats:

$$Q_{\text{total}} = Q_{\text{fusion}} + Q_{\text{heat\_1}} + Q_{\text{vap}} + Q_{\text{heat\_2}}$$

$$Q_{\text{total}} = 6.66\times10^5 + 837{,}200 + 4.52\times10^6 + 167{,}440 = 6{,}190{,}640\ \text{J}$$

Chapter Summary

Heat and Temperature

- Heat is the total internal energy due to molecular motion, and the temperature is the average measure of the kinetic energy for a specific substance.

- The calorie is the amount of heat needed to raise the temperature of one gram of water by one degree Celsius (g/°C).

- Kelvin (K) is the measure of absolute temperature. Absolute zero = 0 K.

- Kelvin and Celsius use the same increments for their degree measurements.

- Specific heat is the ratio of energy absorbed to the temperature rise per unit of mass.

- The heat required to perform a phase change is latent heat (l):

$$Q = m \cdot l$$

- A calorimeter measures the heat of a reaction.
    - The temperature remains constant through a phase change.

- Phase diagram
    - Triple point: all three phases exist simultaneously.
    - Critical point: the substance is always a gas.

- Boltzmann's constant 1.381×10^{-23} J/K: $PV = NkT$

- Coefficient of expansion: the expansion per unit of length of a material for an increase in temperature under constant pressure:

$$\alpha = \frac{\Delta l}{l_i \cdot \Delta T}$$

Heat Transfer

- Conduction requires direct contact.

  Rate of heat transfer (H) by conduction: $H = \frac{kA\Delta T}{L}$

- Convection uses energized particles from one place to another.

- Radiation uses electromagnetic waves.

Laws of Thermodynamics

- Zeroth Law of thermodynamics: two systems in thermal equilibrium with a third system are in thermal equilibrium.

- First Law: energy is always conserved: $\Delta U = Q - W$

  - Work was done *by* a system (−)
  - Work was done *on* a system (+)
  - $Q < 0$ is exothermic, where Q is heat
  - $Q > 0$ is endothermic, where Q is heat

- Second Law: entropy increases or stays constant, never decreases

  - Entropy (S) is a measure of disorder within a system.
  - The natural tendency of all things is to move towards maximum disorder:

    $$\Delta S \geq \frac{Q}{T}$$

  - $\Delta S_{universe} > 0$ spontaneous
  - $\Delta S_{universe} = 0$ at equilibrium
  - $\Delta S_{universe} < 0$ non-spontaneous (reverse process spontaneous)

- Third Law of thermodynamics: as the temperature of a system approaches absolute zero, the entropy (S) approaches a constant value (typically zero).

$$S = k \cdot ln\,(\Omega)$$

where S is the entropy (J/K), k is Boltzmann's constant (1.381×10^{-23} J/K), and Ω is the number of microstates.

- The entropy of a system at absolute zero is zero. However, some systems have more than one minimum energy state.

Enthalpy

- Amount of heat used at constant pressure: $\Delta U = \Delta H - P\Delta V$
- $\Delta H < 0$ exothermic
- $\Delta H > 0$ endothermic
- Hess' Law: $\Delta H_{rxn} = H_{products} - H_{reactants}$

Gibb's Free Energy

- $\Delta G = \Delta H - T\Delta S$
- $G < 0$: spontaneous reaction
- $G = 0$: reaction is at equilibrium
- $G > 0$: nonspontaneous reaction

PV Diagram

- Work is the area under pressure *vs.* volume curve.
- Isothermal - constant temperature: hyperbola
- Adiabatic - no transfer of heat: steeper hyperbola
- Isobaric - constant pressure: horizontal line
- Isochoric - constant volume: vertical line

The Carnot Cycle

- Most efficient heat engine cycle
- Two isothermal processes and two adiabatic processes
- Carnot Efficiency: $\frac{T_H T_C}{T_H} \times 100\%$

The Kinetic Theory of Gases

1. Gases consist of molecules. Gas molecules may be treated as perfect spheres of mass, and the space between each of these masses is many times greater than their diameters.
2. The motion of each gas molecule is random. There is no order or pattern to either the direction or magnitude of the velocity.
3. Molecules follow Newton's Laws of Motion. Each gas molecule moves in a straight line with a constant velocity. If they collide, the gas molecules exert an equal but opposite force on the other.
4. The collisions of molecules are perfectly elastic—they lose no kinetic energy.

The coefficient of Linear Expansion

- The *coefficient of expansion* is a measurement of the expansion or contraction per unit of length of a material that occurs when the temperature is increased or decreased by 1 °C. This term is also expansivity.

$$\alpha_{linear} = \frac{\Delta l}{l_i \cdot \Delta T}$$

where α_{linear} is the coefficient of thermal expansion (K^{-1}), Δl is the change in length (m), l_i is the initial length (m), and ΔT is the change in temperature (K).

Phase diagram: Pressure and Temperature

- The *melting point* of a substance is the temperature at which it changes from a solid to a liquid, and it is the same temperature as the *freezing point*; the reverse reaction occurs at the same temperature.

  The *boiling point* (vaporization) of a substance is the temperature at which it changes phase from a liquid to a gas, and it is the same temperature as the *condensation point*.

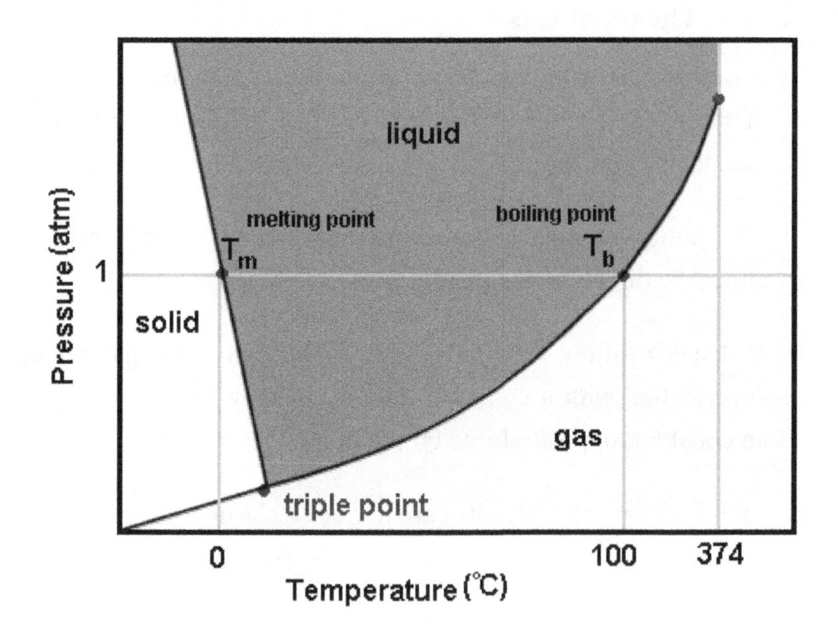

Latent Heat

- The amount of heat required to perform a phase change is *latent heat*. The latent heat is known for most substances.

  The heat (Q) required to convert a substance from one phase to another is:

$$Q = m \cdot l$$

where l is the latent heat (J/kg), and m is the mass of the substance (kg).

Practice Questions

1. A mass of 0.2 kg ethanol, in the liquid state at its melting point of –114.4 °C, is frozen at atmospheric pressure. What is the change in the entropy of the ethanol as it freezes? (Use the heat of fusion of ethanol $L_f = 1.04 \times 10^5$ J/kg)

A. –360 J/K **B.** 54 J/K **C.** –131 J/K **D.** –220 J/K **E.** 285 J/K

2. Isobaric work is

A. $Q - W$ **B.** $P\Delta V$ **C.** $P\Delta T$ **D.** $V\Delta P$ **E.** $Q + W$

3. An adiabatic process is performed on 9 moles of an ideal gas. The initial temperature is 315 K, and the initial volume is 0.70 m^3. The final volume is 0.30 m^3. What is the amount of heat absorbed by the gas? (Use the adiabatic constant for the gas = 1.44)

A. –18 kJ **B.** 32 kJ **C.** 9 kJ **D.** –32 kJ **E.** 0 kJ

4. An 80 g aluminum calorimeter contains 360 g of water at an equilibrium temperature of 20 °C. A 180 g piece of metal, initially at 305 °C, is added to the calorimeter. The final temperature at equilibrium is 35 °C. Assume there is no external heat exchange. What is the specific heat capacity of the metal? (Use the specific heat capacity of aluminum = 910 J/kg·K and the specific heat of water = 4,190 J/kg·K)

A. 260 J/kg·K **B.** 324 J/kg·K **C.** 488 J/kg·K **D.** 410 J/kg·K **E.** 535 J/kg·K

5. A chemist uses 120 g of water that is heated using 65 W of power with 100% efficiency. How much time is required to raise the temperature of the water from 20 °C to 50 °C? (Use the specific heat of water = 4,186 J/g·°C)

A. 136 s **B.** 182 s **C.** 93 s **D.** 232 s **E.** 274 s

6. The Second Law of Thermodynamics leads to the following conclusion:

A. the average temperature of the universe is increasing over time
B. it is theoretically possible to convert heat into work with 100% efficiency
C. disorder in the universe is increasing over time
D. total energy of the universe remains constant
E. entropy of the universe remains constant

7. The statement that 'heat energy cannot be completely transformed into work' is a statement of which thermodynamic law?

A. Third **B.** Second **C.** Zeroth **D.** Fourth **E.** First

8. What is the change in entropy when 20 g of water at 100 °C is turned into steam at 100 °C? (Use latent heat of vaporization of water $L_v = 22.6 \times 10^5$ J/kg)

A. –346 J/K **B.** 346 J/K **C.** –80.8 J/K **D.** 121 J/K **E.** 0 J/K

9. A Carnot-efficiency engine is operated as a heat pump to heat a room in the winter. The heat pump delivers heat to the room at the rate of 32 kJ per second and maintains the room at a temperature of 293 K when the outside temperature is 237 K. The power requirement for the heat pump under these operating conditions is:

A. 6,100 W **B.** 3,400 W **C.** 7,300 W **D.** 14,300 W **E.** 11,450 W

10. Which of the relationships is valid for all types of Carnot heat engines?

I. $\eta = 1 - T_C / T_H$
II. $\eta = 1 - | Q_C / Q_H |$
III. $T_C / T_H = Q_C / Q_H$

A. I only **B.** II only **C.** III only **D.** I, II and III **E.** I and III only

11. What is the change of entropy associated with 8 kg of water freezing to ice at 0 °C? (Use latent heat of fusion $L_f = 80$ kcal/kg)

A. 1.4 kcal/K **C.** –2.3 kcal/K
B. 0 kcal/K **D.** –1.4 kcal/K **E.** 0.8 kcal/K

12. A Carnot-efficiency engine extracts 515 J of heat from a high-temperature reservoir during each cycle and ejects 340 J of heat to a low-temperature reservoir during the same cycle. What is the efficiency of the engine?

A. 67% **B.** 21% **C.** 53% **D.** 17% **E.** 34%

13. A glass beaker of unknown mass contains 65 ml of water. The system absorbs 1,800 cal of heat, and the temperature rises 20 °C. What is the mass of the beaker? (Use specific heat of glass = 0.18 cal/g·°C and specific heat of water = 1 cal/g·°C)

A. 342 g **B.** 139 g **C.** 546 g **D.** 268 g **E.** 782 g

14. A 0.3 kg ice cube at 0 °C has sufficient heat added to result in total melting, and the resulting water is heated to 60 °C. How much total heat is added? (Use the latent heat of fusion for water L_f = 334 kJ/kg, the latent heat of vaporization for water L_v = 2,257 kJ/kg and the specific heat of water = 4,186 kJ/kg·K)

A. 73 kJ **B.** 48 kJ **C.** 176 kJ **D.** 144 kJ **E.** 136 kJ

15. The water flowing over a large dam drops a distance of 60 m. If all the gravitational potential energy is converted to thermal energy, by what temperature does the water rise? (Use the acceleration due to gravity g = 10 m/s^2 and specific heat of water = 4,186 J/kg·K)

A. 0.34 °C **B.** 0.44 °C **C.** 0.09 °C **D.** 0.14 °C **E.** 0.58 °C

Solutions

1. C is correct.

Find the heat from phase change:

$Q = mL_f$

$Q = (0.2 \text{ kg})\cdot(1.04 \times 10^5 \text{ J/kg})$

$Q = 20{,}800 \text{ J}$

Because the ethanol is freezing, Q should be negative due to heat being released.

$Q = -20{,}800 \text{ J}$

Find the change in entropy:

$\Delta S = Q / \text{T}$

$\Delta S = -20{,}800 \text{ J} / (-114.4 \text{ °C} + 273 \text{ K})$

$\Delta S = -131 \text{ J} / \text{K}$

2. B is correct.

$\text{W} = \text{P}\Delta\text{V}$

Isobaric means pressure is constant, and the volume is changing.

3. E is correct.

Adiabatic means that no heat enters or leaves the system.

$Q = 0 \text{ kJ}$

4. C is correct.

$Q = mc\Delta\text{T}$

Find the heat added to aluminum calorimeter:

$Q_A = (0.08 \text{ kg})\cdot(910 \text{ J/kg}\cdot\text{K})\cdot(35 \text{ °C} - 20 \text{ °C})$

$Q_A = 1{,}092 \text{ J}$

Find the heat added to water:

$Q_W = (0.36 \text{ kg})\cdot(4{,}190 \text{ J/kg}\cdot\text{K})\cdot(35 \text{ °C} - 20 \text{ °C})$

$Q_W = 22{,}626 \text{ J}$

Find the total heat added to the system:

$Q_{total} = Q_A + Q_W$

$Q_{total} = 1{,}092 \text{ J} + 22{,}626 \text{ J}$

$Q_{total} = 23{,}718 \text{ J}$

Find the specific heat of the metal:

$Q = mc\Delta T$

$c = Q / m\Delta T$

$c = (23{,}718 \text{ J}) / [(0.18 \text{ kg})\cdot(305 \text{ °C} - 35 \text{ °C})]$

$c = 488 \text{ J/kg}\cdot\text{K}$

5. D is correct.

Watt = 1 J/s

Thermal energy:

$Q = \text{Power} \times \text{time}$

$Q = mc\Delta T$

$P \times t = mc\Delta T$

$t = (mc\Delta T) / P$

$t = [(120 \text{ g})\cdot(4.186 \text{ J/g}\cdot\text{°C})\cdot(50 \text{ °C} - 20 \text{ °C})] / (65 \text{ W})$

$t = 232 \text{ s}$

6. C is correct.

The Second Law of Thermodynamics states that entropy is either constant or increasing over time. A constant entropy process is an idealized process and does not exist. Thus, entropy is always increasing over time.

7. B is correct.

The Second Law of Thermodynamics states that through thermodynamic processes, there is an increase in the sum of entropies of the system, and thus no engine process is 100% efficient.

8. D is correct.

Find the heat from phase change:

$Q = mL_f$

$Q = (0.02\text{ kg})\cdot(22.6 \times 10^5\text{ J/kg})$

$Q = 45{,}200\text{ J}$

Because the water is vaporizing, Q should be positive due to heat being absorbed.

$Q = 45{,}200\text{ J}$

Find the change in entropy:

$\Delta S = Q / \text{T}$

$\Delta S = 45{,}200\text{ J} / (100\text{ °C} + 273\text{ K})$

$\Delta S = 121\text{ J} / \text{K}$

A positive change in entropy indicates that the disorder of the isolated system has increased. When water evaporates into steam, the entropy is positive because the disorder of steam is higher than water.

9. A is correct.

Coefficient of performance assuming an ideal Carnot cycle:

$C_p = Q_H / \text{W}$

$C_p = \text{T}_H / (\text{T}_H - \text{T}_C)$

$Q_H / \text{W} = \text{T}_H / (\text{T}_H - \text{T}_C)$

$\text{W} = [Q_H \cdot (\text{T}_H - \text{T}_C)] / \text{T}_H$

$\text{W} = [(32 \times 10^3\text{ J/s})\cdot(293\text{ K} - 237\text{ K})] / (293\text{ K})$

$\text{W} = 6{,}116\text{ J/s} \approx 6{,}100\text{ W}$

10. D is correct. Carnot efficiency engines can be written as:

$\eta = 1 - T_C / T_H$

$\eta = 1 - | Q_C / Q_H |$

Thus:

$Q_C / Q_H = T_C / T_H$

11. C is correct.

Find the heat from the phase change:

$Q = mL_f$

$Q = (8 \text{ kg})\cdot(80 \text{ kcal/kg})$

$Q = 640 \text{ kcal/kg}$

Because the water is freezing, Q should be negative due to the heat being released.

$Q = -640 \text{ kcal/kg}$

Find the change in entropy:

$\Delta S = Q / T$

$\Delta S = -640 \text{ kcal/kg} / (0 \text{ °C} + 273 \text{ K})$

$\Delta S = -2.3 \text{ kcal/K}$

A negative change in entropy indicates that the disorder of the isolated system has decreased. When water freezes, the entropy is negative because water is more disordered than ice. Thus, the disorder has decreased, and entropy is negative.

12. E is correct.

$Q_H = 515 \text{ J}$

$Q_C = 340 \text{ J}$

Carnot cycle efficiency:

$\eta = (Q_H - Q_C) / Q_H$

$\eta = (515 \text{ J} - 340 \text{ J}) / (515 \text{ J})$

$\eta = 0.34 = 34\%$

13. B is correct.

$Q = (mc\Delta T)_{water} + (mc\Delta T)_{beaker}$

Change in the temperature is the same for both:

$Q = \Delta T[(mc)_{water} + (mc)_{beaker}]$

$1{,}800 \text{ cal} = (20\ °\text{C})\cdot[(65\text{ g})\cdot(1\text{ cal/g}\cdot°\text{C}) + (m_{beaker})\cdot(0.18\text{ cal/g}\cdot°\text{C})]$

$90\text{ cal/}°\text{C} = 65\text{ cal/}°\text{C} + (m_{beaker})\cdot(0.18\text{ cal/g}\cdot°\text{C})$

$25\text{ cal/}°\text{C} / (0.18\text{ cal/g}\cdot°\text{C}) = (m_{beaker})$

$m_{beaker} = 139\text{ g}$

14. C is correct.

Heat to melt the ice cube:

$Q_1 = mL_f$

Heat to raise the temperature:

$Q_2 = mc\Delta T$

Total heat:

$Q_{total} = Q_1 + Q_2$

$Q_{total} = mL_f + mc\Delta T$

$Q_{total} = [(0.3\text{ kg})\cdot(334\text{ kJ/kg})] + [(0.3\text{ kg})\cdot(4.186\text{ kJ/kg}\cdot\text{K})\cdot(60\ °\text{C} - 0\ °\text{C})]$

$Q_{total} = (100.2\text{ kJ}) + [(1.257\text{ kJ/K})\cdot(60\text{ K})]$

$Q_{total} = 175.55\text{ kJ} \approx 176\text{ kJ}$

15. D is correct.

$PE = Q$

$mgh = mc\Delta T$, cancel m from both sides of the expression

$gh = c\Delta T$

$gh / c = \Delta T$

$\Delta T = [(10\text{ m/s}^2)\cdot(60\text{ m})] / 4{,}186\text{ J/kg}\cdot\text{K}$

$\Delta T = 0.14\ °\text{C}$

Chapter 12

Atomic and Nuclear Structure

ATOMIC STRUCTURE AND SPECTRA

- **Emission Spectrum of Hydrogen: Bohr model**
- **Atomic Energy Levels**

ATOMIC NUCLEUS

- **Neutrons, Protons, Isotopes**
- **Atomic Number, Atomic Weight**
- **Nuclear Forces**
- **Radioactive Decay**
- **General Nature of Fission**
- **General Nature of Fusion**
- **Mass Deficit, Energy Liberated, Binding Energy**
- **Mass Spectrometer**

ATOMIC STRUCTURE AND SPECTRA

John Dalton (1755-1844) proposed the *atomic structure* in the early 1800s. Dalton reintroduced atomic theory to explain chemical reactions. His theory centered on five main concepts:

1. All matter is made of indivisible particles of *atoms*.
2. An *element* is made up of identical atoms.
3. Different elements have atoms with different masses.
4. Chemical compounds are made of atoms in specific integer ratios.
5. Atoms are neither created nor destroyed in chemical reactions.

Nobel laureate J. J. Thomson (1856-1940) discovered the *electron* in the late 1800s. By performing cathode ray experiments (two of which are explained below), Thomson discovered that the electron is negatively charged. Thomson also measured the electron's charge-to-mass ratio and identified the electron as a fundamental particle.

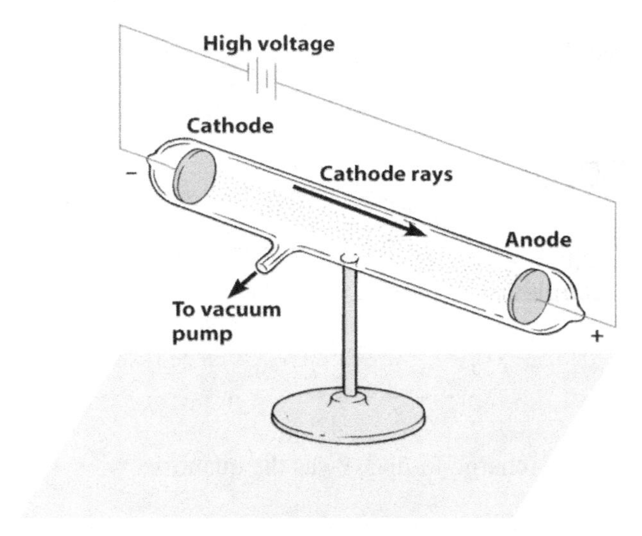

Cathode ray experiment

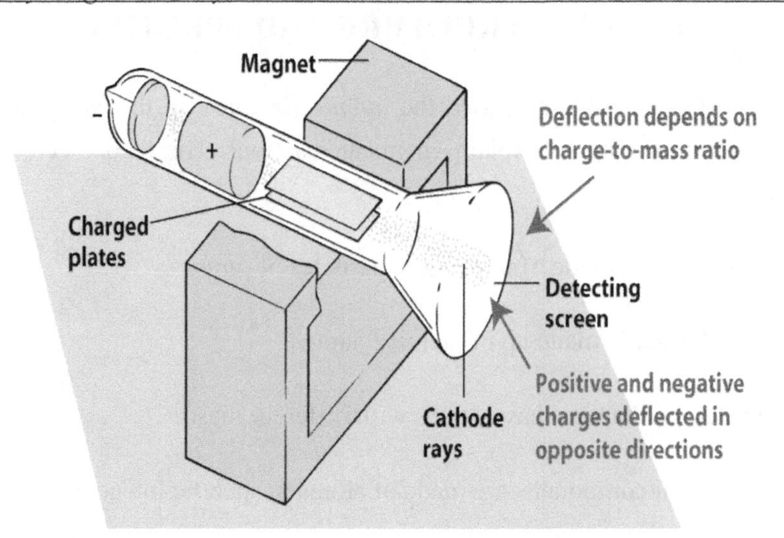

Around 1906, Nobel laureate Robert Millikan (1868-1953) studied minuscule particles, specifically, charged oil droplets in an electric field. Millikan observed that the charge on the oil droplets was a multiple of the electron charge. This discovery, along with Thomson's results, was used to calculate the mass of the electron.

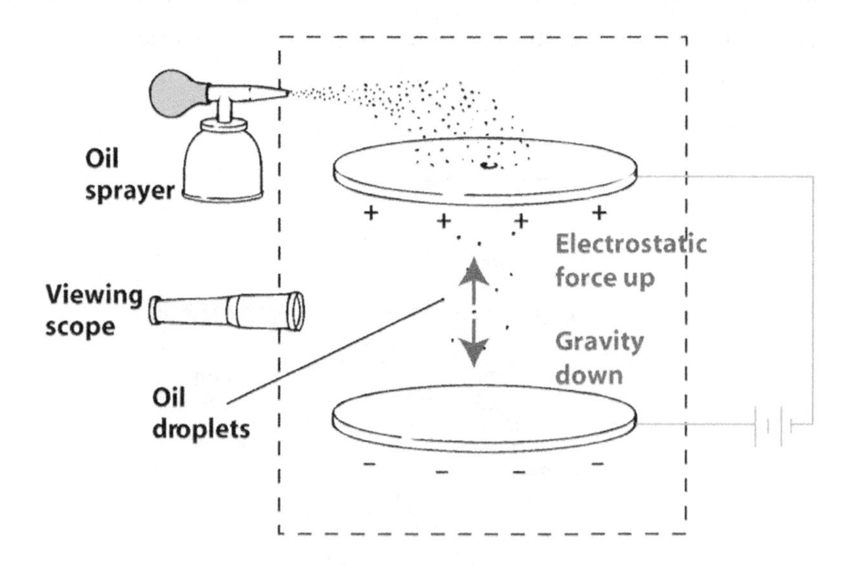

Thomson's result (charge-to-mass) was the quantity:

$$\frac{q}{m} = 1.7584 \times 10^{11} \frac{C}{kg}$$

Millikan's result (q = charge) was:

$$q = 1.60 \times 10^{-19} C$$

Combined, the mass of an electron is:

$$m = \frac{q}{1.7584 \times 10^{11}} = 9.11 \times 10^{-31} kg$$

John Dalton (1766-1844) claimed that atoms were indivisible, that there were no smaller quantities involved. However, Thomson's and Millikan's experiments, along with the discovery of the electron, proved Dalton's statement false.

The electron's mass is small (no measurable volume), but there needs to be a nature of an atom's positive charge. Thomson developed the "*plum pudding*" model. Thomson thought that electrons were embedded in a blob of positively charged matter like "*raisins in plum pudding*."

Nobel laureate Ernest Rutherford (1871-1937), in 1907, conducted an experiment which scattered alpha particles by bouncing them off gold foil.

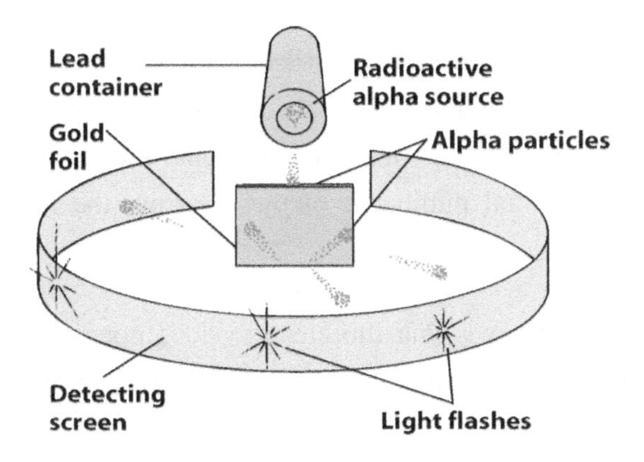

Most particles passed through without significant deflection, but a few scattered at large angles. Rutherford concluded that an atom's positive charge resides in a small nucleus that contained most of the atom's mass (unlike the spread-out blob of matter containing electrons, as the *plum pudding* model suggested). He named these positively charged subatomic particles protons.

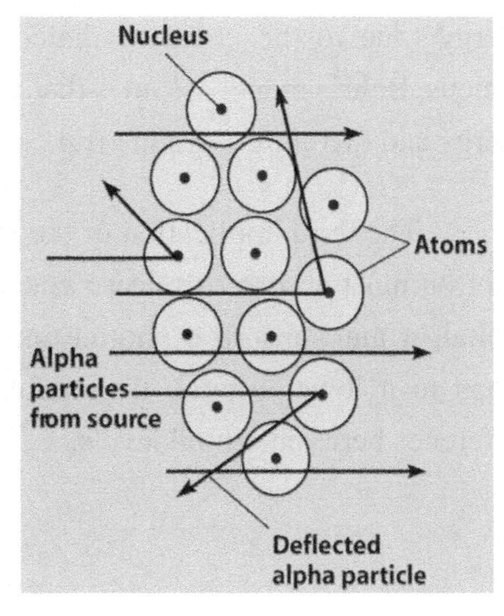

In 1932, Nobel laureate James Chadwick (1891-1974), added the existence of neutral neutrons in the nucleus to the atomic theory.

Emission Spectrum of Hydrogen; Bohr Model

In 1913, Nobel laureate Niels Bohr (1885-1962) proposed a model depicting the atom in which the electron orbits the nucleus as the Earth orbits the Sun. Electrostatic attraction pulls the electron toward the nucleus. However, the electron orbits at high enough speeds to prevent it from crashing into the nucleus. The relationship of the electron orbit with respect to the nucleus leads to several important implications:

1. Electrons only exist in precise allowed orbits.

2. Within an orbit, the electron does not radiate.

3. Radiation is emitted or absorbed when changing orbits.

The first implication of the Bohr model is that electrons occupy specific orbitals which determine their energy. These orbits are characterized by the orbital number n, where $n = 1$ is equal to the ground state and represents the lowest energy level of the electron. The higher the orbital number n of the electron, the higher energy, and the further the electron is from the nucleus.

The second implication is that the electron does not radiate while orbiting the nucleus. As mentioned in earlier chapters, accelerating electric charges emit energy in the form of electromagnetic radiation.

According to classical physics, the electron would, therefore, radiate away its energy (due to the energy radiated from centripetal acceleration) and crash into the nucleus. Bohr's model assumes that within the energy levels, the electron does not radiate energy and therefore stays in orbit.

The third implication of the Bohr model is that when changing energy levels, the electron must absorb or release energy. For a ground state electron to achieve a higher orbital, it must absorb a photon (and thus absorb the photon's energy). If the electron drops to a lower orbital, then the electron must emit a photon equal to the energy difference between orbital levels.

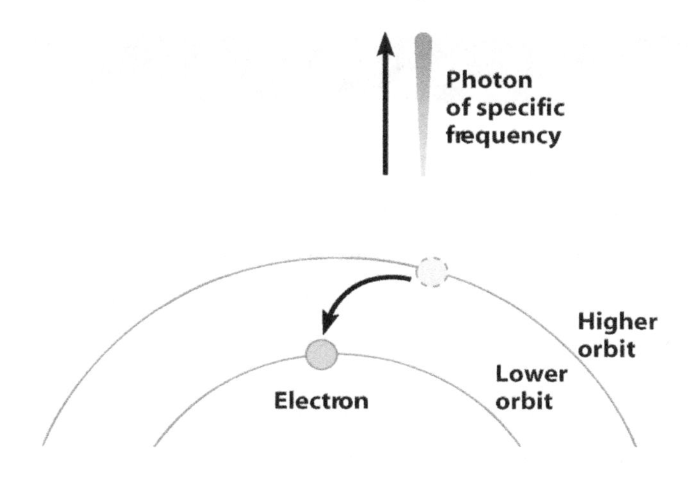

An essential application of the Bohr model is in the identification of elements through their line spectrum. When a high potential difference energizes an element, the electrons are constantly being raised to higher energy levels and then dropping back to lower energies.

The photons released in the process contain specific energies which can be deduced by their frequency. These energies relate the energy differences in the different electron orbitals and allow quantization of the specific orbital energies.

The figure below shows the line spectra of Hydrogen. The different series of light relates to the different orbital transitions of the Hydrogen's electrons.

In 1885, Johann Balmer (1825-1898) proposed the Balmer series for the orbital transitions of electrons from any orbital back to n = 2. Wavelength = $1/\lambda = R(1/2^2 - 1/n2)$

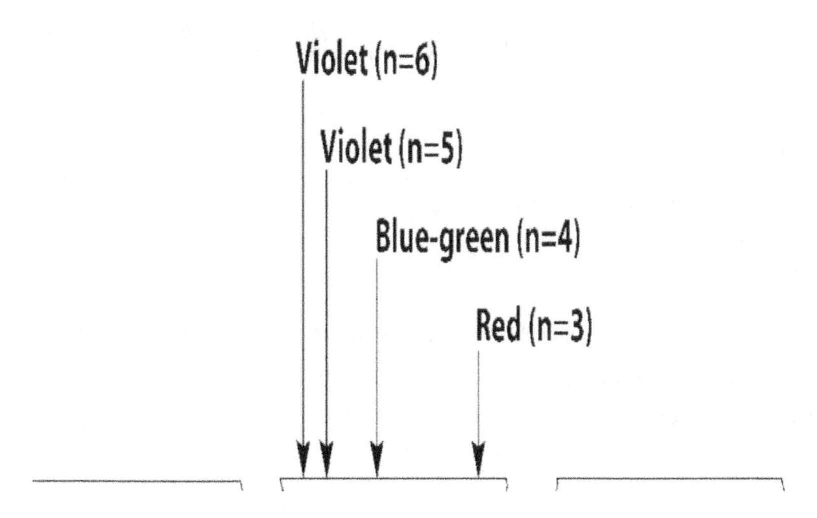

Balmer series to account for electron transitions between orbitals

Atomic Energy Levels

The *quantum theory* of the atom describes the series of energy states that an electron can occupy. The lowest energy state is the "*ground state*," and any higher states are "*excited states*." The energy of the photon (energy) emitted or absorbed during a transition between the specific energy levels equals the difference in state energies.

Below is an example of a hydrogen atom and its energy levels. The violet light represents the higher energy transitions (n = 5 or 6 to n = 2) because violet has the highest energy of the visible spectrum (mnemonic of ROY G BIV).

Red light has lower energy than violet and is the result of a lower energy transition (n = 3 to n = 2).

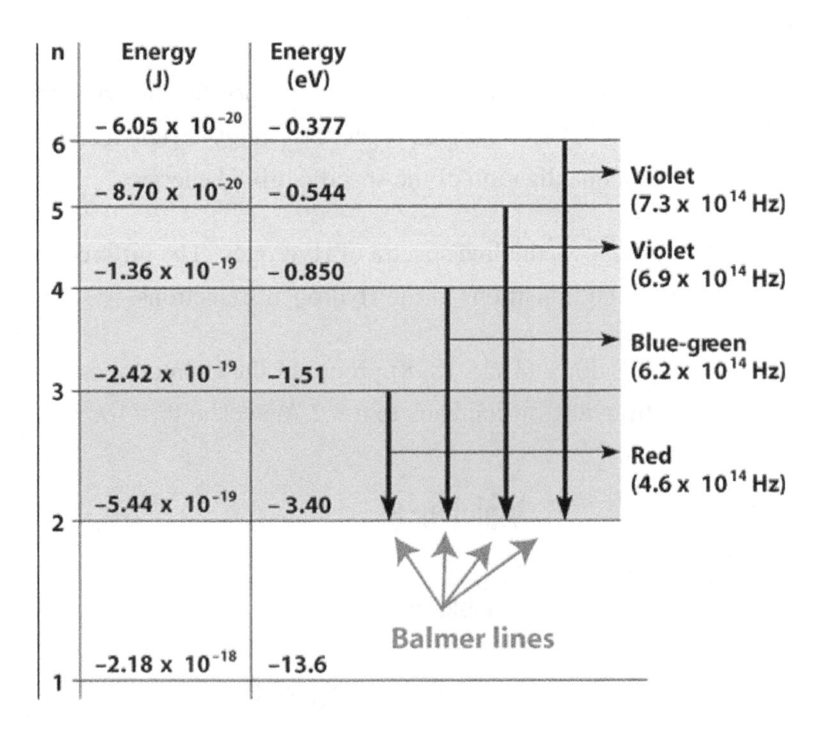

The Bohr theory only holds for the line spectrum of H because there are only two electrons. When more electrons are added, the theory is not able to account for the complications of heavier atoms.

Further experiments have established wave-particle duality of light and matter. Young's two-slit experiment, discussed for introducing the light wave theory, produced interference patterns for both photons and electrons.

In 1932, Nobel laureate Louis de Broglie (1892-1987), proposed the existence of matter waves in regard to the atomic theory. The wavelength is related to momentum, and the matter waves in atoms are standing waves.

Louis de Broglie derived the formula for wavelength:

$$\lambda = \frac{h}{mv}$$

h = Planck's constant = 6.63×10^{-34} J·s

λ = Particle wavelength

m = mass

v = speed

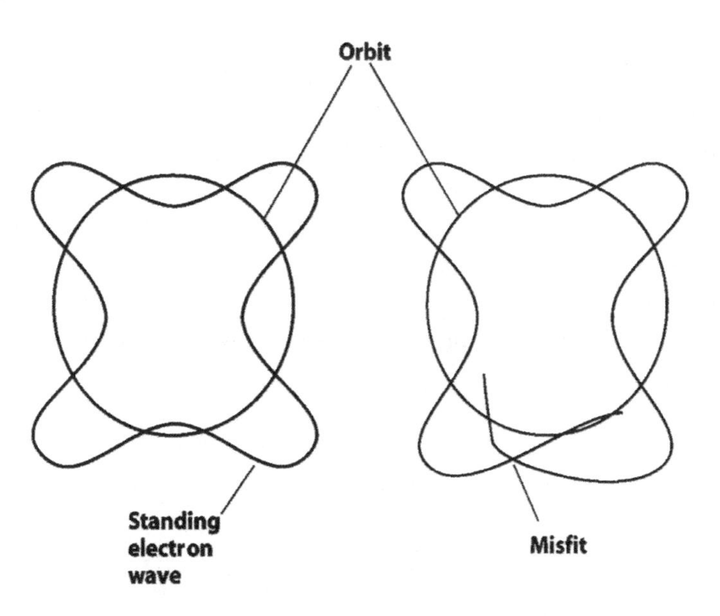

Nobel laureate Erwin Schrödinger (1887-1961) developed the theory of *wave mechanics*. The theory of wave mechanics treats atoms as three-dimensional systems of waves and incorporates the successful ideas of the Bohr model, along with other foundational principles.

Schrödinger's wave mechanics theory describes the hydrogen atom and other atoms with many electrons, as well as providing for the fundamental understanding of chemistry. The *quantum mechanics* model is a visualization of wave functions and probability distributions.

Quantum numbers specify electronic quantum states, and that an electron is delocalized.

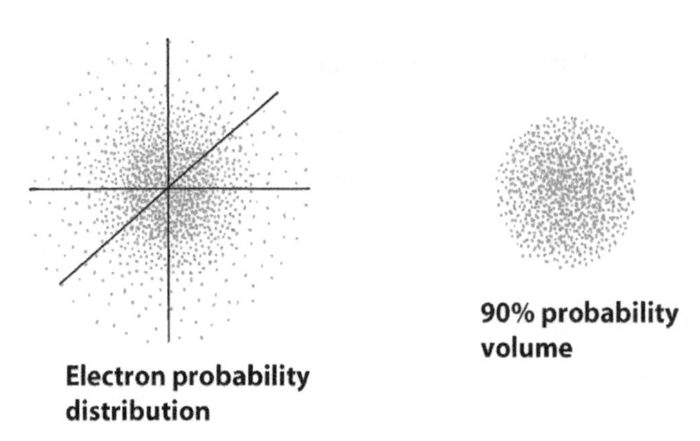

Quantum numbers define and describe a particular electron in an atom.

The *principal quantum number*, given by the lowercase *n*, states the energy level (i.e., *shells*) in which the electron can most likely be found. The principal quantum number also represents an average distance from the nucleus.

The *angular momentum quantum number* is the second quantum number and is given by the lowercase letter *l*.

The angular momentum quantum number describes the spatial distribution of the particular orbits, which are labeled *s*, *p*, *d*, *f*, *g*, *h*, etc. These are *subshells*.

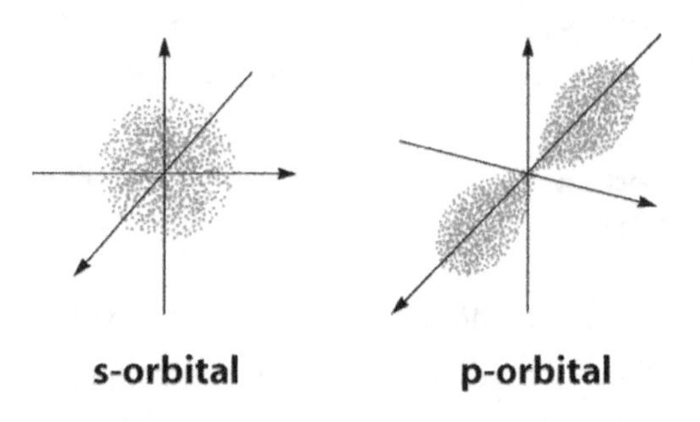

The *magnetic quantum number* (m) is the third quantum number. It states the spatial orientation of the orbit within the energy (n) and the shape (l).

The magnetic quantum number (m_l) divides the subshell even more into specific orbitals that hold the electron.

The number of orbitals is given by $2l + 1$. The m_l has values of $-l$ to l, with increments of one including zero ($-l$, .., 0, ..., $+l$).

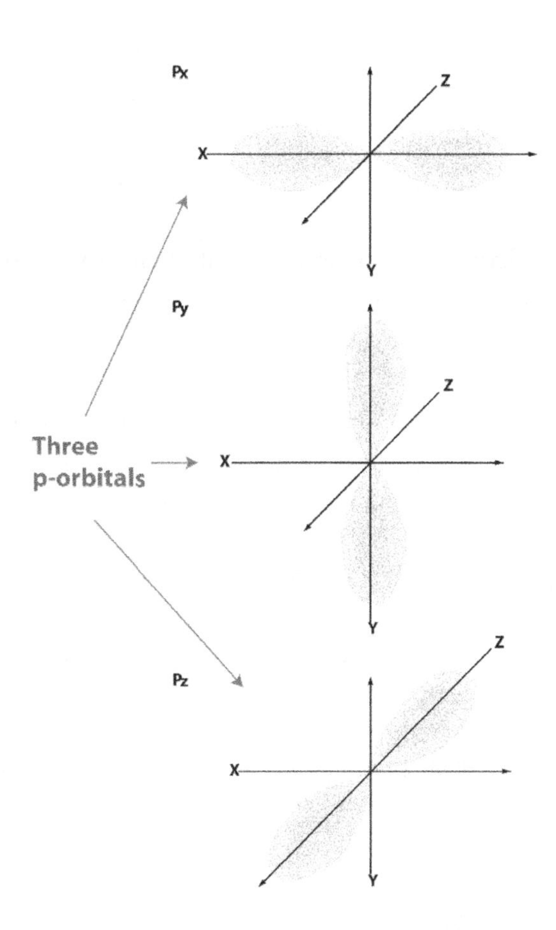

p orbitals orient along the x-, y- and z-axis and can accommodate 6 electrons

The *spin quantum number* is the fourth quantum number. It is either $+\frac{1}{2}$ or $-\frac{1}{2}$ and represents the electron spin orientation. There are only two values for the spin quantum number because an electron can only spin in one of two directions ($+\frac{1}{2}$ or $-\frac{1}{2}$).

Pauli's exclusion principle states that no two electrons in the same atom can have the same set of quantum numbers. Therefore, each orbital can only contain two electrons, and the electrons must have opposite spins.

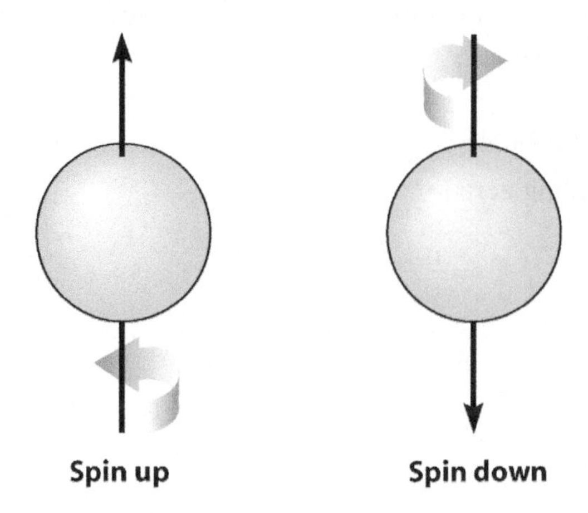

An *electron configuration* is an arrangement of electrons into atomic orbitals and specifies the atom's quantum state. The chemical properties of an atom are often determined by examining the atom's electronic structure.

When writing electron configurations, it is imperative to follow the Pauli exclusion principle. Each electron has unique quantum numbers, a maximum of two electrons per orbital, one spins up, and one spins down.

Electrons fill available orbitals in the order of increasing energy (n).

Each shell can only hold a finite number of subshells and the corresponding total number of electrons. The capacity of electrons in each subshell of an orbital is as follows:

s = 2 electrons in 1 subshell

p = 6 electrons in 3 subshells

d = 10 electrons in 5 subshells

f = 14 electrons in 7 subshells

Each orbital contains a maximum of two electrons. It is possible to fill half an orbital, but it must be in the outer (valence) shell. For example, strontium has 38 electrons.

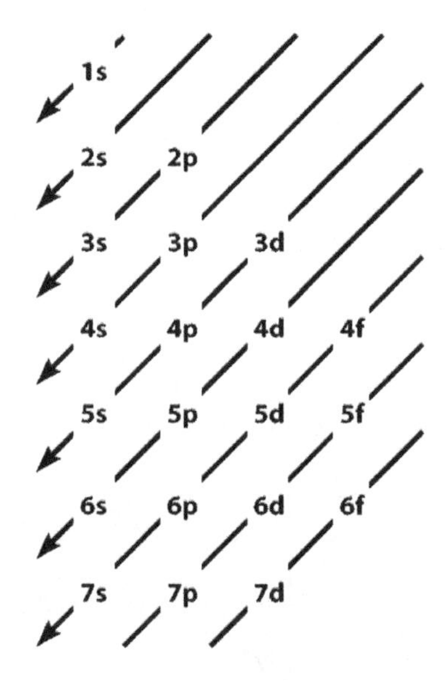

$$1s^2 2s^2 2p^6 3s^2 3p^6 4s^2 3d^{10} 4p^6 5s^2$$

The chemical properties of the elements can be understood by looking at the periodic table. Most chemical reactions follow the rules aligned with their *valence electron configuration.* Valence electrons are those in the outer orbits that determine chemical properties for the atom.

Elements are more stable with full outer (valence) shells, and thus donate their extra electrons or attract electrons from another element; both actions attempt to complete the valence shell. Which of these occurs (donate or attract electrons) generally depends on how many valence electrons the element has.

In the periodic table, rows are *periods,* and columns are *families* or *groups*. The families are:

Alkali metals (IA)

Alkaline earth (IIA)

Halogens (VIIA)

Noble gases (VIIIA)

A-group elements are the main group or the representative elements.

B-group elements are the transition elements or metals.

Key: 1 — Atomic number; H — Symbol; 1.01 — Atomic weight (rounded value)

IA	IIA	IIIB	IVB	VB	VIB	VIIB	VIIIB	VIIIB	VIIIB	IB	IIB	IIIA	IVA	VA	VIA	VIIA	VIIIA
1 H 1.01																	2 He 4.0
3 Li 6.94	4 Be 9.01											5 B 10.8	6 C 12.0	7 N 14.0	8 O 16.0	9 F 19.0	10 Ne 20.2
11 Na 23.0	12 Mg 24.3											13 Al 27.0	14 Si 28.1	15 P 31.0	16 S 32.1	17 Cl 35.5	18 Ar 39.9
19 K 39.1	20 Ca 40.1	21 Sc 45.0	22 Ti 47.9	23 V 50.9	24 Cr 52.0	25 Mn 54.9	26 Fe 55.8	27 Co 58.9	28 Ni 58.7	29 Cu 63.5	30 Zn 65.4	31 Ga 69.7	32 Ge 72.6	33 As 74.9	34 Se 79.0	35 Br 79.9	36 Kr 83.8
37 Rb 85.5	38 Sr 87.6	39 Y 88.9	40 Zr 91.2	41 Nb 92.9	42 Mo 95.9	43 Tc 98.9	44 Ru 101.1	45 Rh 102.9	46 Pd 106.4	47 Ag 107.9	48 Cd 112.4	49 In 114.8	50 Sn 118.7	51 Sb 121.8	52 Te 127.6	53 I 126.9	54 Xe 131.3
55 Cs 132.9	56 Ba 137.3	57 La 138.9	72 Hf 168.5	73 Ta 180.9	74 W 183.9	75 Re 186.2	76 Os 190.2	77 Ir 192.2	78 Pt 195.1	79 Au 197.0	80 Hg 200.6	81 Tl 204.4	82 Pb 207.2	83 Bi 209.0	84 Po (210)	85 At (210)	86 Rn (222)
87 Fr (223)	88 Ra 226.0	89 Ac (227)	104 Rf (261)	105 Db (262)	106 Sg (266)	107 Bh (264)	108 Hs (269)	109 Mt (268)	110 (269)	111 (272)	112 (277)		114 (285)		116 (289)		118 (293)

58 Ce 140.1	59 Pr 140.9	60 Nd 144.2	61 Pm (145)	62 Sm 150.4	63 Eu 152.0	64 Gd 157.3	65 Tb 158.9	66 Dy 162.5	67 Ho 164.9	68 Er 167.3	69 Tm 168.9	70 Yb 173.0	71 Lu 175.0
90 Th 232.0	91 Pa (231)	92 U 238.0	93 Np (237)	94 Pu (244)	95 Am (243)	96 Cm (247)	97 Bk (247)	98 Cf (251)	99 Es (252)	100 Fm (257)	101 Md (258)	102 No (259)	103 Lr (260)

( ) represents an isotope

Noble gases (VIIIA) have completely filled valence shells. They are inert or do not react readily with other elements.

Elements with 1 to 3 outer electrons will *lose their valence electrons* to become positive ions (such as metals).

Elements with 5 to 7 outer electrons tend *to gain electrons* and form negative ions (such as nonmetals).

Metalloids are chemical elements that have a mixture of properties of metals and nonmetals. There is no standard definition of a metalloid and no complete agreement on which elements belong to this category.

The six commonly recognized metalloids are: boron, silicon, germanium, arsenic, antimony, and tellurium. Five elements that are less frequently classified as metalloids are: carbon, aluminium, selenium, polonium, and astatine. On the Periodic Table, all eleven metalloid elements are located in a diagonal are of the *p*-block, extending from boron at the upper left to astatine at lower right.

Quantized energy levels for electrons

In 1900, Nobel laureate Max Planck (1858-1947) introduced *quantized energy*. In 1905, Albert Einstein (1879-1955) discovered that light was made up of quantized photons. He also found that higher frequency photons have more energetic photons. He expressed this with the equation:

$$E = hf$$

h = Planck's constant = 6.63×10^{-34} J·s

E = Photon energy

f = Photon frequency

The distinct lines of the emission spectrum prove that electron energy is quantized into discrete energy levels. If electron energy is not quantized, then a continuous spectrum would be observed.

The energy of a particular level is calculated using the quantum number, and given by the equation:

$$E_n = -\frac{13.6}{n^2}(eV)$$

The equation is negative, so all energies are negative.

Negative energies mean that energy contributes to the "stability" of the system — the electron's *binding energy*. The more negative (lower) the energy, the more stable the orbit, and the more difficult it is to liberate the electron. The less negative (higher) the energy, the less stable the orbit, and the easier it is to remove the electron. At the highest energy, 0 eV, there is no binding energy, so the electron dissociates.

For atoms other than hydrogen, the shape of the energy level curve stays the same. However, the numerator is a constant other than 13.6 eV. The precise relationship for atoms other than hydrogen is:

$$E = -\frac{Z^2 R_E}{n^2}$$

where Z is the atomic number.

Higher Z values give more negative binding energy (more stable) because of greater charge, the more electrostatic attraction.

Calculation of energy emitted or absorbed when an electron changes energy level

The *Rydberg formula* governs the wavelength of the emitted or absorbed radiation:

$$\frac{1}{\lambda} = R\left(\frac{1}{n_f^2} - \frac{1}{n_i^2}\right)$$

where λ is the wavelength, n_f is the final energy level, n_i is the initial energy level, and R is the Rydberg constant of value $1.0973 \times 10^7\ \mathrm{m}^{-1}$.

The energy of the emitted or absorbed radiation is:

$$E = hf = h\nu = h\frac{c}{\lambda}$$

where E is energy, f and ν both represent frequency, and c is the speed of light.

Energy is emitted from transitions to lower energy levels ($n_f < n_i$) and absorbed from transitions to higher energy levels ($n_f > n_i$).

The photon energy can also be computed using the equation:

$$hf = E_H - E_L$$

Neutrons, Protons, Isotopes

The *nucleus* is made of protons and neutrons.

Protons have a positive charge and a mass of:

$$m_p = 1.67262 \times 10^{-27} \text{ kg}$$

Neutrons are electrically neutral and slightly more massive than protons:

$$m_n = 1.67493 \times 10^{-27} \text{ kg}$$

Neutrons and protons are collectively *nucleons* (as they reside in the nucleus).

The different nuclei are *nuclides*.

The number of protons defines an element and refers to the neutral atom (having no charge). If two atoms have the same number of protons, they are the same element.

An *isotope* is an atom that has the same atomic number as its counterpart on the periodic table, but which has a different number of neutrons. Isotopes often have similar chemical properties, but different stabilities. For instance, some isotopes decay and give off radiation particles, while others are stable and do not decay. For many elements, several different isotopes exist in nature.

Natural abundance is the percentage of an element that consists of a particular isotope in nature.

$^{14}_{7}N$ **99.63%** **0.9963(14.00307 u)**

Natural abundances **Sum = atomic weight = 14.0067 u**

$^{15}_{7}N$ **0.37%** **0.0037(15.00011 u)**

Atomic Particles			
Name	Mass (amu)	Charge	Location
Proton	1	+1	In the nucleus
Neutron	1	0	In the nucleus
Electron	0	−1	Surrounding the nucleus

Atomic Number, Atomic Weight

The number of protons determines an element's *atomic number* (Z). A total of 118 elements have been identified. The number of nucleons is equal to the *atomic mass* number (A). The atomic mass number is found by adding the number of protons and neutrons. The value of neutrons is found by subtracting the atomic number (Z) from the atomic mass number (A). The expression $N = A - Z$ gives the neutron number.

The *atomic weight* is the weighted average of *atomic mass* for all isotopes of a given atom and is used for an element. The atomic mass is used for an isotope.

A and Z are sufficient to specify a nuclide. Nuclides are symbolized as follows:

$${}^{A}_{Z}X$$

X is the chemical symbol for the element; it contains the same information as Z but in a more recognizable form.

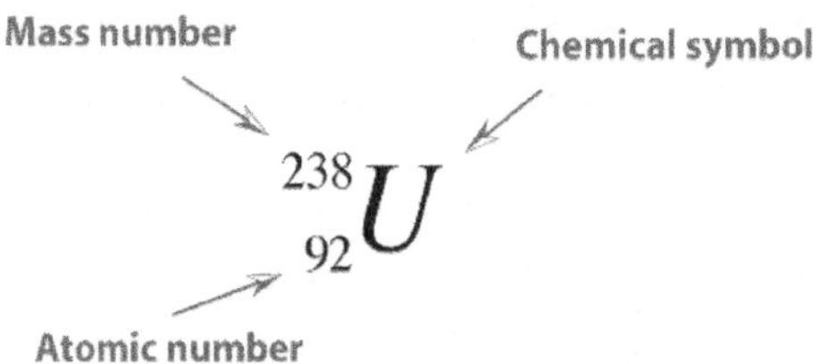

Because of wave-particle duality, the size of the nucleus is somewhat fuzzy.

Measurements of high-energy electron scattering yield the radius to be:

$$r \approx (1.2 \times 10^{-15}\ m)(A^{\frac{1}{3}})$$

Masses of atoms are measured with reference to the carbon-12 (^{12}C) atom, which is assigned a mass of exactly 12 amu (or Daltons).

$$1\text{amu} = 1.6605 \times 10^{-27}\ \text{kg} = 931.5\ \text{MeV}/c^2$$

From the table, the electron is considerably less massive than a nucleon.

Rest Masses in Kilograms, Unified Atomics Mass Units, and MeV/c^2			
	Mass		
Object	**kg**	**u**	**MeV/c^2**
Electron	9.1094×10^{-31}	0.00054858	0.51100
Proton	1.67262×10^{-27}	1.007276	938.27
${}^{1}_{1}H$ atom	1.67353×10^{-27}	1.007825	938.78
Neutron	1.67493×10^{-27}	1.008665	939.57

Nuclear Forces

The two forces in the nucleus are the *strong nuclear force* and the *electromagnetic force.*

The *strong nuclear force* binds the nucleons together and therefore contributes to the binding energy.

The *electromagnetic force* is due to electrostatic repulsion in the group of positively charged protons in the nucleus (like-charges repel). The nucleus stays together because the strong nuclear force is much stronger than the electromagnetic repulsion.

The strong force is short-ranged, less than 10^{-15} m.

The repulsion between proton particles that it overcomes is the *proton-proton Coulomb repulsion.*

Proton-proton chain reaction

$$^{1}_{1}H + ^{1}_{1}H \rightarrow ^{2}_{1}H + ^{0}_{1}e$$

$$^{2}_{1}H + ^{2}_{1}H \rightarrow ^{3}_{2}He + ^{1}_{0}n$$

$$^{3}_{2}He + ^{3}_{2}He \rightarrow ^{4}_{2}He + 2^{1}_{1}H$$

Coulomb's Law (F) relates the repulsion force on two charged particles depending on their charge (q) and their distance (r) from each other. The force (F) is significant when the distance is small.

$$F_{Coulomb} = \frac{kq_1q_2}{r^2}$$

To compare how tightly bound different nuclei are, divide the binding energy by the mass number A to calculate the binding energy per nucleon.

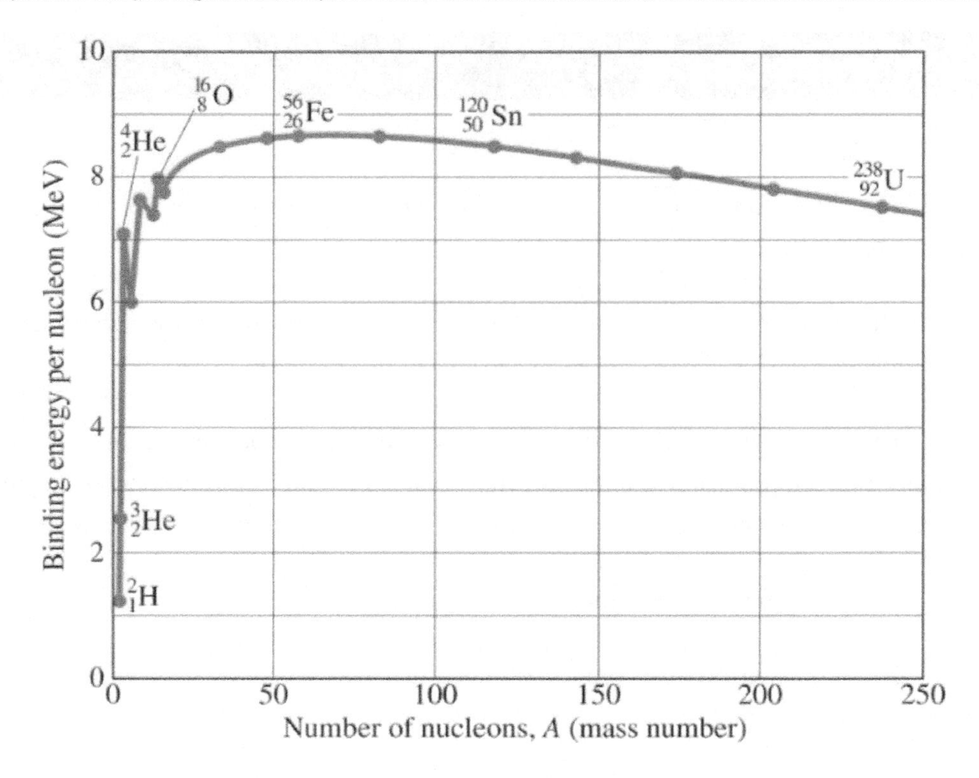

Chemical Reactions

For chemical reactions, compounds and chemical change must first be explained. The atom is the smallest elemental unit.

A *molecule* is the smallest particle retaining the characteristic chemical properties of a substance; they are combinations of atoms.

Examples of molecules include oxygen and hydrogen gas, which are diatomic molecules (made of two atoms). Ozone is a triatomic oxygen molecule (made of three atoms). The noble gases such as helium and neon are monatomic molecules (made of one atom, as they do not bond).

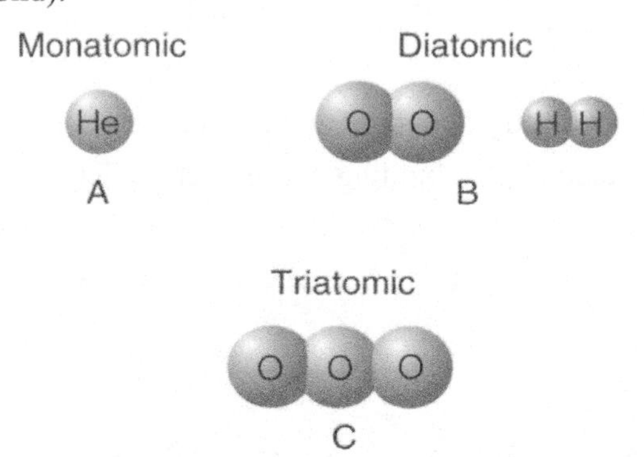

During chemical reactions, there is a formation or breaking of chemical bonds that form new molecules (*products*) from the original (*reactants*).

Chemical energy is the internal bonding potential energy, and the chemical equation is a symbolic summary of the chemical reaction. Some chemical reaction cannot be reversed.

Firewood burned to create carbon dioxide cannot be turned back into wood.

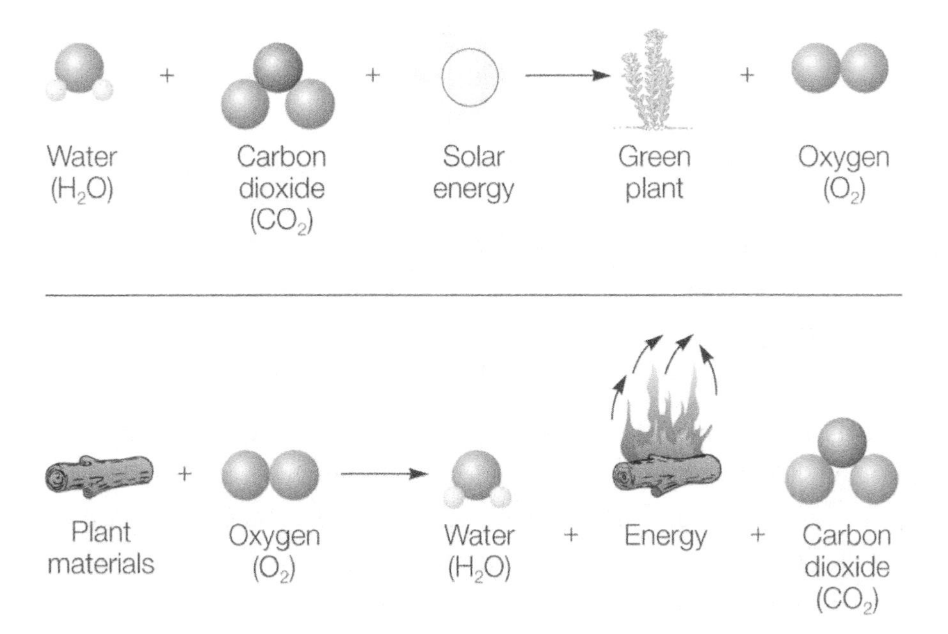

Valence electrons and ions (charged atoms) are essential for chemical reactions. The outer electrons (valence electrons) determine the chemical properties of an atom.

In 1916, Gilbert Newton Lewis proposed the *Lewis dot notation* to show the number of electrons an element has in its outer (valence) shell. Gilbert Lewis was nominated for the Nobel prize 41 times. However, despite his several seminal contributions to chemistry and his influence on many others that were awarded the recognition, he never received the Nobel prize.

From the periodic table below, moving from the left to right has each group (vertical column) has one more electron in its outer shell.

In the diagram, the transition metals are absent. Transition elements do not generally follow this trend of increasing valence electrons when moving across the table.

The elements shown below are the main elements used when assigning the Lewis dot notation for reactions.

H							He
Li	Be	B	C	N	O	F	Ne
Na	Mg	Al	Si	P	S	Cl	Ar
K	Ca	Ga	Ge	As	Se	Br	Kr
Rb	Sr	In	Sn	Sb	Te	I	Xe
Cs	Ba	Tl	Pb	Bi	Po	At	Rn
Fr	Ra						

Atoms attempt to acquire a full outer shell of eight electrons. Electrons may be gained, lost, or shared in the process of an element reacting with another element. When electrons are transferred, they must follow a few rules. The number of gained electrons must equal the number of lost electrons, and electrons are either lost or gained to form a stable form with complete *octets*.

Diatomic molecules will not form unless both octets are satisfied.

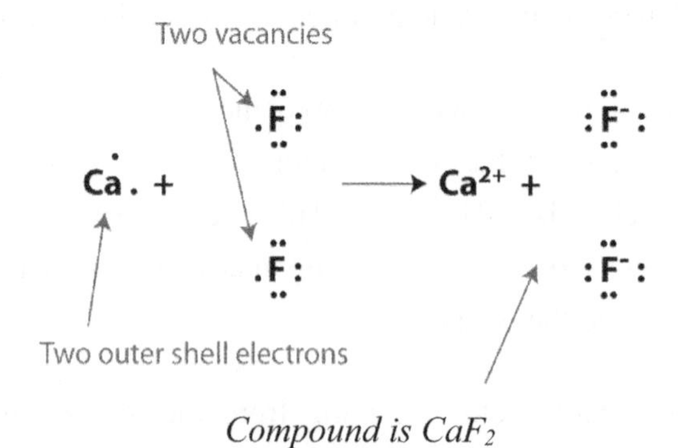

Compound is CaF_2

There are a few exceptions to this rule. Lithium and hydrogen are both most stable when they fill their first orbital, and therefore only have two electrons. Aluminum and boron both function with only six valence electrons. They are more stable with eight electrons but form stable compounds without eight electrons. Some compounds will have more than an octet—these are mostly found in the halogen family.

Another exception to the rule is for a noble gas. For instance, xenon (Xe) reacts with six fluorine (F) atoms to create xenon hexafluoride (XeF_6).

Depicted below is sodium (Na) reacting with another element and losing its one outer electron. Sodium is now stable because it has eight electrons in its outer shell.

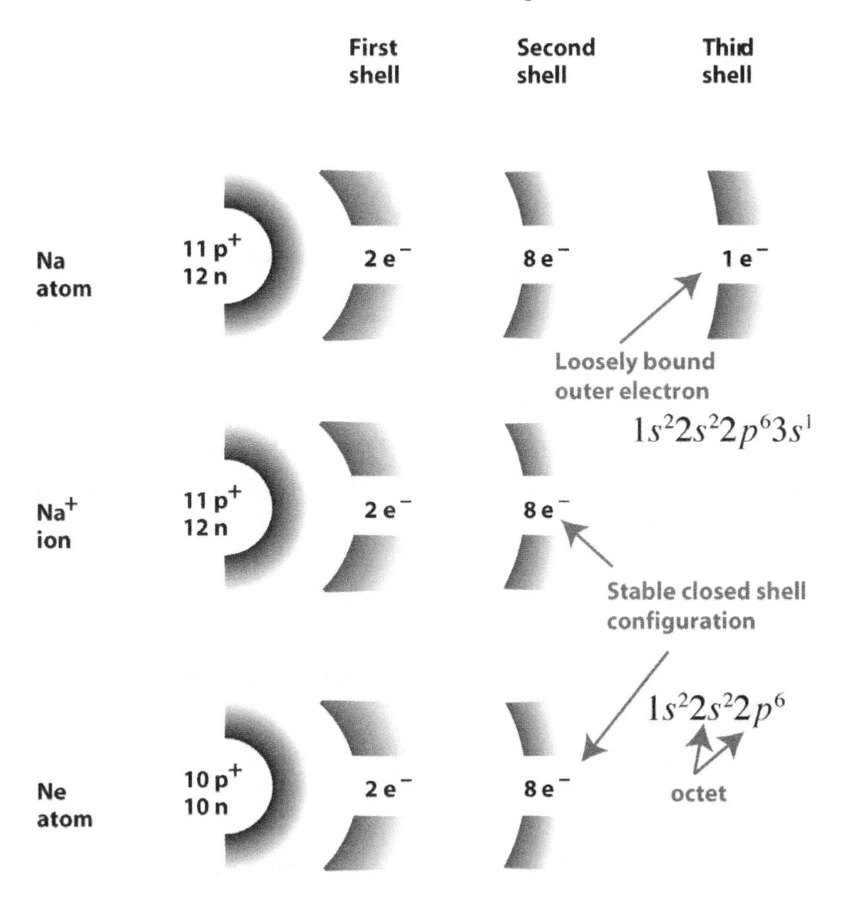

Chemical bonds are attractive forces that hold atoms together in compounds. They can be described regarding molecular (delocalized) or atomic (localized) orbitals. There are three types of chemical bonds: metallic, ionic, and covalent.

Metallic bonds occur between metallic elements, and the outer electrons can move freely throughout the metal. The electrons combine to form an "*electron gas*" that fills spaces within a rigid crystalline lattice of metal atoms. This allows the efficient conduction of heat and electricity.

When these types of bonds are formed, there is not just one molecule. Instead, these bonds create an orderly geometric structure, as depicted below in the formation of sodium chloride (NaCl). Sodium loses an electron while chlorine gains one—both elements then have eight valence electrons.

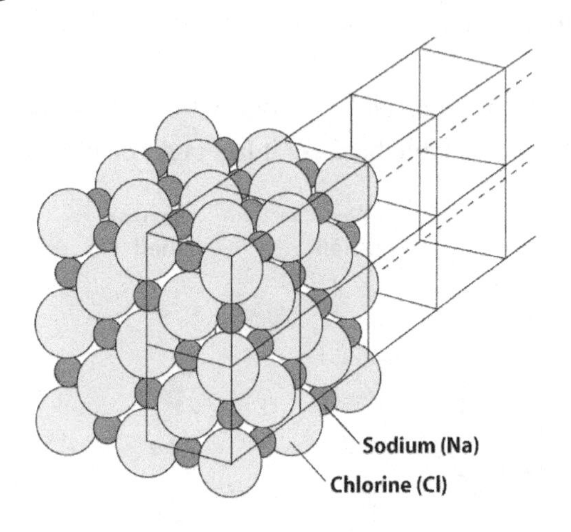

$$Na^{+} + Cl^{-} \rightarrow NaCl$$

A *covalent bond* is when electrons are shared between electrons, so both atoms achieve octets in their outermost orbital. This overlap of shared electron clouds between nuclei yields net attraction.

Atoms within covalent compounds are electrically neutral, or nearly so.

Typically, this bonding occurs between nonmetallic elements.

A *covalent compound* is held together by covalent bonds, which are represented by electron dot diagrams.

Bonding pairs are shared electrons, while lone pairs (non-bonding electrons) are not shared.

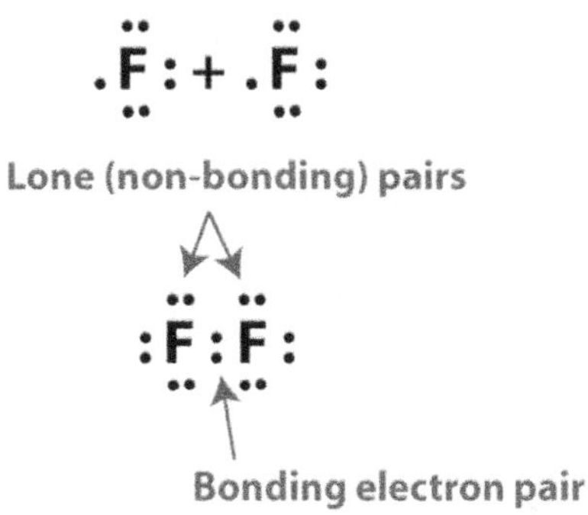

.F: + .F:

Lone (non-bonding) pairs

:F:F:

Bonding electron pair

Fluorine molecule, F_2 with Lewis dot structure representation

Structures and compounds of nonmetallic elements combined with hydrogen:

Nonmetallic Elements	Element (E Represents Any Element of Family)	Compound
Family IVA: C, Si, Ge	·Ė·	H H:E:H H
Family VA: N, P, As, Sb	·E·	H:E:H H
Family VIA: O, S, Se, Te	·E·	H:E:H
Family VIIA: F, Cl, Br, I	·E:	H:E:

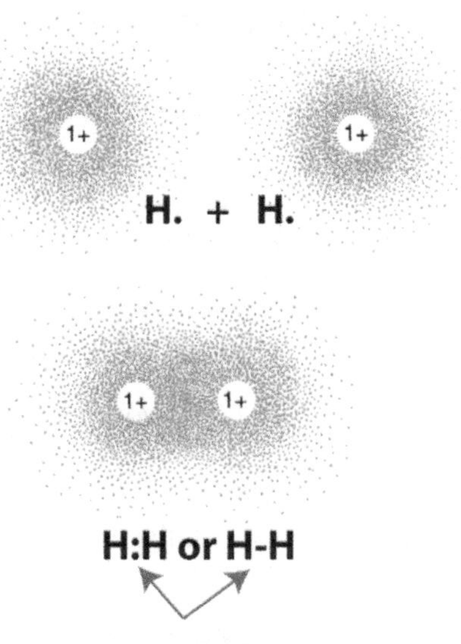

Hydrogen nuclei bonding - lines represent the shared pairs of electrons

Some covalent bonds require the sharing of more than one electron pair. When this occurs, they are double or triple bonds.

These bonds are symbolized by a single bar and represent two electrons.

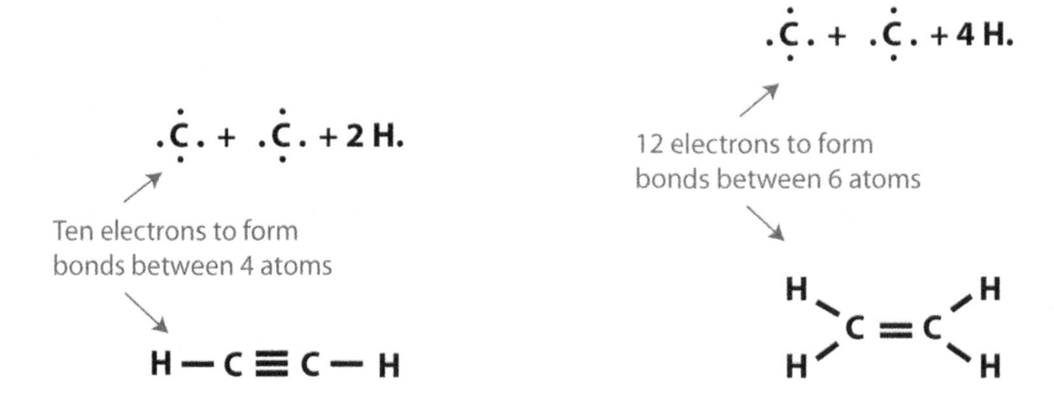

An *ionic bond* occurs when electrons are transferred between atoms. The only force involved in this bonding is the electrostatic force. This electrostatic force binds the atoms. The reaction energy released during the reaction is equal to the heat of formation.

Each ionic reaction can be conceptually divided into half-reactions.

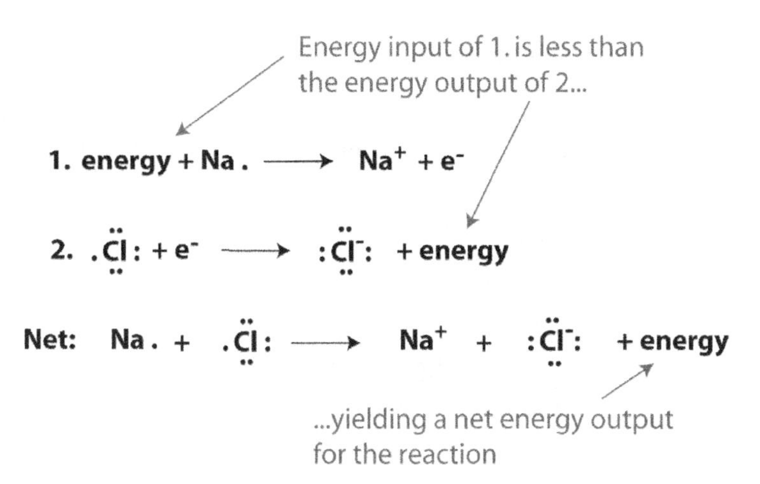

Ionic bonds characterize ionic compounds. They are commonly in the form of white, crystalline solids that are soluble in water.

Families IA and IIA lose electrons and form positive ions

Families VIA and VIIA gain electrons to form negative ions.

Polyatomic ions are made up of more than one element.

Common ions of some representative elements		
Element	**Symbol**	**Ion**
Lithium	Li	1^+
Sodium	Na	1^+
Potassium	K	1^+
Magnesium	Mg	2^+
Calcium	Ca	2^+
Barium	Ba	2^+
Aluminum	Al	3^+
Oxygen	O	2^-
Sulfur	S	2^-
Hydrogen	H	1^+, 1^-
Fluorine	F	1^-
Chlorine	Cl	1^-
Bromine	Br	1^-
Iodine	I	1^-

Common ions of some transition elements		
Single-Charge Ions		
Element	**Symbol**	**Charge**
Zinc	Zn	2^+
Tungsten	W	6^+
Silver	Ag	1^+
Cadmium	Cd	2^+
Variable-Charge Ions		
Chromium	Cr	$2^+, 3^+, \ldots$
Manganese	Mn	$2^+, 4^+, \ldots$
Iron	Fe	$2^+, 3^+$
Cobalt	Co	$2^+, 3^+$
Nickel	Ni	$2^+, 3^+$
Copper	Cu	$1^+, 2^+$
Tin	Sn	$2^+, 4^+$
Gold	Au	$1^+, 3^+$
Mercury	Hg	$1^+, 2^+$
Lead	Pb	$2^+, 4^+$

Some common polyatomic ions

Acetate	$(C_2H_3O_2)^-$
Ammonium	$(NH_4)^+$
Borate	$(BO_3)^{3-}$
Carbonate	$(CO_3)^{2-}$
Chlorate	$(ClO_3)^-$
Chromate	$(CrO_4)^{2-}$
Cyanide	$(CN)^-$
Dichromate	$(Cr_2O_7)^{2-}$
Hydrogen carbonate (or bicarbonate)	$(HCO_3)^-$
Hydrogen sulfate (or bisulfate)	$(HSO_4)^-$
Hydroxide	$(OH)^-$
Hypochlorite	$(ClO)^-$
Nitrate	$(NO_3)^-$
Nitrite	$(NO_2)^-$
Perchlorate	$(ClO_4)^-$
Permanganate	$(MnO_4)^-$
Phosphate	$(PO_4)^{3-}$
Phosphite	$(PO_3)^{3-}$
Sulfate	$(SO_4)^{2-}$
Sulfite	$(SO_3)^{2-}$

When writing a chemical formula for an ionic compound, list the elements in the compound and their proportions (subscripts). The proportions are decided by the amount of electron gain or loss.

Always write the symbol for a positive ion first, followed by a negative ion symbol. Then, assign subscripts to ensure the compound is electrically neutral.

For example, when magnesium and chloride react, the chemical formula for the compound is $MgCl_2$. This is because magnesium has two electrons to donate (Mg^{2-}), but chlorine only needs one electron to fill its valence orbital (Cl^{1+}).

There must be two chlorines, and each will take one of magnesium's two extra electrons.

$$Ca \rightarrow Ca^{2+} \qquad Cl \rightarrow Cl^{-}$$

Forms +2 ion   Forms -1 ion

$$CaCl_2$$

+2+2(-1) = neutral compound

When compounds are formed, there is an unequal sharing of electrons. This is because of the difference in electronegativity in the elements involved in the reaction.

Electronegativity is the measure of an atom's ability to attract electrons.

The amount of difference reveals what type of bond is between the two atoms.

The meaning of absolute differences in electronegativity		
Absolute Difference	→	**Type of Bond Expected**
1.7 or greater	means	ionic bond
between 0.5 and 1.7	means	polar covalent bond
0.5 or less	means	covalent bond

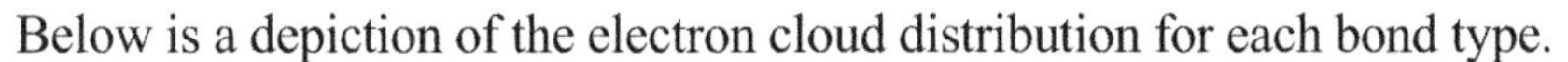

Below is a depiction of the electron cloud distribution for each bond type.

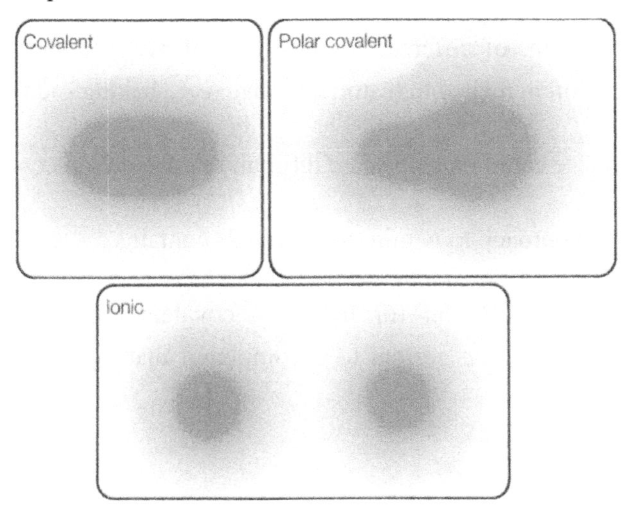

The electronegativity of each element can be found from the table:

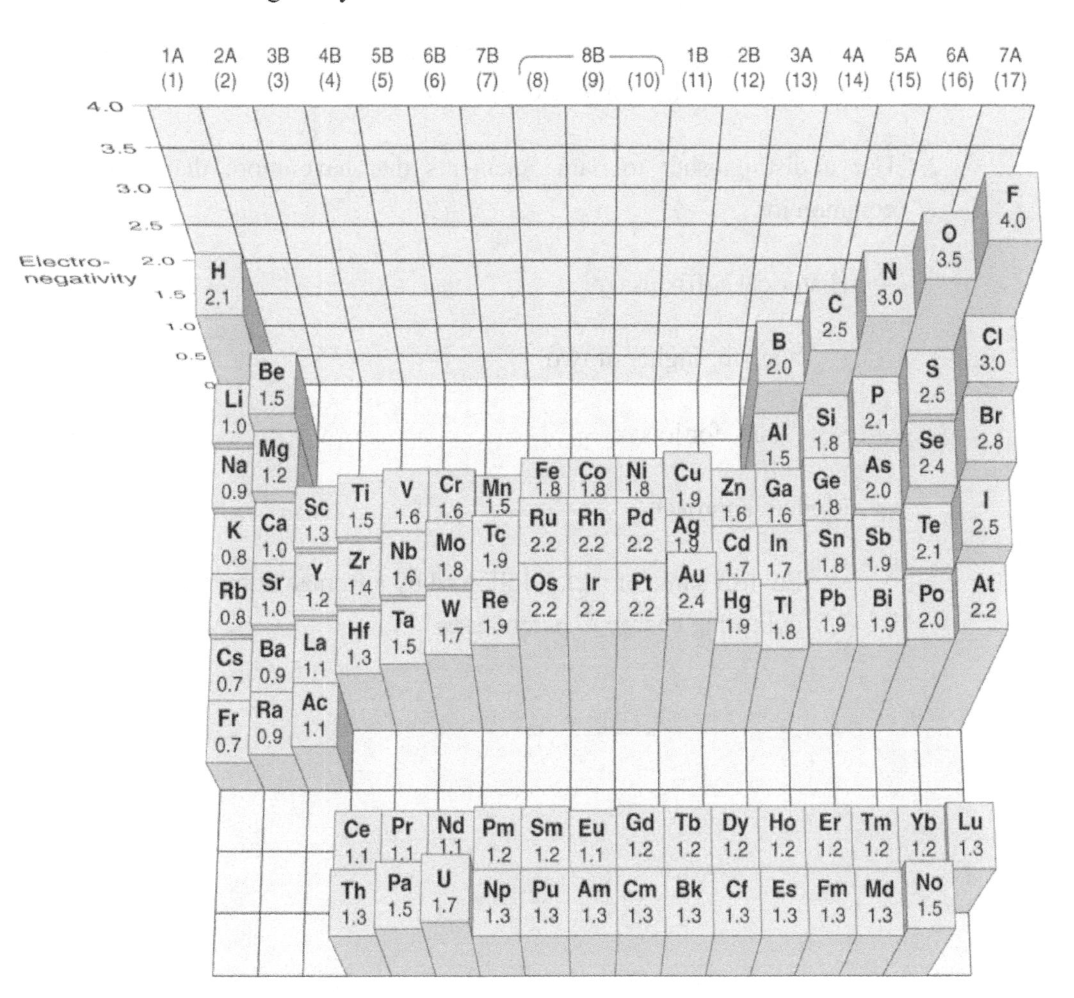

Naming Compounds

There are millions of different combinations of over 90 elements. Many of these have common names often related to historical usage (e.g., baking soda or washing soda).

However, these common names are difficult to relate to molecular composition.

The modern approach to naming compounds contains a systematic set of rules.

These rules are different for ionic *vs.* covalent compounds, but there is a universal rule: the suffix "*-ide*" refers to a compound that contains only two different elements.

Ionic Compounds

1. Start with the name of the metal ion (the positive one) first, then add the nonmetal ion (the negative one).

2. Use a distinguisher to name elements that have more than one common ion

   – Historical suffix usage

     - "-ic" for higher of two

     - "-ous" for lower

   – Modern approach

     - English name of metal, followed by a Roman numeral indicating charge in parentheses

Modern names of some variable-charge ions	
Ion	**Name of Ion**
Fe^{2+}	Iron(II) ion
Fe^{3+}	Iron(III) ion
Cu^{+}	Copper(I) ion
Cu^{2+}	Copper(II) ion
Pb^{2+}	Lead(II) ion
Pb^{4+}	Lead(IV) ion
Sn^{2+}	Tin(II) ion
Sn^{4+}	Tin(IV) ion
Cr^{2+}	Chromium(II) ion
Cr^{3+}	Chromium(III) ion
Cr^{6+}	Chromium(VI) ion

Covalent compounds

Covalent compounds are molecular or composed of two or more nonmetals. The same elements can combine to form several different compounds, depending on the concentration of each element involved.

When naming covalent compounds, there are two rules to follow. Use the table below for the prefix and stem names:

1. The first element given in the formula is named first.
   - A Greek prefix indicates the subscript or concentration of the element
2. Only the stem name of the second element comes next.
   - a Greek prefix indicates concentration
   - the suffix "*-ide*" is added to the end

Prefixes and element stem names			
Prefix	*Meaning*	*Element*	*Stem*
Mono-	1	Hydrogen	Hydr-
Di-	2	Carbon	Carb-
Tri-	3	Nitrogen	Nitr-
Tetra-	4	Oxygen	Ox-
Penta-	5	Fluorine	Fluor-
Hexa-	6	Phosphorus	Phosph-
Hepta-	7	Sulfur	Sulf-
Octa-	8	Chlorine	Chlor-
Nona-	9	Bromine	Brom-
Deca-	10	Iodine	Iod-

Note: the *a* or *o* ending on the prefix if often dropped if the stem name begins with a vowel (e.g., "*tetroxide*," not "*tetraoxide*").

Examples: carbon dioxide (CO_2) and carbon tetrachloride (CCl_4).

Radioactive Decay

Natural radioactivity is a spontaneous emission of particles or energy from an unstable nucleus. At the end of the 19th century, minerals were found to darken a photographic plate even in the absence of light. This phenomenon is radioactivity.

When an atom is unstable, it will undergo *radioactive decay.* The maximum stability for a nucleon number is 2, 8, 20, 28, 50, 82, or 126. The higher the binding energy per nucleon, the more stable the nucleus.

More massive nuclei require extra neutrons to overcome the Coulomb repulsion of the protons to be stable. There is a band of stability that depends on the ratio of protons to neutrons that makes an atom stable. The Coulomb force is long-range; therefore, extra neutrons are needed for stability in high-Z nuclei:

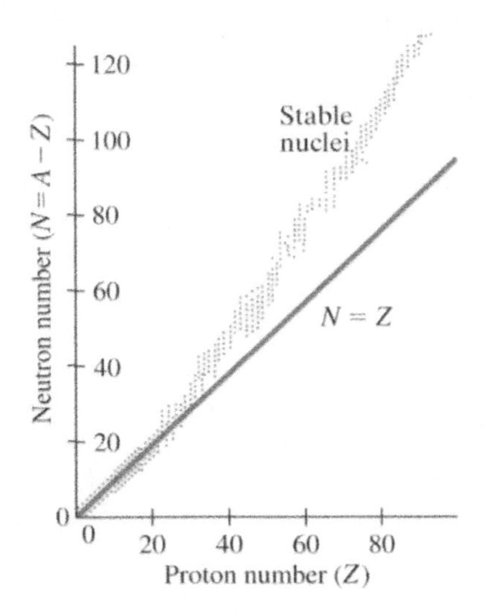

Pairs of protons and pairs of neutrons add stability to the atom.

Conversely, an odd number of both protons and neutrons makes a less stable atom. For added stability, specific neutron-to-proton ratios are desired: 1:1 for isotopes with up to 20 protons and (1+1):1 for increasingly heavy isotopes. Nuclei that are unstable undergo decay; many such decays are governed by another force called the weak nuclear force. There are three types of decay: alpha, beta, and gamma decay.

All these forms of radiation are *ionizing radiation* because they ionize material that they go through.

Alpha emission is an expulsion of a helium nucleus. Alpha emission is the least penetrating and can even be stopped by a piece of paper.

Beta emission is an expulsion of an electron. Beta emission is more penetrating than alpha decay and can be stopped by 1 cm of aluminum.

Gamma decay is an emission of a high energy photon (electromagnetic radiation). Gamma decay is the most penetrating emission, and can only be stopped by a thick, dense shield (e.g., approximately 13.5 feet of water, 6.5 feet of concrete or about 1.3 feet of lead).

Radioactive Decay			
Unstable Condition	**Type of Decay**	**Emitted**	**Product Nucleus**
More than 83 protons	Alpha emission	${}^{4}_{2}\alpha\ ({}^{4}_{2}He)$	Lost 2 protons and 2 neutrons
Neutron-to-proton ratio too large	Beta emission	${}^{0}_{-1}\beta\ ({}^{0}_{-1}e)$	Gained 1 proton, no mass change
Excited Nucleus	Gamma emission	${}^{0}_{0}\gamma$	No change
Neutron-to-proton ratio too small	Other emissions	${}^{0}_{1}e$	Lost 1 proton, no mass change

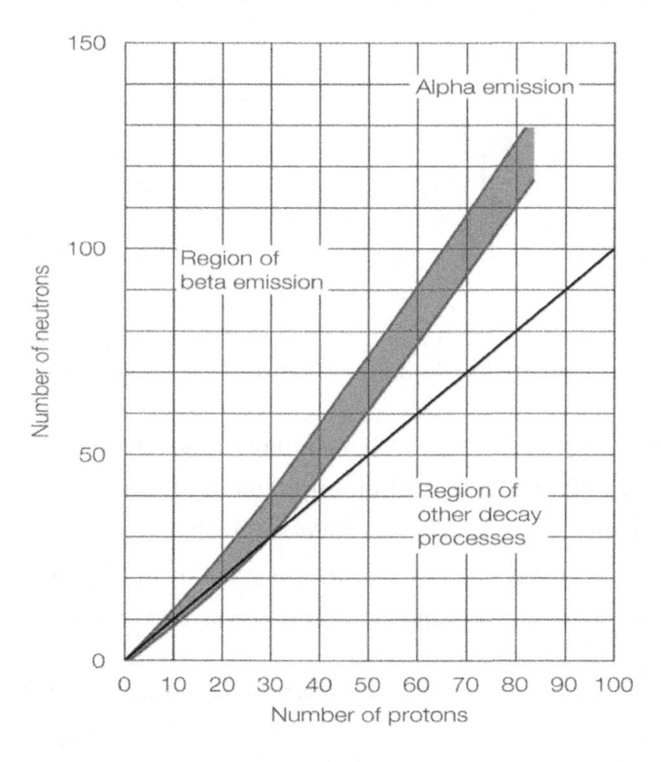

Alpha and beta rays are bent in opposite directions in a magnetic field, while gamma rays are not bent at all.

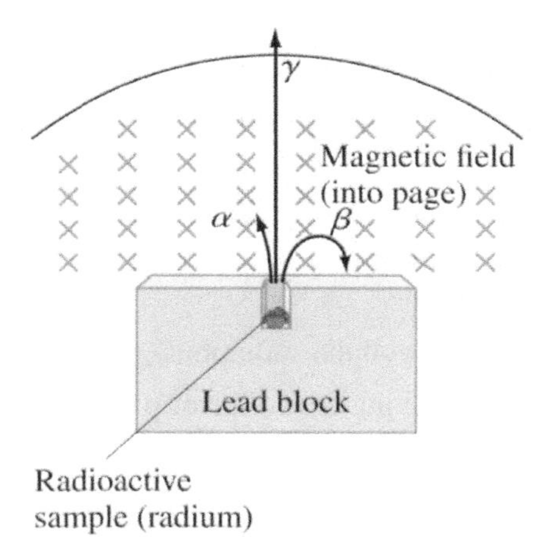

The *conservation of mass* dictates that the total atomic weight before the decay must equal the total atomic weight after. The *conservation of charge* dictates that the total atomic number before the decay must equal the total atomic number after. A new conservation law that is evident by studying radioactive decay is that the total number of nucleons cannot change.

Do not get thrown off by unfamiliar particles. If they have a weight and a charge, incorporate these numbers into calculations. Problems of identifying decay products are just math work. The atomic number (bottom number) determines what element it is.

Alpha Decay

An example of alpha decay is radium-227 (^{227}Ra) decaying into radon-222 (^{222}Rn):

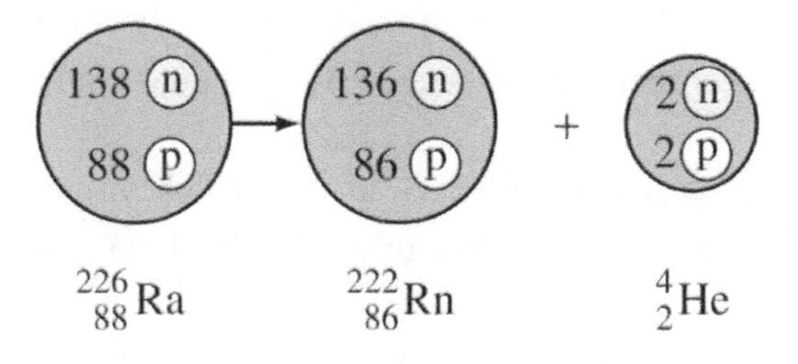

In general, alpha decay can be written:

$$^{A}_{Z}N \rightarrow \, ^{A-4}_{Z-2}N' + \, ^{4}_{2}He$$

Alpha decay occurs when the strong nuclear force cannot hold a large nucleus together.

The mass of the parent nucleus is greater than the sum of the masses of the daughter nucleus and the alpha particle; this difference is the *disintegration energy*.

Alpha decay is much more likely than other forms of nuclear disintegration because the alpha particle itself is stable.

Some smoke detectors use alpha radiation—the presence of smoke absorbs the alpha rays and keep them from striking the collector plate.

Beta Decay

Beta-decay occurs when a nucleus emits an electron. An example is the decay of carbon-14 (^{14}C):

$$^{14}_{6}C \rightarrow {}^{14}_{7}N + e^{-} + \text{neutrino}$$

The nucleus has 14 nucleons, but it has one more proton and one fewer neutron. This decay is an example of an interaction that proceeds via the weak nuclear force.

The electron in beta decay is not an orbital electron; it is created in the decay.

The fundamental process is a neutron decaying to a proton, electron, and neutrino:

$$n \rightarrow p + e^{-} + neutrino$$

The need for a particle such as a *neutrino* was discovered through analysis of energy and momentum conservation in beta decay—it could not be a two-particle decay.

Neutrinos are notoriously difficult to detect. They interact only weakly, and direct evidence for their existence was not available for a long time.

The symbol for the neutrino is the Greek letter nu (ν); the beta decay of carbon-14 is written as (the bar over the neutrino means that it is an antineutrino):

$$^{14}_{6}C \rightarrow {}^{14}_{7}N + e^{-} + \overline{\nu}$$

Beta decay can occur where the nucleus emits a positron rather than an electron:

$$^{19}_{10}Ne \rightarrow {}^{19}_{9}F + e^{+} + \overline{\boldsymbol{v}}$$

A nucleus can capture one of its inner electrons:

$$^{7}_{4}Be + e^{-} \rightarrow {}^{7}_{3}Li + v$$

Gamma Decay

Gamma rays are high-energy photons. They are emitted when a nucleus decays from an excited state to a lower state, just as electrons returning to a lower state emits photons.

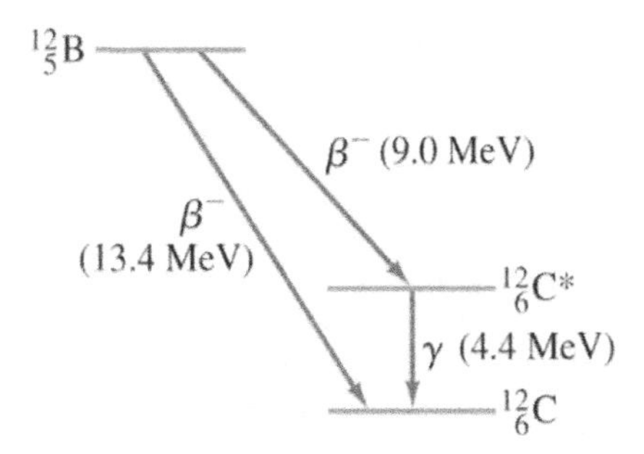

Half-life, stability, exponential decay, semi-log plots

An element's *half-life* is the amount of time required for ½ of a radioactive sample to decay.

For example, a 1 kg sample of an unstable isotope with a one-day half-life:

After 1 day - 500 g remain

After 2 days - 250 g remain

After 3 days - 125 g remain

Half-lives of some radioactive isotopes		
Isotope	**Half-Life**	**Mode of Decay**
${}_{1}^{3}H$ (tritium)	12.26 years	Beta
${}_{6}^{14}C$	5,730 years	Beta
${}_{38}^{90}Sr$	28 years	Beta
${}_{53}^{131}I$	8 days	Beta
${}_{54}^{133}Xe$	5.27 days	Beta
${}_{92}^{238}U$	4.51×10^9 years	Alpha
${}_{94}^{242}Pu$	3.79×10^5 years	Alpha
${}_{94}^{240}Pu$	6,760 years	Alpha
${}_{94}^{239}Pu$	24,360 years	Alpha
${}_{19}^{40}K$	1.3×10^9 years	Alpha

Half-life graphs are shown with decreasing exponential:

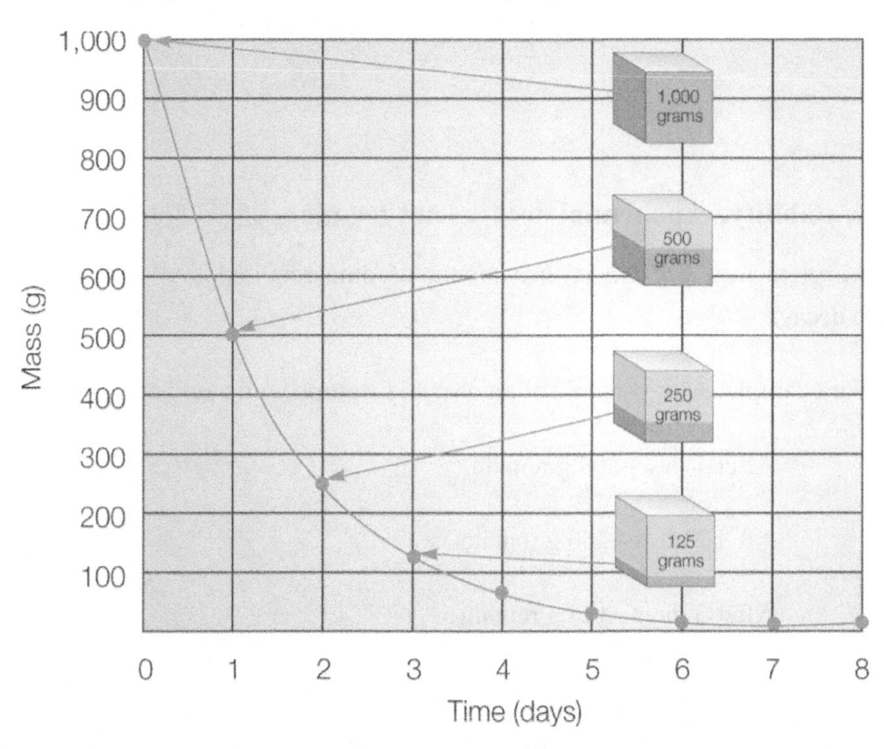

Nuclear decay is a random process; the decay of another nucleus does not influence the decay of one nucleus.

Additionally, the decay rate is unaffected by temperature, pressure, volume, or any other environmental factor.

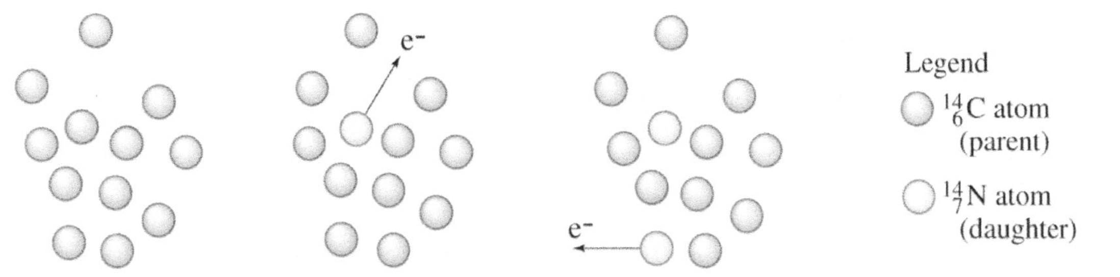

Therefore, the number of decays in a short time interval is proportional to the number of nuclei present and to the time:

$$\Delta N = -\lambda N \Delta t$$

Here, λ is a constant characteristic of that nuclide, the decay constant.

This equation is using calculus, for N as a function of time:

$$N = N_0 e^{-\lambda t}$$

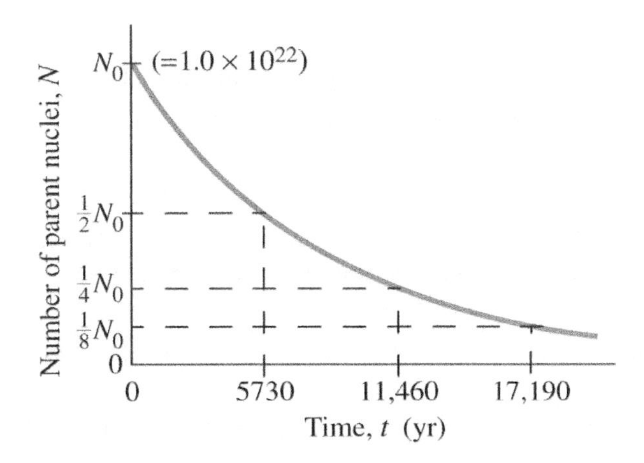

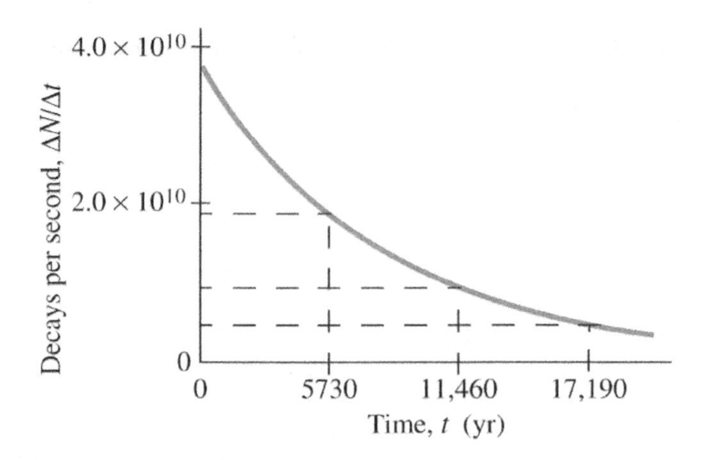

The half-life can be related to the decay constant:

$$T_{\frac{1}{2}} = \frac{\ln 2}{\lambda} = \frac{0.693}{\lambda}$$

It can also be written:

$$N_t = N_{t=0} \cdot \left(\frac{1}{2}\right)^{\#half-lives} = N_{t=0} \cdot \left(\frac{1}{2}\right)^{\frac{t}{half-life}}$$

where $N_{t=0}$ is the amount the original starting material, N_t is the amount of the original material that is left, and t is time.

Although the above is the official half-life equation, it may be easier to multiply rather than to divide. Therefore, a more user-friendly equation is:

$$N_{t=0} = N_t \cdot 2^{\#half-lives} = N_t \cdot 2^{\frac{t}{half-life}}$$

Semi-log plots convert exponential curves into straight lines.

- Something that curves up becomes a straight line with a positive slope.
- Something that curves down becomes a straight line with a negative slope.
- For exponential decay, a semi-log plot graphs the log of amount vs. time.
- For exponential decay, a semi-log plot is a straight line with a negative slope.
- The semi-log plot intercepts the x-axis, where the original y-value is 1.

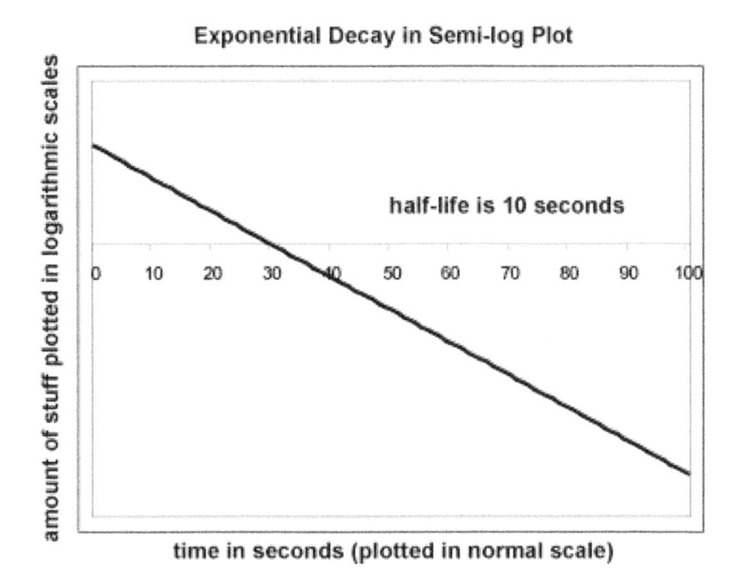

A decay series occurs when one radioactive isotope decays to another radioactive isotope, which decays to another and so on. This decay series allows for the creation of nuclei that otherwise would not exist in nature.

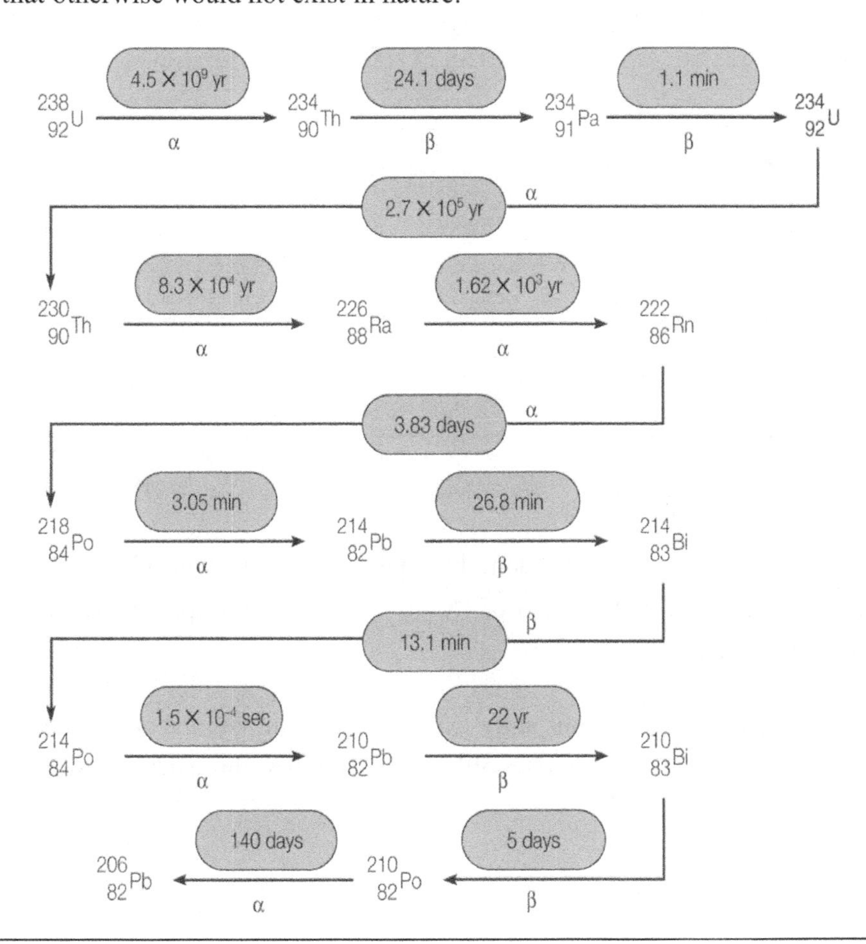

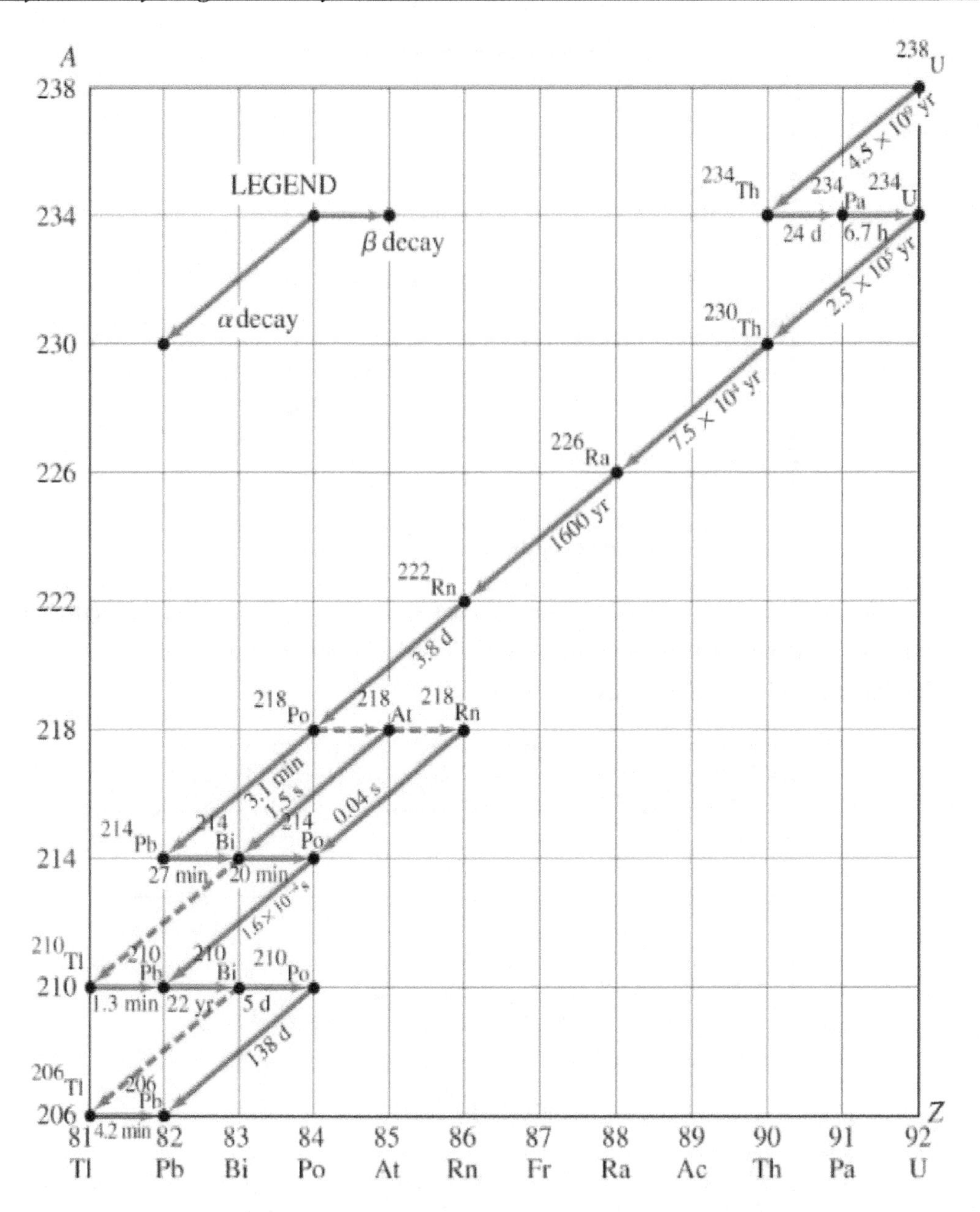

Radioactive dating can be done by analyzing the fraction of carbon in organic material that is carbon-14. The ratio of ^{14}C to ^{12}C in the atmosphere has been roughly constant over thousands of years.

A living plant or tree constantly exchanges carbon with the atmosphere and will have the same carbon ratio in its tissue. When the plant dies, this exchange stops.

^{14}Carbon has a half-life of about 5,730 years; it gradually decays and becomes a smaller and smaller fraction of the total carbon in the plant tissue. This fraction can be measured, and the age of the tissue deduced.

Objects older than about 60,000 years cannot be dated this way—there is little ^{14}carbon remaining. Other isotopes are useful for geologic time scale dating, Uranium-238 (^{238}U) has a half-life of 4.5×10^9 years and has been used to date the oldest rocks on Earth as about 4 billion years old.

When a nucleus decays through alpha emission, energy is released. Why is it that these nuclei do not decay immediately?

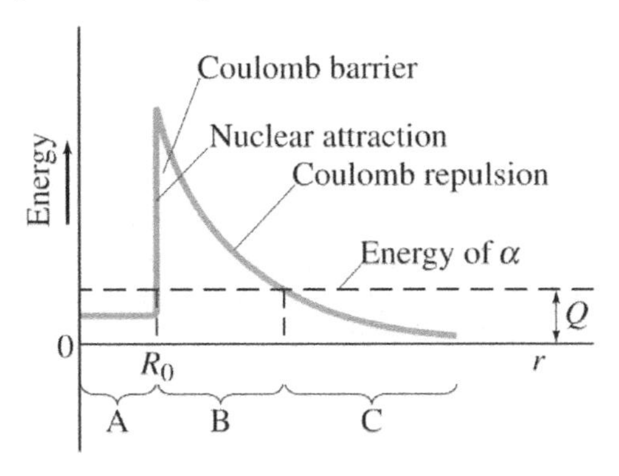

Although energy is released in the decay, there is an energy barrier. The alpha particle escapes through a quantum mechanical phenomenon of tunneling. For the *Heisenberg uncertainty principle*, energy conservation can be violated if the violation does not last long:

$$Q = M_p c^2 - (M_D + m_\alpha)c^2$$

The higher the energy barrier, the less time the alpha particle has to get through it, and the less likely it is to happen. This requirement accounts for the extensive variation in half-lives for alpha decay.

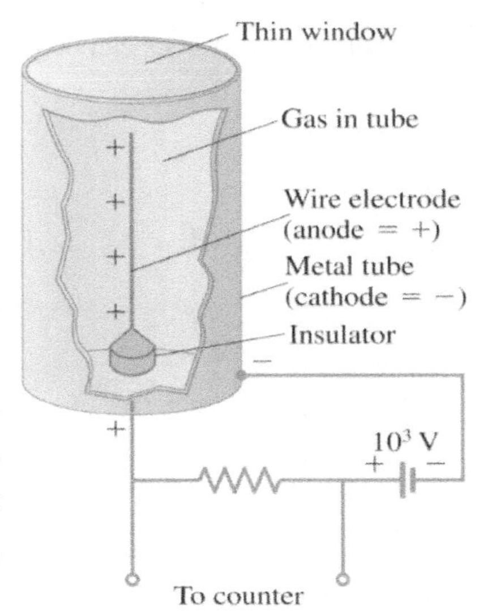

There are a few methods to take measurements of decay. An ionization counter is a device that detects ions produced by radiation, such as a *Geiger counter*. The Geiger counter is a gas-filled tube with a wire in the center. The wire is at high voltage; the case is grounded. When a charged particle passes through, it ionizes the gas. The ions cascade onto the wire, producing a pulse.

A *scintillation counter* uses a scintillator. A scintillator is a material that emits light when a charged particle passes through it. The scintillator is made light-tight, and the light flashes are viewed with a photomultiplier tube, which has a photocathode that emits an electron when struck by a photon, and then a series of amplifiers.

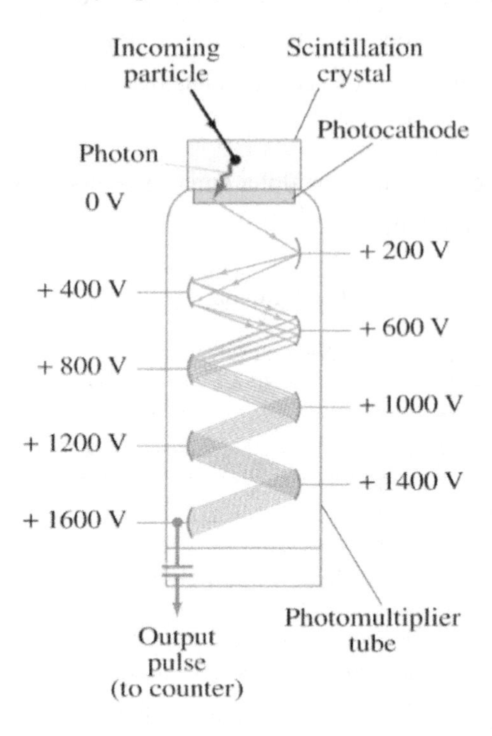

A *cloud chamber* contains a supercooled gas; when a charged particle goes through it, droplets form along its track.

Similarly, a *bubble chamber* contains a superheated liquid, which forms bubbles. In either case, the tracks can be photographed and measured.

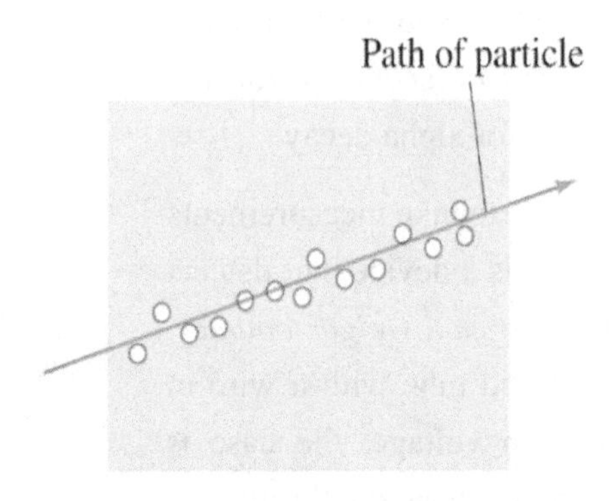

A *wire drift chamber* is similar but vastly more sophisticated than a Geiger counter. Many wires are present, some at high voltage, and some grounded. In addition to the presence of a signal, the time it takes the pulse to arrive at the wire is measured, allowing precise measurement of position. Radiation can be measured at the source. The activity is the number of disintegrations per unit time.

There are several units of radioactivity:

Names, symbols, and conversion factors for radioactivity			
Name	**Symbol**	**To Obtain**	**Multiply by**
Becquerel	Bq	Ci	2.7×10^{-11}
gray	Gy	rad	100
sievert	Sv	rem	100
curie	Ci	Bq	3.7×10^{10}
rem	rem	Sv	0.01
millirem	mrem	rem	0.001
rem	rem	millirem	1,000

Radiation is also measured where it is absorbed. Human exposure is measured in rem, but the SI unit is the millisievert. Another measurement is the absorbed dose—the effect the radiation has on the absorbing material.

The rad, a unit of dosage, is the amount of radiation that deposits energy at a rate of 1.00×10^{-2} J/kg in any material. The SI unit for dose is the gray (Gy). The dosage is related to effects on organisms.

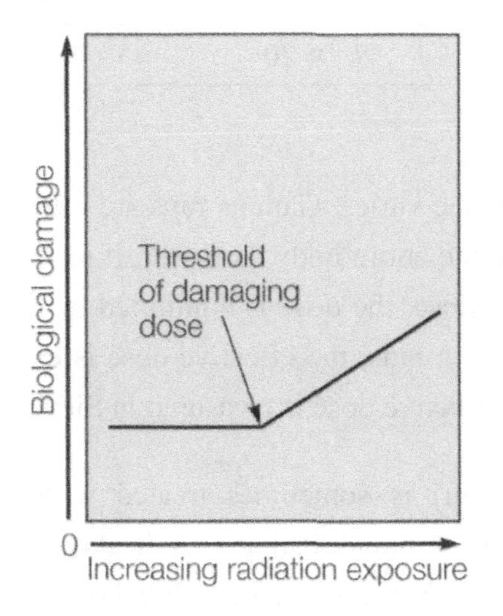

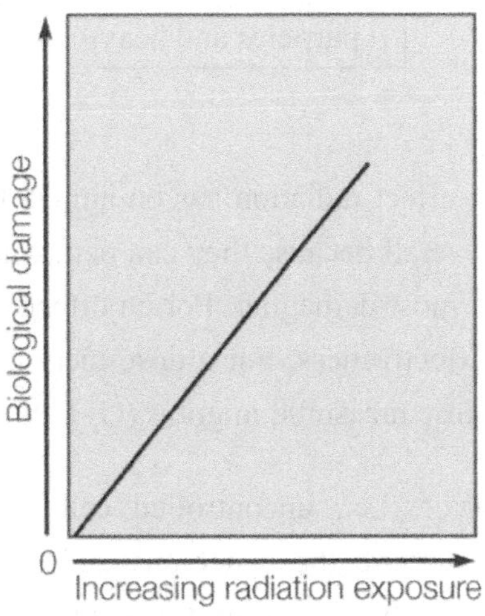

An approximate single dose, whole-body effects of radiation exposure	
Level	**Comment**
0.130 rem	Average annual exposure to natural background radiation
0.500 rem	The upper limit of annual exposure to the public
25.0 rem	The threshold for observable effects such as reduced blood cell count
100.0 rem	Fatigue and other symptoms of radiation sickness
200.0 rem	Definite radiation sickness, bone marrow damage, the possibility of developing leukemia
500.0 rem	The lethal dose for 50 percent of individuals
1,000.0 rem	The lethal dose for all

Radiation damages biological tissue, but it can also be used to treat cancer and other diseases. It is crucial to be able to measure the amount, or dose, of radiation, received.

Relative Biological Effectiveness (RBE)	
Type	**RBE**
X- and γ rays	1
β (electrons)	1
Protons	2
Slow Neutrons	5
Fast Neutrons	≈ 10
α particles and heavy ions	≈ 20

The effect radiation has on human tissue varies. Gamma rays are often the most dangerous overall because they can penetrate the entire body; however, if ingested, alpha rays are the most damaging. For an effective dose, the dose is multiplied by the relative biological effectiveness. For a dose measured in rads, the effective dose is expressed in rem. For a dose measured in grays (Gy), the effective dose is measured in Sieverts (Sv).

Cancer (i.e., uncontrolled cell growth) is sometimes treated with *radiation therapy* to destroy the cancerous cells. For minimal damage to healthy tissue, the radiation source is rotated, so it penetrates through different parts of the body as it targets the tumor (i.e., a mass of abhorrent cells).

Radioactive isotopes are widely used in medicine for diagnostic purposes. They can be used as non-invasive scans or tools to check for unusual concentrations that could signal a tumor or other problem. The radiation is detected with a gamma-ray detector.

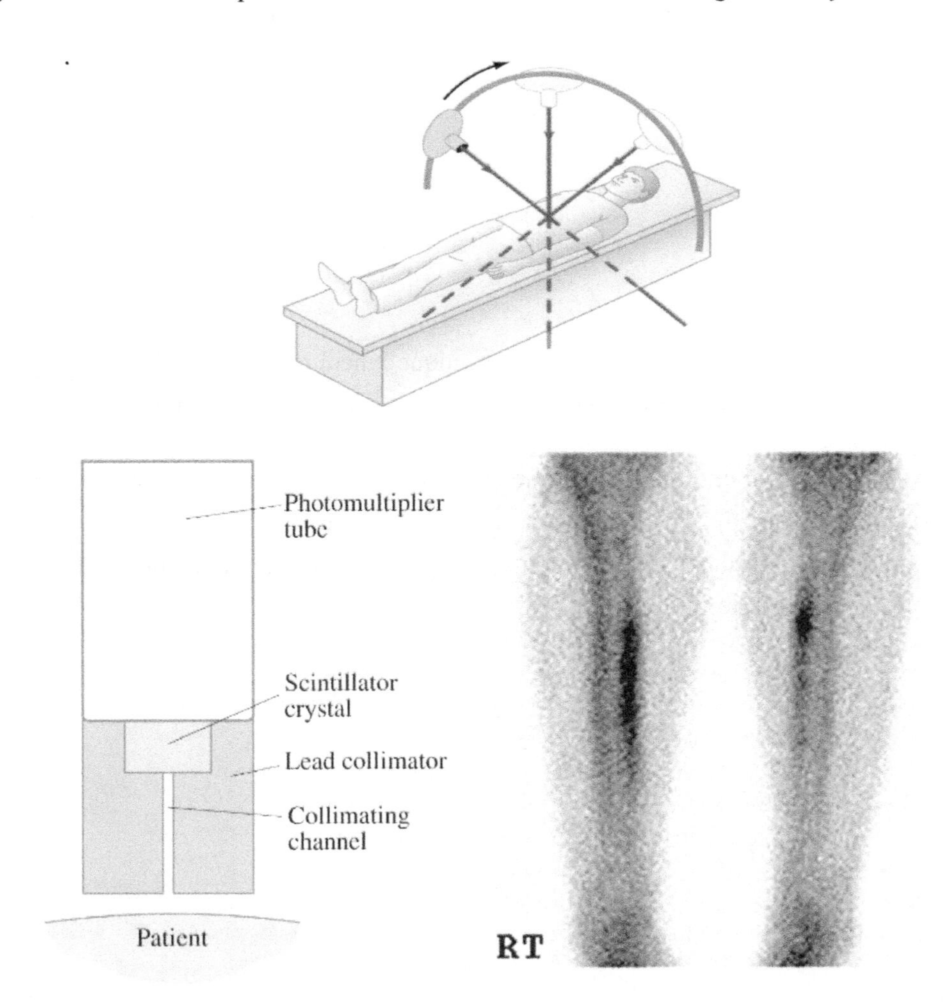

Radioactive tracers can also be detected using tomographic techniques, where a three-dimensional image is gradually built up through successive scans.

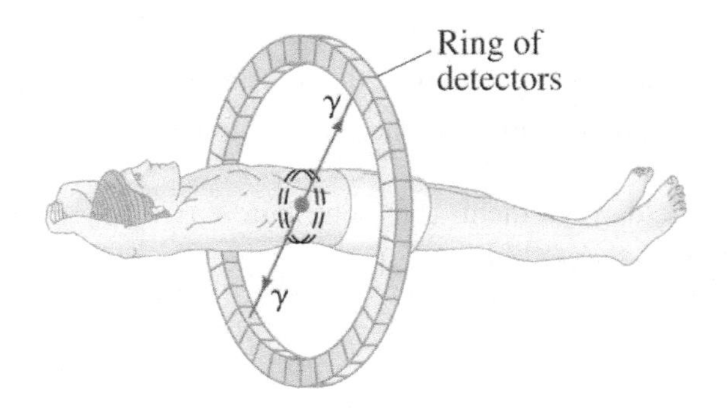

A proton in a magnetic field has its spin either parallel or antiparallel to the field.

The field splits the energy levels slightly; the energy difference is proportional to the field magnitude, where the vector B represents the magnetic field.

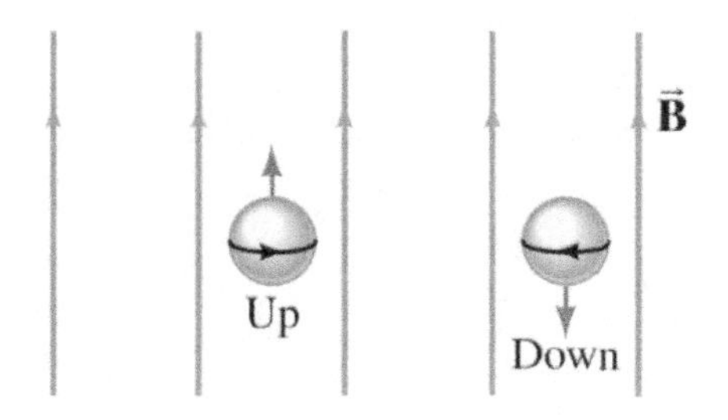

Nuclear magnetic resonance is a technique where the examined object is placed in a static magnetic field, and radio frequency (RF) electromagnetic radiation is applied.

When the radiation has the proper energy to excite the spin-flip transition, many photons are absorbed. The value of the field depends on the local molecular neighborhood. This value of spin-flip transitions allows information about the structure of the molecules to be determined.

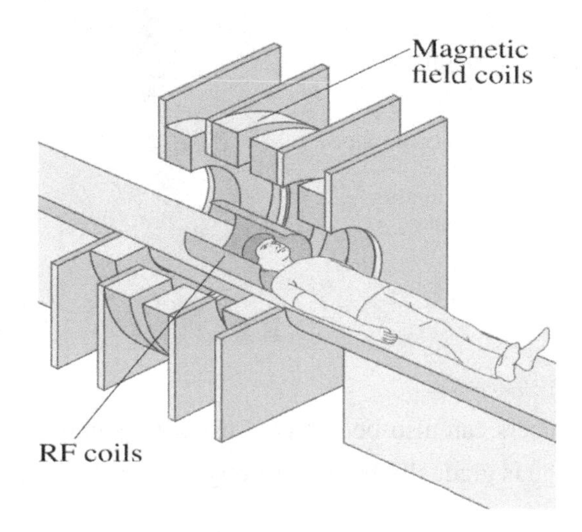

Magnetic resonance imaging works the same way; the transition is excited in hydrogen atoms, which are the most common in the human body.

Giving the field a gradient can contribute to image accuracy, as it allows determining the origin of a particular signal.

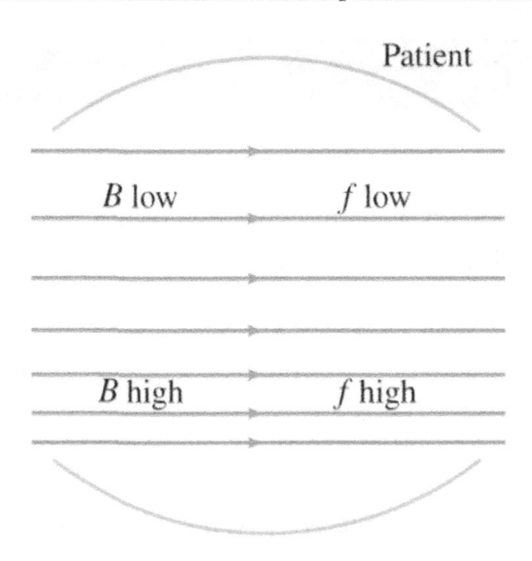

Technique	Optimal Resolution
Conventional X-ray	½ mm
CT scan, X-ray	½ mm
Nuclear medicine (tracers)	1 cm
SPECT (single photon emission)	1 cm
PET (positron emission)	2 – 5 mm
MRI (NMR)	½ – 1 mm
Ultrasound	0.3 – 2 mm

General Nature of Fission

There are two types of nuclear reactions: fission and fusion. Nuclear fission is the process whereby the nucleus of an atom is split into two smaller fragments. A nuclear reaction takes place when another nucleus or particle strikes a nucleus.

Transmutation is when the original nucleus is transformed into another:

$$^{4}_{2}\text{He} + ^{14}_{7}\text{N} \rightarrow ^{17}_{8}\text{O} + ^{1}_{1}\text{H}$$

Energy and momentum must be conserved in nuclear reactions.

$$\text{a} + \text{X} \rightarrow \text{Y} + \text{b}$$

The reaction energy, or Q-value, is the sum of the initial masses less the sum of the final masses, multiplied by c^2:

$$Q = (M_a + M_X - M_b - M_Y)c^2$$

If Q is positive, the reaction is exothermic and occurs regardless of how small the initial kinetic energy is.

If Q is negative, there is a minimum initial kinetic energy that must be available before the reaction can take place.

Neutrons are effective in nuclear reactions, as they have no charge and therefore are not repelled by the nucleus.

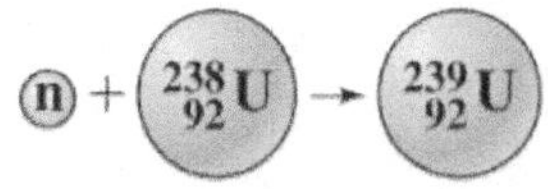

$$\text{n} + ^{238}_{92}\text{U} \rightarrow ^{239}_{92}\text{U}$$

Neutron captured by $^{238}_{92}\text{U}$.

$$^{239}_{92}\text{U} \rightarrow ^{239}_{93}\text{Np} + e^{-} + \bar{\nu}$$

$^{239}_{92}\text{U}$ decays by β decay to neptunium-239.

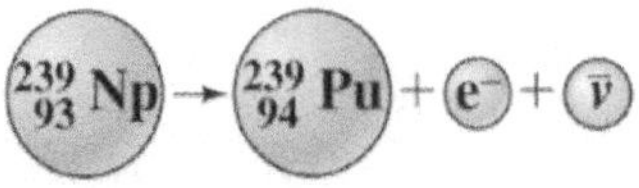

$$^{239}_{93}\text{Np} \rightarrow ^{239}_{94}\text{Pu} + e^{-} + \bar{\nu}$$

$^{239}_{93}\text{Np}$ itself decays by β decay to produce plutonium-239.

Nuclear fission occurs when a nucleus splits apart. The energy released in a fission reaction is tremendous. Also, since smaller nuclei are stable with fewer neutrons, several neutrons emerge from each fission. These neutrons can be used to induce fission in other nuclei, causing a chain reaction.

This occurs when Uranium undergoes fission when struck by a free neutron. The critical mass determines if the nuclei have sufficient mass and concentration to produce a chain reaction. The mass distribution of the fragments shows that the first two pieces are substantial but usually unequal in size.

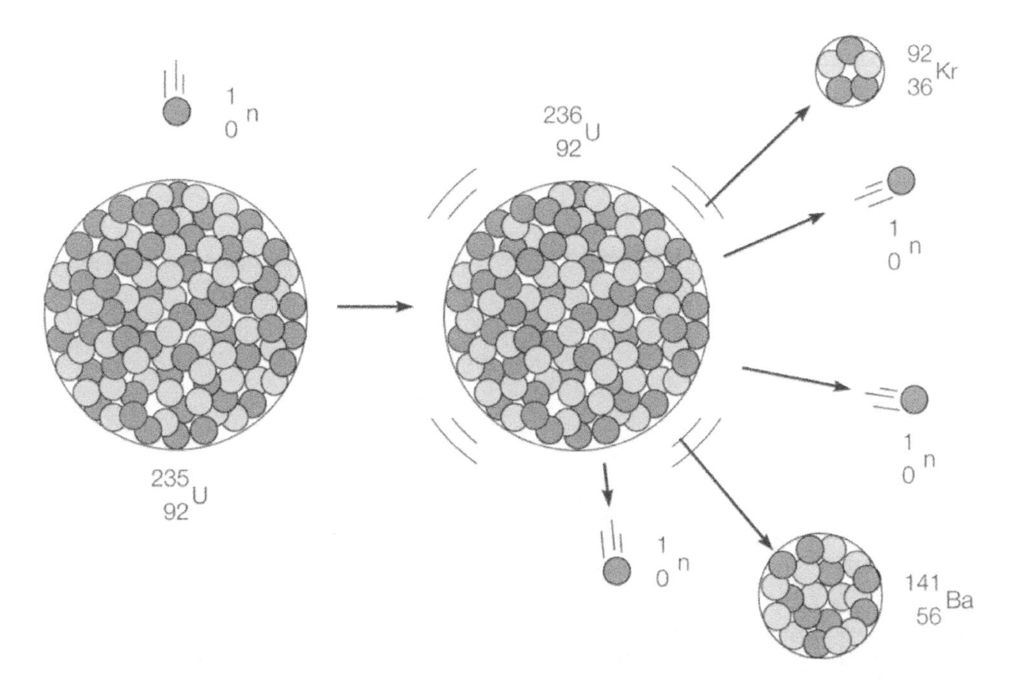

Nuclear energy is an interconversion of mass and energy. Nuclear energy is ultimately connected to the origins of the universe and the life cycles of the starts.

The *Big Bang theory* states that incredibly hot, dense, primordial plasma cooled, creating protons and neutrons. Continued cooling leads to hydrogen atoms, which collapsed gravitationally into 1st generation stars.

The ultimate source of nuclear energy is a gravitational attraction.

The mass deficit is the difference between the masses of the reactants and products.

The binding energy is the energy required to break a nucleus into individual protons and neutrons. The ratio of binding energy to the nucleon number indicates how stable a nucleus is. Iron-56 is the most stable nucleus:

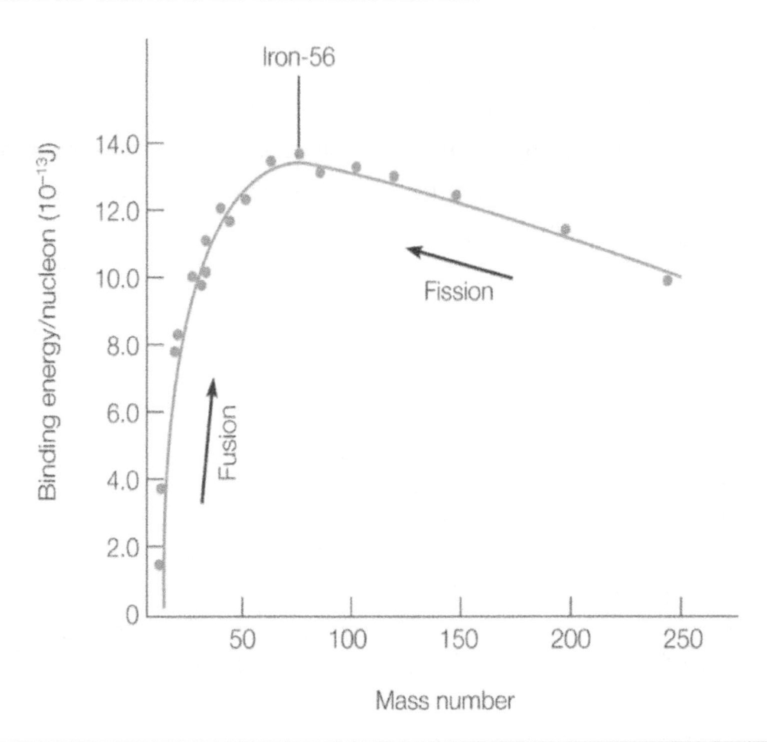

Isotope	Major Mode of Decay	Half-Life
Tritium	Beta	12.26 years
Carbon-14	Beta	5,930 years
Argon-41	Beta, gamma	1.83 hours
Iron-55	Electron capture	2.7 years
Colbalt-58	Beta, gamma	71 days
Colbalt-60	Beta, gamma	5.26 years
Nickel-63	Beta	92 years
Krypton-85	Beta, gamma	10.76 years
Strontium-89	Beta	5.4 days
Strontium-90	Beta	28 years
Yttrium-91	Beta	59 days
Zirconium-93	Beta	9.5×10^5 years
Zirconium-95	Beta, gamma	65 days
Niobium-95	Beta, gamma	35 days

Isotope	Major Mode of Decay	Half-Life
Technetium-99	Beta	2.1×10^5 years
Ruthenium-106	Beta	1 year
Iodine-129	Beta	1.6×10^7 years
Iodine-131	Beta, gamma	8 days
Xenon-133	Beta, gamma	5.27 days
Cesium-134	Beta, gamma	2.1 years
Cesium-135	Beta	2×10^6 years
Cesium-137	Beta	30 years
Cerium-141	Beta	32.5 days
Cerium-144	Beta, gamma	285 days
Promethium-147	Beta	2.6 years
Samarium-151	Beta	90 years
Europium-154	Beta, gamma	16 years
Lead-210	Beta	22 years
Radon-222	Alpha	3.8 days
Radium-226	Alpha, gamma	1,620 years
Thorium-229	Alpha	7,300 years
Thorium-230	Alpha	26,000 years
Uranium-234	Alpha	2.48×10^5 years
Uranium-235	Alpha, gamma	7.13×10^8 years
Uranium-238	Alpha	4.51×10^9 years
Neptunium-237	Alpha	2.14×10^6 years
Plutonium-238	Alpha	89 years
Plutonium-239	Alpha	24,360 years
Plutonium-240	Alpha	6,760 years
Plutonium-241	Beta	13 years
Plutonium-242	Alpha	3.79×10^5 years
Americium-241	Alpha	458 years
Americium-243	Alpha	7,650 years
Curium-242	Alpha	163 days
Curium-244	Alpha	18 years
Thorium-229	Alpha	7,300 years

For a nuclear reactor, shown below, the chain reaction needs to be self-sustaining—it continues indefinitely—but is controlled.

A *moderator* is needed to slow the neutrons; otherwise, their probability of interacting is too small.

Common moderators are heavy water and graphite. Unless the moderator is heavy water, the fraction of fissionable nuclei in natural uranium is too small to sustain a chain reaction, about 0.7%. It needs to be enriched to about 2–3%.

Neutrons that escape from the uranium do not contribute to fission.

There is a critical mass below which a chain reaction does not occur because too many neutrons escape.

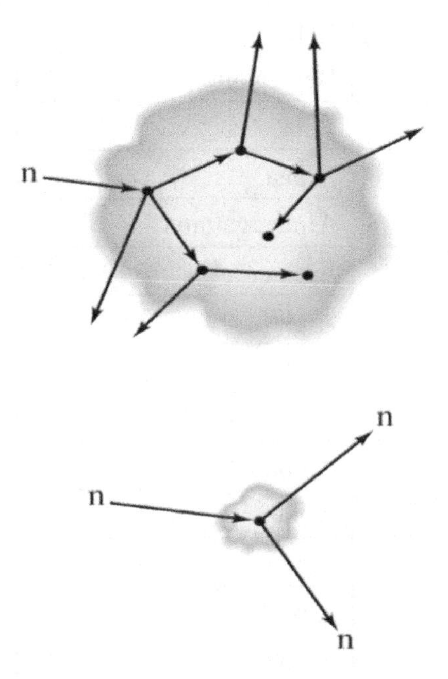

Finally, there are *control rods*, usually cadmium or boron, which absorb neutrons and can be used for fine control of the reaction, to keep it just barely critical.

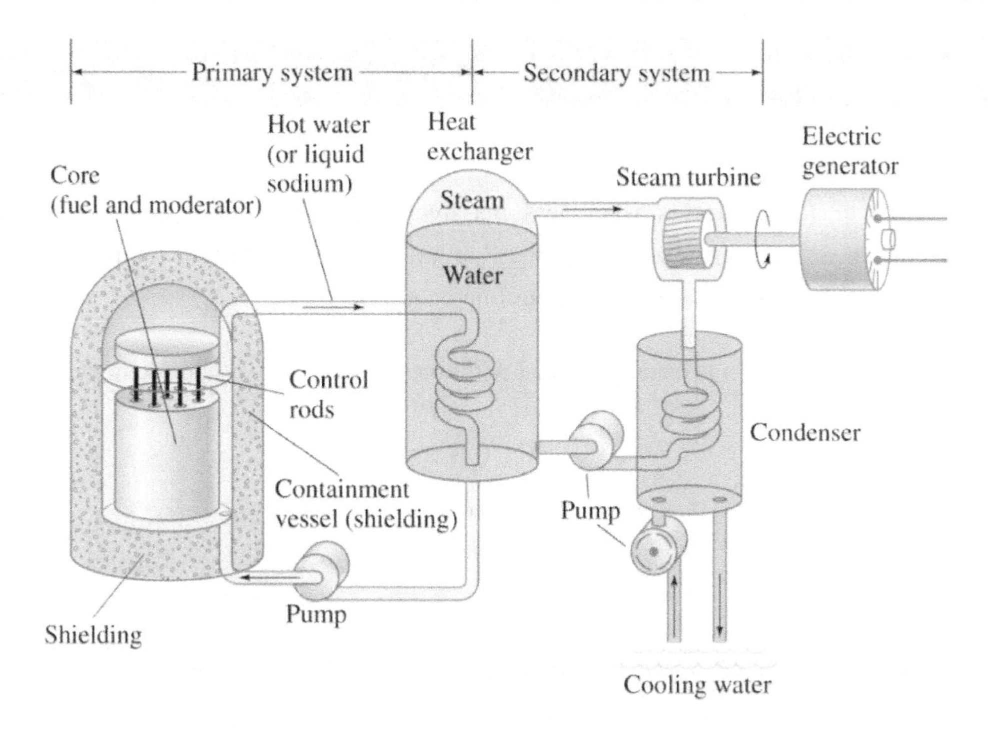

Some problems associated with nuclear reactors include the disposal of radioactive waste and the possibility of an accidental release of radiation (e.g., Chernobyl, Ukraine in 1986 and Fukushima, Japan in 2011).

The atomic bomb also uses fission, but the core is deliberately designed to undergo a massive uncontrolled chain reaction when the uranium is formed into a critical mass during the detonation process.

General Nature of Fusion

Nuclear fusion occurs when less massive nuclei form more massive nuclei. Nuclear fusion is the energy source for the Sun and other stars. Fusion requires a high temperature, high density, and sufficient confinement time.

Controlled fusion is usually studied with magnetic or inertial confinement.

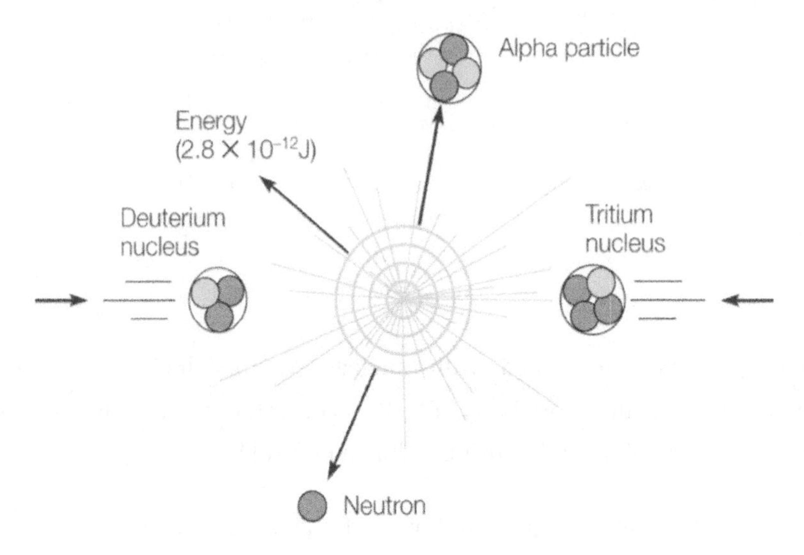

When light nuclei fuse to form heavier nuclei, they release energy in the process.

An example is the sequence of fusion processes that change hydrogen in the Sun into helium.

The fusion reactions are listed with the energy released for each process:

$${}_1^1H + {}_1^1H \rightarrow {}_1^2H + e^+ + v \qquad (0.42\ \text{MeV})$$

$${}_1^1H + {}_1^2H \rightarrow {}_2^3He + \gamma \qquad (5.49\ \text{MeV})$$

$${}_2^3He + {}_2^3He \rightarrow {}_2^4He + {}_1^1H + {}_1^1H \qquad (12.86\ \text{MeV})$$

The net effect of the fusion reaction is to transform four protons into a helium nucleus plus two positrons, two neutrinos, and two gamma rays.

$$4{}_1^1H \rightarrow {}_2^4He + 2e^+ + 2v + 2\gamma$$

More massive stars can fuse heavier elements in their cores, all the way up to iron, the most stable nucleus.

Three fusion reactions are being considered for power reactors:

$$ {}_1^2H + {}_1^2H \rightarrow {}_1^3H + {}_1^1H \qquad (4.03 \ \text{MeV}) $$

$$ {}_1^2H + {}_1^2H \rightarrow {}_2^3He + n \qquad (3.27 \ \text{MeV}) $$

$$ {}_1^2H + {}_1^3H \rightarrow {}_2^4He + n \qquad (17.59 \ \text{MeV}) $$

These fusion reactions use standard fuels—deuterium or tritium—and release much more energy per nucleon than does a fission reaction.

A successful fusion reactor has not yet been achieved, but fusion bombs (thermonuclear bombs) have been built.

Several geometries for the containment of the incredibly hot plasma that results in a fusion reactor have been developed.

Mass Deficit, Energy Liberated, Binding Energy

Balanced equations present nuclear reactions. The charge is always conserved, as is the mass number.

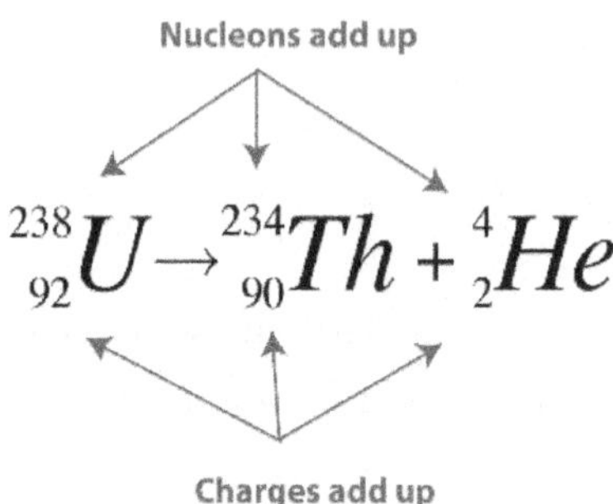

Names, symbols, and properties of particles in nuclear equations			
Name	Symbol	Mass Number	Charge
Proton	^1_1H (or ^1_1p)	1	1+
Electron	$^{0}_{-1}e$ (or $^{0}_{-1}\beta$)	0	1–
Neutron	^1_0n	1	0
Gamma Photon	$^0_0\gamma$	0	0

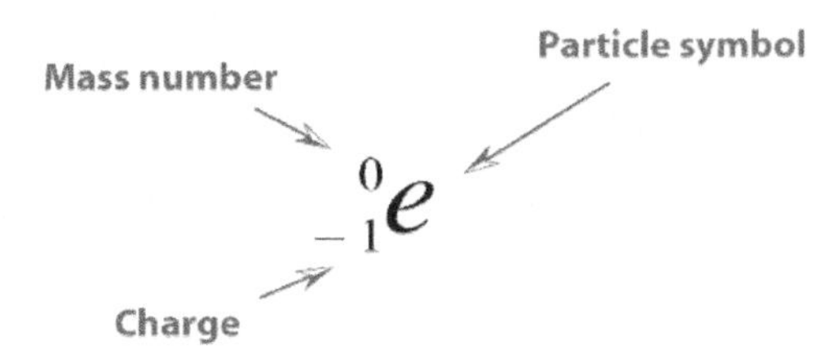

The difference in mass before and after a reaction is the mass deficit or mass defect.

If the total mass before the reaction is different from the total mass after the reaction, then the difference in mass is made up for by energy.

Energy is liberated when mass is lost during a reaction because mass and energy are always conserved. The total mass and energy before a reaction are the same as the total mass and energy after the reaction.

The energy that makes up for the mass deficit is calculated by:

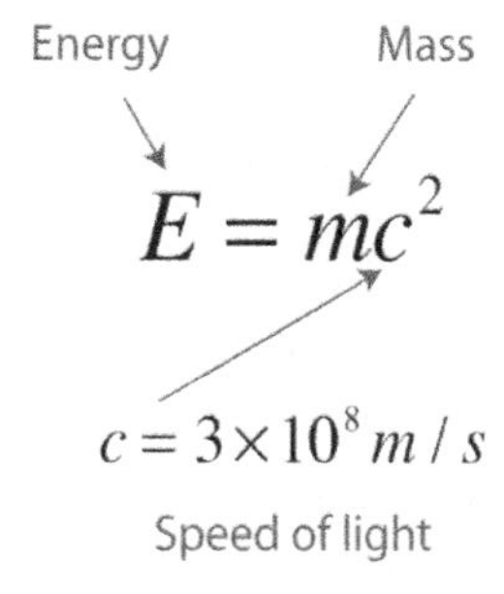

The mass of the nucleons is always higher than the mass of the atom. The mass has become energy, such as radiation or kinetic energy, released during the formation of the nucleus. The difference between the mass of the constituents and the mass of the nucleus is the binding energy of the nucleus.

The energy liberated is equal to the binding energy. The binding energy is used when converting ΔM into its equivalent in energy (ΔMc^2).

$$M_{nucleons} = M_{atom} + \text{binding energy}/c^2$$

Binding energy commonly refers to nuclear binding energy (the energy that binds the nucleons together).

Binding energy per nucleon is strongest for Iron (^{56}Fe), and weakest for Deuterium (the 2-nucleon isotope of hydrogen).

Less commonly used is the electron binding energy. This is because electron binding energy is commonly referred to as the ionization energy.

$$m_{nucleons} - m_{atom} = \text{mass deficit (also mass defect)} = \Delta m$$

Reactant masses

$$2(1.0078u) + 2(1.0087u) = 4.0330u$$

$$2\,{}_1^1H + 2\,{}_0^1n \rightarrow {}_2^4He$$

$4.0026u$

Mass defect = binding energy **Helium mass**

$$\Delta m = 4.0330\ u - 4.0026\ u = 0.0304\ u = 28.3\ \text{Mev}$$

Mass Spectrometer

Mass spectrometry is a technique that identifies the chemicals in a substance. A mass spectrometer measures the mass-to-charge ratio of a sample, as well as the number of ions in gas-phase. This measurement produces a mass spectrum.

A magnetic field can deflect charged particles. A mass spectrometer works by turning atoms into ions, increasing their acceleration, so they all have the same kinetic energy, and deflecting them with a magnetic field. It creates ions by removing electrons from the atom—even if an atom forms typically a negative ion, a mass spectrometer usually works with positive ions.

The particles' paths are deflected, and their mass can be calculated because the force of deflection is known. The curve of their deflection path separates the atoms. The lighter the ion, the more it is deflected.

The higher the positive charge on an ion, the more the magnetic field deflects it. A mass spectrum is then produced, which resembles a bar graph (shown below). The mass spectrum gives the concentration of the atom, organized by their mass-to-charge ratio.

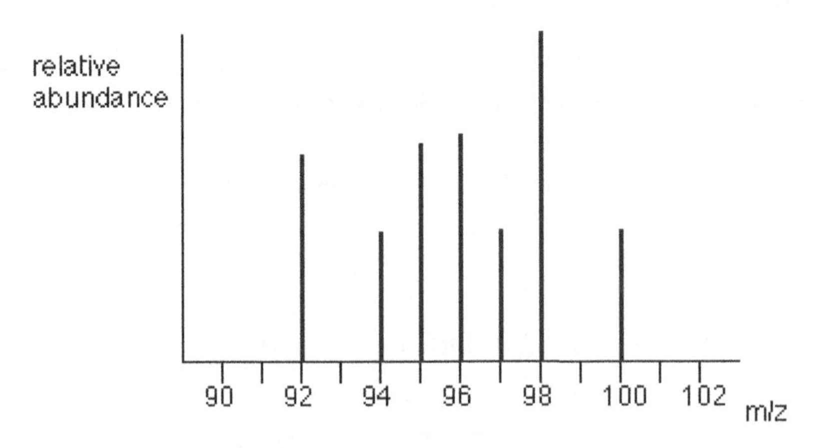

Mass spectrum plotting relative abundance against the mass-charge ratio

Chapter Summary

Atoms

- Nuclei contain protons and neutrons—nucleons.
- The total number of nucleons, A, is the atomic mass number.
- The number of protons, Z, is the atomic number.
- Isotope notation: ${}^{A}_{Z}X$
- Nuclear masses are measured in u; carbon-12 is defined as having a mass of 12 amu.
- 1 amu = 1.6605×10^{-27} kg = 931.5 MeV/c^2
- The difference between the mass of the nucleus and the mass of its constituents is the binding energy.

Decay

- Unstable nuclei decay through alpha, beta or gamma emission.
- An alpha particle is a helium nucleus; a beta particle is an electron or positron; a gamma-ray is a highly energetic photon.
- Nuclei are held together by the strong nuclear force; the weak nuclear force is responsible for beta decay.
- Electric charge, linear and angular momentum, mass-energy, and nucleon number are all conserved.
- Radioactive decay is a statistical process.
- The number of decays per unit time is proportional to the number of nuclei present: $N = N_0e^{-\lambda t}$
- The half-life is the time it takes for half the nuclei to decay.

Nuclear Reactions

- A nuclear reaction occurs when nuclei collide, and different nuclei are produced.

- Reaction energy or Q-value:

$$Q = M_p c^2 - (M_D + m_\alpha)c^2$$

- Fission is when a heavy nucleus splits into two smaller nuclei.
  - Chain reactions occurred when neutrons emitted in a fission trigger more reactions.
  - Critical mass is the minimum mass needed to sustain a chain reaction.
  - A moderator is used to slow the neutrons, so collisions are more likely.

- Fusion is when small nuclei combine to form larger nuclei.
  - Sun's energy comes from fusion reactions.
  - A useful fusion reactor has not yet been built.
  - Radiation damage is measured using dosimetry.
  - The effect of the absorbed dose depends on the type of radiation.

Practice Questions

1. In a nuclear reaction, the mass of the products is less than the mass of the reactants. Why is this not observed in a chemical reaction?

A. In chemical reactions, the mass is held constant by the nucleus
B. In chemical reactions, the mass deficit is balanced by a mass surplus
C. The mass deficit in chemical reactions is too small to be observed
D. The mass does not convert to energy in chemical reactions
E. None of the above are true

2. The isotope ${}^{13}_{7}N$ decays by positron emission to what isotope?

A. ${}^{14}_{6}C$ **B.** ${}^{11}_{7}N$ **C.** ${}^{13}_{6}C$ **D.** ${}^{12}_{6}C$ **E.** ${}^{12}_{7}N$

3. The radioactive gas radon is:

I. more hazardous to smokers than nonsmokers
II. the single greatest source of human radiation exposure
III. a product of the radioactive decay series of uranium

A. I only **B.** II only **C.** III only **D.** I and III only **E.** I, II and III

4. Which of the statements best describes the strong nuclear force?

A. The strength of the force increases with distance
B. The force is powerful and is effective over a large range of distances
C. The electrical force is stronger than the nuclear force
D. The force is powerful but is effective only within a short range of distances
E. None of the above are correct statements

5. Natural line broadening can be understood regarding the:

A. Schrodinger wave equation
B. Pauli exclusion principle
C. de Broglie wavelength
D. uncertainty principle
E. quantum numbers

6. A blue photon has a:

A. longer wavelength than a red photon and travels with a higher speed
B. longer wavelength than a red photon and travels with the same speed
C. shorter wavelength than a red photon and travels with a higher speed
D. longer wavelength than a red photon and travels with a lower speed
E. shorter wavelength than a red photon and travels with the same speed

7. Which type of nuclear radiation is powerful light energy that is *not* deflected as it passes between electrically charged plates?

A. Gamma
C. Alpha
B. Beta
D. Nuclide
E. None of the above

8. The main reason that there is a limit to the size of a stable nucleus is the:

A. weakness of the electrostatic force
B. weakness of the gravitational force
C. short-range effect of the strong nuclear force
D. limited range of the gravitational force
E. long-distance effect of the strong nuclear force

9. Elements combine in fixed mass ratios to form compounds. This fixed mass ratio requires those elements:

A. have unambiguous atomic numbers
B. are always chemically active
C. are composed of continuous matter without subunits
D. are composed of discrete subunits called atoms
E. none of the above are required

10. What is rem?

A. A unit for measuring rapid electron motion
B. The number of radiation particles absorbed per second
C. The number of radiation particles emitted per second
D. The maximum exposure limit for occupational safety
E. A unit for measuring radiation exposure

11. The intensity of X-rays, gamma rays, or any other radiation is:

A. inversely proportional to the square of the distance from the source
B. inversely proportional to the distance from the source
C. directly proportional to the square of the distance from the source
D. directly proportional to the distance from the source
E. inversely proportional to the cube of the distance from the source

12. Ionizing radiation is:

A. a neutron that has acquired a charge, resulting in the formation of an ion
B. high-energy radiation that removes electrons from atoms or molecules
C. radiation that only interacts with ions
D. equivalent to a proton
E. released by ions and reacts with nuclei

13. What nucleus results when ^{55}Ni decays by positron emission?

A. ^{55}Ca **B.** ^{55}Ni **C.** ^{55}Co **D.** ^{55}Fe **E.** None of the above

14. Which of the following types of radiation has the highest energy?

A. γ rays
B. Visible light rays
C. α particles
D. β particles
E. All have the same energy

15. If a star has a peak intensity at 580 nm, what is its temperature? (Use Wien's displacement constant $b = 2.9 \times 10^{-3}$ K·m)

A. 5,000 °C
B. 2,000 °C
C. 5,000 °F
D. 2,000 K
E. 5,000 K

Solutions

1. D is correct.

Mass is conserved in a chemical reaction; no particles are created or destroyed. The atoms are rearranged to form products from reactants.

In nuclear reactions, the mass difference is due to mass converting into energy.

2. C is correct.

Positron emission occurs during β^+.

The decay equation:

$$^{13}_{7}\text{N} \rightarrow {}^{13}_{6}\text{C} + e^+ + \nu_e$$

A proton converts into a neutron, and a positron is ejected with an electron neutrino. The daughter nuclide is ^{13}C.

3. E is correct. Radon gas is a natural decay product of uranium that accounts for the greatest source of yearly radiation exposure in humans. Radon is more hazardous to smokers due to the combined carcinogenic effects of smoking and radiation exposure.

4. D is correct. The strong nuclear force is the strongest of the four fundamental forces. However, it only acts within a small range of distance (about the diameter of the nucleus).

5. D is correct. Natural line broadening is the extension of a spectral line over a range of frequencies. This occurs due in part to the uncertainty principle, which relates the time in which an atom is excited to the energy of its emitted photon.

$$\Delta E \Delta t > \hbar / 2$$

where $\hbar$ is the symbol for the reduced Planck's constant

Because energy is related to frequency by:

$$E = hf$$

where h is Planck's constant

The range of frequencies observed (broadening) is due to the uncertainty in energy outlined by the uncertainty principle.

6. E is correct.

Speed is constant for all electromagnetic waves and only changes due to the transmission medium, not the frequency or wavelength of the wave.

Blue photons have higher energy and thus higher frequencies than red light due to:

$$E = hf$$

However, the frequency is inversely proportional to wavelength:

$$\lambda = c / f$$

Thus, blue photons have shorter wavelengths than red photons due to their higher frequencies.

7. A is correct.

Gamma radiation is a high energy electromagnetic wave, and as such, it has no charge and will not deflect within an electric field.

8. C is correct.

Nuclei with atomic numbers greater than 83 are inherently unstable and thus radioactive. This limit in size is because the strong nuclear force has a short-range, and as the nucleus gets larger, the strong nuclear force cannot overcome the Coulomb repulsion from the protons within the nucleus.

9. D is correct.

Its molar mass can measure a mass of an element or compound. The molar mass relates the mass of the element or compound to a discrete number of subunits (atoms for elements, molecules for compounds).

10. E is correct.

The rem is short for the Roentgen equivalent in man and is designed to measure the biological damage of ionizing radiation. It does not measure the number of particles absorbed or emitted, nor is it the maximum occupational safety exposure limit for radiation.

11. A is correct.

The intensity of electromagnetic radiation concerning distance from the point source:

$$I = S / 4\pi r^2$$

where I = intensity, S = point source strength, and r = radial distance from the point source

The intensity of the radiation from the point source is inversely proportional to the square of the distance from the point source.

12. B is correct. Ionizing radiation can be high energy charged particles (alpha, beta) or high energy electromagnetic waves (X-rays, gamma rays).

All are termed ionizing because they possess enough energy to remove electrons from atoms or molecules and ionize them.

13. C is correct.

Nuclear reaction:

$$^{55}_{28}\text{Ni} \rightarrow {}^{55}_{27}\text{Co} + e^+ + \nu_e$$

where Co = product, e^+ = positron and ν_e = electron neutrino

In positron emission (β^+ decay), a proton in the nucleus converts to a neutron while releasing a positron and an electron neutrino.

The atomic number decreases by one, but the mass number stays constant.

14. A is correct. Gamma rays are high energy electromagnetic waves. They have the highest energy of all radiation (e.g., alpha, beta, and electromagnetic spectrum).

15. E is correct.

Use Wien's Displacement Law:

$$\lambda_{max} = b / T$$

$$T = (2.9 \times 10^{-3}\ \text{K·m}) / (580 \times 10^{-9}\ \text{m})$$

$$T = 5{,}000\ \text{K}$$

Appendix

COMMON PHYSICS FORMULAS & CONVERSIONS

Constants and Conversion Factors

1 unified atomic mass unit	$1\ u = 1.66 \times 10^{-27}$ kg
	$1\ u = 931$ MeV/c^2
Proton mass	$m_p = 1.67 \times 10^{-27}$ kg
Neutron mass	$m_n = 1.67 \times 10^{-27}$ kg
Electron mass	$m_e = 9.11 \times 10^{-31}$ kg
Electron charge magnitude	$e = 1.60 \times 10^{-19}$ C
Avogadro's number	$N_0 = 6.02 \times 10^{23}$ mol^{-1}
Universal gas constant	R = 8.31 J/(mol·K)
Boltzmann's constant	$k_B = 1.38 \times 10^{-23}$ J/K
Speed of light	$c = 3.00 \times 10^8$ m/s
Planck's constant	$h = 6.63 \times 10^{-34}$ J·s
	$h = 4.14 \times 10^{-15}$ eV·s
	$hc = 1.99 \times 10^{-25}$ J·m
	$hc = 1.24 \times 10^3$ eV·nm
Vacuum permittivity	$\varepsilon_0 = 8.85 \times 10^{-12}$ C^2/N·m^2
Coulomb's Law constant	$k = 1/4\pi\varepsilon_0 = 9.0 \times 10^9$ N·m^2/C^2
Vacuum permittivity	$\mu_0 = 4\pi \times 10^{-7}$ (T·m)/A
Magnetic constant	$k' = \mu_0/4\pi = 10^{-7}$ (T·m)/A
Universal gravitational constant	$G = 6.67 \times 10^{-11}$ m^3/kg·s^2
Acceleration due to gravity at Earth's surface	g = 9.8 m/s^2
1 atmosphere pressure	1 atm = 1.0×10^5 N/m^2
	1 atm = 1.0×10^5 Pa
1 electron volt	1 eV = 1.60×10^{-19} J
Balmer constant	$B = 3.645 \times 10^{-7}$ m
Rydberg constant	$R = 1.097 \times 10^7$ m^{-1}
Stefan constant	$\sigma = 5.67 \times 10^{-8}$ W/m^{2}K^4

Units

Name	Symbol
meter	m
kilogram	kg
second	s
ampere	A
kelvin	K
mole	mol
hertz	Hz
newton	N
pascal	Pa
joule	J
watt	W
coulomb	C
volt	V
ohm	Ω
henry	H
farad	F
tesla	T
degree Celsius	°C
electronvolt	eV

Prefixes

Factor	Prefix	Symbol
10^{12}	tera	T
10^{9}	giga	G
10^{6}	mega	M
10^{3}	kilo	k
10^{-2}	centi	c
10^{-3}	milli	m
10^{-6}	micro	μ
10^{-9}	nano	n
10^{-12}	pico	p

Values of Trigonometric Functions for Common Angles

θ	$\sin\theta$	$\cos\theta$	$\tan\theta$
0°	0	1	0
30°	1/2	$\sqrt{3}/2$	$\sqrt{3}/3$
37°	3/5	4/5	3/4
45°	$\sqrt{2}/2$	$\sqrt{2}/2$	1
53°	4/5	3/5	4/3
60°	$\sqrt{3}/2$	1/2	$\sqrt{3}$
90°	1	0	∞

Newtonian Mechanics

Translational Motion	$v = v_0 + a\Delta t$ $x = x_0 + v_0\Delta t + \frac{1}{2}a\Delta t^2$ $v^2 = v_0^2 + 2a\Delta x$ $\vec{a} = \frac{\sum \vec{F}}{m} = \frac{\vec{F}_{net}}{m}$
Rotational Motion	$\omega = \omega_0 + \alpha t$ $\theta = \theta_0 + \omega_0 t + \frac{1}{2}\alpha t^2$ $\omega^2 = \omega_0^2 + 2\alpha\Delta\theta$ $\vec{\alpha} = \frac{\sum \vec{\tau}}{I} = \frac{\vec{\tau}_{net}}{I}$
Force of Friction	$\lvert\vec{F}_f\rvert \leq \mu\lvert\vec{F}_n\rvert$
Centripetal Acceleration	$a_c = \frac{v^2}{r}$
Torque	$\tau = r_{\perp}F = rF\sin\theta$
Momentum	$\Delta\vec{p} = m\vec{v}$
Impulse	$\vec{J} = \Delta\vec{p} = \vec{F}\Delta t$
Kinetic Energy	$K = \frac{1}{2}mv^2$
Potential Energy	$\Delta U_g = mg\Delta y$
Work	$\Delta E = W = F_{\parallel}d$ $= Fd\cos\theta$

a = acceleration

A = amplitude

E = energy

F = force

f = frequency

h = height

I = rotational inertia

J = impulse

K = kinetic energy

k = spring constant

ℓ = length

m = mass

N = normal force

P = power

p = momentum

L= angular momentum

r = radius of distance

T = period

t = time

U = potential energy

v = velocity or speed

W = work done on a system

Power	$P = \frac{\Delta E}{\Delta t} = \frac{\Delta W}{\Delta t}$	x = position
		y = height
Simple Harmonic Motion	$x = A\cos(\omega t) = A\cos(2\pi f t)$	α = angular acceleration
Center of Mass	$x_{cm} = \frac{\sum m_i x_i}{\sum m_i}$	
Function of Time Function of Time Angular Momentum	$L = I\omega$	μ = coefficient of friction
Angular Impulse	$\Delta L = \tau \Delta t$	θ = angle
		τ = torque
Angular Kinetic Energy	$K = \frac{1}{2} I \omega^2$	ω = angular speed
Work	$W = F\Delta r \cos\theta$	
Power	$P = Fv\cos\theta$	
Spring Force	$\lvert\vec{F}_s\rvert = k\lvert\vec{x}\rvert$	
Spring Potential Energy	$U_s = \frac{1}{2}kx^2$	
Period of Spring Oscillator	$T_s = 2\pi\sqrt{m/k}$	
Period of Simple Pendulum	$T_p = 2\pi\sqrt{\ell/g}$	
Period	$T = \frac{2\pi}{\omega} = \frac{1}{f}$	
Gravitational Body Force	$\lvert\vec{F}_g\rvert = G\frac{m_1 m_2}{r^2}$	
Gravitational Potential Energy of Two Masses	$U_G = -\frac{G m_1 m_2}{r}$	

Electricity and Magnetism

Quantity	Formula
Electric Field	$\vec{E} = \frac{\vec{F}_E}{q}$
Electric Field Strength	$\lvert\vec{E}\rvert = \frac{1}{4\pi\varepsilon_0}\frac{\lvert q\rvert}{r^2}$
Electric Field Strength	$\lvert\vec{E}\rvert = \frac{\lvert\Delta V\rvert}{\lvert\Delta r\rvert}$
Electrostatic Force Between Charged Particles	$\lvert\vec{F}_E\rvert = \frac{1}{4\pi\varepsilon_0}\frac{\lvert q_1 q_2\rvert}{r^2}$
Electric Potential Energy	$\Delta U_E = q\Delta V$
Electrostatic Potential due to a Charge	$V = \frac{1}{4\pi\varepsilon_0}\frac{q}{r}$
Capacitor Voltage	$V = \frac{Q}{C}$
Capacitance of a Parallel Plate Capacitor	$C = \kappa\varepsilon_0\frac{A}{d}$
Electric Field Inside a Parallel Plate Capacitor	$E = \frac{Q}{\varepsilon_0 A}$
Capacitor Potential Energy	$U_C = \frac{1}{2}Q\Delta V = \frac{1}{2}C(\Delta V)^2$
Current	$I = \frac{\Delta Q}{\Delta t}$
Resistance	$R = \frac{\rho l}{A}$

A = area

B = magnetic field

C = capacitance

d = distance

E = electric field

ϵ = emf

F = force

I = current

l = length

P = power

Q = charge

q = point charge

R = resistance

r = separation

t = time

U = potential energy

V = electric potential

v = speed

κ = dielectric constant

ρ = resistivity

θ = angle

Φ = flux

Power	$P = I\Delta V$
Current	$I = \frac{\Delta V}{R}$
Resistors in Series	$R_s = \sum_i R_i$
Resistors in Parallel	$\frac{1}{R_p} = \sum_i \frac{1}{R_i}$
Capacitors in Parallel	$C_p = \sum_i C_i$
Capacitors in Series	$\frac{1}{C_s} = \sum_i \frac{1}{C_i}$
Magnetic Field Strength	$B = \frac{\mu_0 I}{2\pi r}$
Magnetic Force	$\vec{F}_M = q\vec{v} \times \vec{B}$ $\vec{F}_M = \lvert q\vec{v}\rvert \lvert sin\,\theta\rvert \lvert\vec{B}\rvert$ $\vec{F}_M = I\vec{l} \times \vec{B}$ $\vec{F}_M = \lvert I\vec{l}\rvert \lvert sin\,\theta\rvert \lvert\vec{B}\rvert$
Magnetic Flux	$\Phi_B = \vec{B} \cdot \vec{A}$ $\Phi_B = \lvert\vec{B}\rvert\, cos\,\theta\, \lvert\vec{A}\rvert$
Electromagnetic Induction	$\epsilon = \frac{-\Delta\Phi_B}{\Delta t}$ $\epsilon = Blv$

Fluid Mechanics and Thermal Physics

Density	$\rho = \frac{m}{V}$	
Pressure	$P = \frac{F}{A}$	
Absolute Pressure	$P = P_0 + \rho g h$	
Buoyant Force	$F_b = \rho V g$	
Fluid Continuity Equation	$A_1 v_1 = A_2 v_2$	
Bernoulli's Equation	$P_1 + \rho g y_1 + \frac{1}{2}\rho v_1^2 = P_2 + \rho g y_2 + \frac{1}{2}\rho v_2^2$	
Heat Conduction	$\frac{Q}{\Delta t} = \frac{kA\Delta T}{d}$	
Thermal Radiation	$P = e\sigma A(T^4 - T_C^4)$	
Ideal Gas Law	$PV = nRT = Nk_B T$	
Average Energy	$K = \frac{3}{2}k_B T$	
Work	$W = -P\Delta V$	
Conservation of Energy	$\Delta E = Q + W$	
Linear Expansion	$\Delta l = a l_o \Delta T$	
Heat Engine Efficiency	$n_c = \lvert W/Q_H \rvert$	
Carnot Heat Engine Efficiency	$n_c = \frac{T_H - T_C}{T_H}$	
Energy of Temperature Change	$Q = mc\Delta T$	
Energy of Phase Change	$Q = mL$	

A = area
c = specific heat
d = thickness
e = emissivity
F = force
h = depth
k = thermal conductivity
K = kinetic energy
l = length
L = latent heat
m = mass
n = number of moles
n_c *= efficiency*
N = number of molecules
P = pressure or power
Q = energy transferred to a system by heating
T = temperature
t = time
E = internal energy
V = volume
v = speed
W = work done on a system
y = height
σ = Stefan constant
α = coefficient of linear expansion
ρ = density

Optics

Wavelength to Frequency	$\lambda = \frac{v}{f}$	
Index of Refraction	$n = \frac{c}{v}$	
Snell's Law	$n_1 \sin\theta_1 = n_2 \sin\theta_2$	
Thin Lens Equation	$\frac{1}{s_i} + \frac{1}{s_0} = \frac{1}{f}$	
Magnification Equation	$\lvert M \rvert = \left\lvert \frac{h_i}{h_o} \right\rvert = \left\lvert \frac{s_i}{s_o} \right\rvert$	
Double Slit Diffraction	$d \sin\theta = m\lambda$	
	$\Delta L = m\lambda$	
Critical Angle	$\sin\theta_c = \frac{n_2}{n_1}$	
Focal Length of Spherical Mirror	$f = \frac{R}{2}$	

d = separation

f = frequency or focal length

h = height

L = distance

M = magnification

m = an integer

n = index of refraction

R = radius of curvature

s = distance

υ = speed

x = position

λ = wavelength

θ = angle

Acoustics

Standing Wave/ Open Pipe Harmonics	$\lambda = \frac{2L}{n}$	
Closed Pipe Harmonics	$\lambda = \frac{4L}{n}$	
Harmonic Frequencies	$f_n = nf_1$	
Speed of Sound in Ideal Gas	$v_{sound} = \sqrt{\frac{yRT}{M}}$	
Speed of Wave Through Wire	$v = \sqrt{\frac{T}{m/L}}$	
Doppler Effect (Approaching Stationary Observer)	$f_{observed} = (\frac{v}{v - v_{source}})f_{source}$	
Doppler Effect (Receding Stationary Observer)	$f_{observed} = (\frac{v}{v + v_{source}})f_{source}$	
Doppler Effect (Observer Moving towards Source)	$f_{observed} = (1 + \frac{v_{observer}}{v})f_{source}$	
Doppler Effect (Observer Moving away from Source)	$f_{observed} = (1 - \frac{v_{observer}}{v})f_{source}$	

f = frequency

L = length

m = mass

M = molecular mass

n = harmonic number

R = gas constant

T = tension

v = velocity

y = adiabatic constant

λ = wavelength

Modern Physics

Photon Energy	$E = hf$	B = *Balmer constant*
		c = *speed of light*
Photoelectric Electron Energy	$K_{max} = hf - \phi$	E = *energy*
		f = *frequency*
Electron Wavelength	$\lambda = \frac{h}{p}$	K = *kinetic energy*
		m = *mass*
Energy Mass Relationship	$E = mc^2$	p = *momentum*
Rydberg Formula	$\frac{1}{\lambda} = R(\frac{1}{n_f^2} - \frac{1}{n_i^2})$	R = *Rydberg constant*
		v = *velocity*
Balmer Formula	$\lambda = B(\frac{n^2}{n^2 - 2^2})$	λ = *wavelength*
		ϕ = *work function*
Lorentz Factor	$\gamma = \frac{1}{\sqrt{1 - \frac{v^2}{c^2}}}$	γ = *Lorentz factor*

Geometry and Trigonometry

Rectangle	$A = bh$	A = *area*
		C = *circumference*
Triangle	$A = \frac{1}{2}bh$	V = *volume*
		S = *surface area*
Circle	$A = \pi r^2$	b = *base*
	$C = 2\pi r$	h = *height*
		l = *length*
Rectangular Solid	$V = lwh$	w = *width*
Cylinder	$V = \pi r^2 l$	r = *radius*
	$S = 2\pi rl + 2\pi r^2$	θ = angle
Sphere	$V = \frac{4}{3}\pi r^3$	
	$S = 4\pi r^2$	
Right Triangle	$a^2 + b^2 = c^2$	
	$\sin\theta = \frac{a}{c}$	
	$\cos\theta = \frac{b}{c}$	
	$\tan\theta = \frac{a}{b}$	

Glossary

A

Absolute humidity (or saturation value) – the maximum amount of water vapor that could be present in 1 m^3 of the air at any given temperature.

Absolute magnitude – a classification scheme which compensates for the differences in the distance to stars; calculates the brightness that stars would appear to have if they were all at a defined, standard distance of 10 parsecs.

Absolute scale – temperature scale set so that zero is the theoretically lowest temperature possible (this would occur when all random motion of molecules has ceased).

Absolute zero – the theoretically lowest temperature possible, at which temperature the molecular motion vanishes; –273.16 °C or 0 K.

Absorptance – the ratio of the total absorbed radiation to the total incident radiation.

Acceleration – the rate of change of velocity of a moving object with respect to time; the SI units are m/s^2; by definition, this change in velocity can result from a change in speed, a change in direction, or a combination of changes in both speed and direction.

Acceleration due to gravity – the acceleration produced in a body due to the Earth's attraction; denoted by the letter g (SI unit – m/s^2); on the surface of the Earth, its average value is 9.8 m/s^2; increases when going towards the poles from the equator; decreases with altitude and with depth inside the Earth; the value of g at the center of the earth is zero.

Achromatic – capable of transmitting light without decomposing it into its constituent colors.

Acoustics – the science of the production, transmission, and effects of sound.

Acoustic shielding – a sound barrier that prevents the transmission of acoustic energy.

Adiabatic – any change in which there is no gain or loss of heat.

Adiabatic cooling – the decrease in temperature of an expanding gas that involves no additional heat flowing out of the gas; the cooling from the energy lost by expansion.

Adiabatic heating – the increase in temperature of the compressed gas that involves no additional heat flowing into the gas; the heating from the energy gained by compression.

Afocal lens – a lens of zero convergent power whose focal points are infinitely distant.

Air mass – a large, uniform body of air with the nearly same temperature and moisture conditions throughout.

Albedo – the fraction of the total light incident on a reflecting surface, especially a celestial body, which is reflected in all directions.

Allotropic forms – elements that can have several different structures with different physical properties (e.g., graphite and diamond).

Alpha (α) particle – the nucleus of a helium atom (two protons and two neutrons) emitted as radiation from a decaying heavy nucleus (α-decay).

Alternating current – an electric current that first moves in one direction, then in the opposite direction with a regular frequency.

Amorphous – term that describes solids that have neither definite form nor structure.

Amp – unit of electric current; equivalent to coulomb/second.

Ampere – the full name of the unit amp; the SI unit of electric current; one ampere is the flow of one coulomb of charge per second.

Amplitude – the maximum absolute value attained by the disturbance of a wave or by any quantity that varies periodically.

Amplitude (of an oscillation) – the maximum displacement of a body from its mean position during an oscillatory motion.

Amplitude (of waves) – the maximum displacement of particles of the medium from their mean positions during the propagation of a wave.

Angle of contact – the angle between tangents to the liquid surface and the solid surface inside the liquid; both the tangents are drawn at the point of contact.

Angle of incidence – the angle of an incident (arriving) ray or particle to a surface; measured from a line perpendicular to the surface (the normal).

Angle of reflection – the angle of a reflected ray or particle from a surface; measured from a line perpendicular to the surface (the normal).

Angle of refraction – the angle between the refracted ray and the normal.

Angle of repose – the angle of inclination of a plane with the horizontal such that a body placed on the plane is on the verge of sliding but does not.

Angstrom – a unit of length; 1 = 10^{-10} m.

Angular acceleration – the rate of change of angular velocity of a body moving along a circular path; denoted by a.

Angular displacement – the angle described at the center of the circle by a moving body along a circular path. It is measured in radians.

Angular momentum – also called a moment of momentum; the cross-product of position vector and momentum.

Angular momentum quantum number – from the quantum mechanics model of the atom, one of four descriptions of the energy state of an electron wave; describes the energy sublevels of electrons within the principal energy levels of an atom.

Angular velocity – the rate of change of angular displacement per unit of time.

Annihilation – a process in which a particle and an antiparticle combine and release their rest energies in other particles.

Antineutrino – the antiparticle of neutrino; has zero mass and spin ½.

Archimedes principle – a body immersed in a fluid experiences an apparent loss of weight which is equal to the weight of the fluid displaced by the body.

Astronomical unit – the radius of the Earth's orbit is defined as one astronomical unit (A.U.).

Atom – the smallest unit of an element that can exist alone or in combination with other elements.

Atomic mass unit – relative mass unit (amu) of an isotope based on the standard of the carbon-12 isotope; one atomic mass unit (1 amu) = 1/12 the mass of a ^{12}C atom = 1.66×10^{-27} Kg.

Atomic number – the number of protons in the nucleus of an atom.

Atomic weight – weighted average of the masses of stable isotopes of an element as they occur in nature; based on the abundance of each isotope of the element and the atomic mass of the isotope compared to carbon-12.

Avogadro's number – the number of carbon-12 atoms in exactly 12.00 g of C that is 6.02×10^{23} atoms or other chemical units; the number of chemical units in one mole of a substance.

Avogadro's Law – under the same conditions of temperature and pressure, equal volumes of all gases contain an equal number of molecules.

Axis – the imaginary line about which a planet or other object rotates.

B

Background radiation – ionizing radiation (e.g., alpha, beta, gamma rays) from natural sources.

Balanced forces – when some forces act on a body and the resultant force is zero; see *Resultant forces*.

Balmer lines – lines in the spectrum of the hydrogen atom in the visible range; produced by the transition between n 2 and n = 2, with n being the principal quantum number.

Balmer series – a set of four-line spectra; narrow lines of color emitted by hydrogen atom electrons as they drop from excited states to the ground state.

Bar – a unit of pressure; equal to 10^5 Pascal.

Barometer – an instrument that measures atmospheric pressure; used in weather forecasting and determining elevation above sea level.

Baryon – subatomic particle composed of three quarks.

Beat – a phenomenon of the periodic variation in the intensity of sound due to the superposition of waves differing slightly in frequency; rhythmic increases and decreases of volume from constructive and destructive interference between two sound waves of slightly different frequencies.

Bernoulli's theorem – states that the total energy per unit volume of a non-viscous, incompressible fluid in a streamline flow will remain constant.

Beta (β) particle – high-energy electron emitted as ionizing radiation from a decaying nucleus (β-decay); also a beta ray.

Big bang theory – the current model of galactic evolution in which the universe is assumed to have been created by an intense and brilliant explosion from a primeval fireball.

Binding energy – the net energy required to break a nucleus into its constituent protons and neutrons; also, the energy equivalent released when a nucleus is formed.

Black body – an ideal body which would absorb all incident radiation and reflect none.

Black body radiation – electromagnetic radiation emitted by an ideal material (the black body) that perfectly absorbs and perfectly emits radiation.

Black hole – the remaining theoretical core of a supernova that is so dense that even light cannot escape.

Bohr model – model of the structure of the atom that attempted to correct the deficiencies of the solar system model and account for the Balmer series.

Boiling point – the temperature at which a phase change of liquid to gas takes place through boiling; the same temperature as the condensation point.

Boundary – the division between two regions of differing physical properties.

Boyle's Law – for a given mass of a gas at a constant temperature, the volume of the gas is inversely proportional to the pressure.

Brewster's Law – states that the refractive index of a material is equal to the tangent of the polarizing angle for the material.

British thermal unit (Btu) – the amount of energy or heat needed to increase the temperature of one pound of water one-degree Fahrenheit.

Brownian motion – the continuous random motion of solid microscopic particles when suspended in a fluid medium due to their ongoing bombardment by atoms and molecules.

Bulk modulus of elasticity – the ratio of normal stress to the volumetric strain produced in a body.

Buoyant force – the upward force on an object immersed in a fluid.

C

Calorie – a unit of heat; 1 Calorie = 4.186 joule.

Candela – the SI unit of luminous intensity defined as the luminous intensity in each direction of a source that emits monochromatic photons of frequency 540×10^{12} Hz and has a radiant intensity in that direction of 1/683 W/sr.

Capacitance – the ratio of the charge stored per increase in potential difference.

Capacitor – an electrical device used to store charge and energy in the electrical field.

Capillarity – the rise or fall of a liquid in a tube of a fine bore.

Carnot's theorem – no engine operating between two temperatures can be more efficient than a reversible engine working between the same two temperatures.

Cathode rays – negatively charged particles (electrons) that are emitted from a negative terminal in an evacuated glass tube.

Celsius scale of temperature – the ice-point is taken as the lower fixed point (0 °C), and the steam-point is taken as the upper fixed point (100 °C); the interval between the ice-point and the steam-point is divided into 100 equal divisions; the unit division on this scale is 1 °C; previously called the centigrade scale; the relationship relates the temperatures on the Celsius scale and the Fahrenheit scale, $C/100 = (F - 32) / 180$; the temperature of a healthy person is 37 °C or 98.6 °F.

Centrifugal force – an apparent outward force on an object in circular motion; a consequence of the third law of motion.

Centripetal force – the radial force required to keep an object moving in a circular path.

Chain reaction – a self-sustaining reaction where some of the products can produce more reactions of the same kind (e.g., in a nuclear chain reaction, neutrons are the products that produce more nuclear reactions in a self-sustaining series).

Charles' Law – for a given mass of a gas at constant pressure, the volume is directly proportional to the temperature.

Chromatic aberration – an optical lens defect causing color fringes due to the lens bringing different colors of light to focus at different points.

Circular motion – the motion of a body along a circular path.

Closed system – the system which cannot exchange heat or matter with the surroundings.

Coefficient of areal expansion – the fractional change in surface area per degree of temperature change; see *Coefficient of thermal expansion.*

Coefficient of linear expansion – the fractional change in length per degree of temperature change; see *Coefficient of thermal expansion.*

Coefficient of thermal expansion – the fractional change in the size of an object per degree of change in temperature at constant pressure; the SI unit is K^{-1}.

Coefficient of volumetric expansion – the fractional change in volume per degree of temperature change; see *Coefficient of thermal expansion.*

Coherent source – a source in which there is a constant phase difference between waves emitted from different parts of the source.

Compression – a part of a longitudinal wave in which the density of the particles of the medium is higher than the typical density.

Compressive stress – a force that tends to compress the surface as the Earth's plates move into each other.

Condensation (sound) – a compression of gas molecules; a pulse of increased density and pressure that moves through the air at the speed of sound.

Condensation (water vapor) – where more vapor or gas molecules are returning to the liquid state than are evaporating.

Condensation nuclei – tiny particles such as tiny dust, smoke, soot, or salt crystals suspended in the air on which water condenses.

Condensation point – the temperature at which a gas or vapor changes back to liquid; see *Boiling point.*

Conduction – the transfer of heat from a region of higher temperature to a region of lower temperature by increased kinetic energy moving from molecule to molecule.

Constructive interference – the condition in which two waves are arriving at the same place at the same time and in phase add amplitudes to create a new wave.

Control rods – rods inserted between fuel rods in a nuclear reactor to absorb neutrons and control the rate of the nuclear chain reaction.

Convection – transfer of heat from a region of higher temperature to a region of lower temperature by the displacement of high-energy molecules (e.g., the displacement of warmer, less dense air (higher kinetic energy) by cooler, denser air (lower kinetic energy)).

Conventional current – the opposite of electron current; considers an electric current to consist of a drift of positive charges that flow from the positive terminal to the negative terminal of a battery.

Coulomb – a unit used to measure the quantity of electric charge; equivalent to the charge resulting from the transfer of 6.24 billion particles such as the electron.

Coulomb's Law – the relationship between charge, distance, and magnitude of the electrical force between two bodies; the force between any two charges is directly proportional to the product of charges and inversely proportional to the square of the distance between the charges.

Covalent bond – a chemical bond formed by the sharing of a pair of electrons.

Covalent compound – chemical compound held together by a covalent bond or bonds.

Crest – the point of maximum positive displacement on a transverse wave.

Critical angle – the limit to the angle of incidence when all light rays are reflected internally.

Critical mass – the mass of fissionable material needed to sustain a chain reaction.

Curvilinear motion – the motion of a body along a curved path.

Cycle – a complete vibration.

Cyclotron – a device used to accelerate the charged particles.

D

De-acceleration – negative acceleration when the velocity of a body decreases with time.

Decibel – unit of the sound level; if P1 & P2 are two amounts of power, the first is said to be n decibels greater, where n = 10 log10 (P1/P2).

Decibel scale – a nonlinear scale of loudness based on the ratio of the intensity level of a sound to the intensity at the threshold of hearing.

Density – the mass of a substance per unit volume.

Destructive interference – the condition in which two waves arriving at the same point at the same time out of phase add amplitudes that cancel to create zero total disturbance; see *Constructive interference*.

Dewpoint temperature – the temperature at which condensation begins.

Dew – condensation of water vapor into droplets of liquid on surfaces.

Diffraction – the bending of light around the edge of an opaque object.

Diffuse reflection – light rays reflected in many random directions, as opposed to the parallel rays reflected from a perfectly smooth surface such as a mirror.

Diopter – unit of measure of the refractive power of a lens.

Direct current – an electrical current that always flows in one direction.

Direct proportion – when two variables increase or decrease together in the same ratio (at the same rate).

Dispersion – the splitting of white light into its component colors of the spectrum.

Displacement – a vector quantity for the change in the position of an object as it moves in a particular direction; also the shortest distance between the initial position and the final position of a moving body.

Distance – a scalar quantity for the length of the path traveled by a body irrespective of the direction it goes in.

Doppler effect – an apparent change in the frequency of sound or light due to the relative motion between the source of the sound or light and the observer.

E

Echo – a reflected sound that can be distinguished from the original sound, usually arriving 0.1 s or more after the original sound.

Einstein mass-energy relation – $E = mc^2$; E is the energy released, m is the mass defect, and c is the speed of light.

Elastic potential energy – the potential energy of a body by its configuration (i.e., shape).

Elastic strain – an adjustment to stress in which materials recover their original shape after stress is released.

Electric circuit – consists of a voltage source that maintains an electrical potential, a continuous conducting path for a current to follow, and a device where the electrical potential does work; a switch in the circuit is used to complete or interrupt the conducting path.

Electric current – the flow of electric charge; the electric force field produced by an electrical charge.

Electric field line – an imaginary curve tangent to which at any given point gives the direction of the electric field at that point.

Electric field lines – a map of an electric field representing the direction of the force that a test charge would experience; the direction of an electric field shown by lines of force.

Electric generator – a mechanical device that uses wire loops rotating in a magnetic field to produce electromagnetic induction to generate electricity.

Electric potential energy – potential energy due to the position of a charge near other charges.

Electrical conductors – materials that have electrons that are free to move throughout the material (e.g., metals); allows electric current to flow through the material.

Electrical energy – a form of energy from electromagnetic interactions.

Electric force – a fundamental force that results from the interaction of electrical charges; it is the most powerful force in the universe.

Electrical insulators – electrical nonconductors, or materials that obstruct the flow of electric current.

Electrical nonconductors – materials that have electrons that do not move easily within the material (e.g., rubber); also called electrical insulators.

Electrical resistance – the property of opposing or reducing electric current.

Electrolyte – water solution of ionic substances that will conduct an electric current.

Electromagnet – a magnet formed by a solenoid that can be turned on and off by turning the current on and off.

Electromagnetic force – one of four fundamental forces; the force of attraction or repulsion between two charged particles.

Electromagnetic induction – the process in which current is induced in a coil whenever there is a change in the magnetic flux linked with the coil.

Electromagnetic waves – the waves which are due to oscillating electrical and magnetic fields and do not need any material medium for their propagation; can travel through a material medium (e.g., light waves and radio waves); travel in a vacuum with a speed of 3.0×10^8 m/s.

Electron – a subatomic particle that has the smallest negative charge possible; usually found in an orbital of an atom but is gained or lost when atoms become ions.

Electron configuration – the arrangement of electrons in orbits and sub-orbits about the nucleus of an atom.

Electron current – the opposite of conventional current; considers electric current to consist of a drift of negative charges that flows from the negative terminal to the positive terminal of a battery.

Electron pair – a pair of electrons with different spin quantum numbers that may occupy an orbital.

Electron volt – the energy gained by an electron moving across a potential difference of one volt; equal to 1.60×10^{-19} Joules.

Electronegativity – the comparative ability of atoms of an element to attract bonding electrons.

Electrostatic charge – an accumulated electric charge on an object from a surplus of electrons or a deficiency of electrons.

Element – a pure chemical substance that cannot be broken down into anything simpler by chemical or physical means; there are over 100 known elements, the fundamental materials of which all matter is made.

Endothermic process – the process in which heat is absorbed.

Energy – the capacity of a body to do work; a scalar quantity; the SI unit is the Joule; there are five forms: mechanical, chemical, radiant, electrical, and nuclear.

Escape velocity – the minimum velocity with which an object must be thrown upward to overcome the gravitational pull and escape into space; the escape velocity depends on the mass and radius of the planet/star, but not on the mass of the body being thrown upward.

Evaporation – a process of more molecules leaving a liquid for the gaseous state than returning from the gas to the liquid; can occur at any given temperature from the surface of a liquid; takes place only from the surface of the liquid; causes cooling; faster if the surface of the liquid is large, the temperature is higher, and the surrounding atmosphere does not contain a large amount of vapor of the liquid.

Exothermic process – the process in which heat is evolved.

F

Fahrenheit scale of temperature – the ice-point (lower fixed point) is taken as 32 °F, and the steam-point (upper fixed point) is taken as 212 °F; the interval between these two points is divided into 180 equal divisions; the unit division on the Fahrenheit scale is 1 °F; the relationship relates the temperatures on the Celsius scale and the Fahrenheit scale, C/100 = (°F – 32) / 180; the temperature of a healthy person is 37 °C or 98.6 °F.

Farad – the SI unit of capacitance; the capacitance of a capacitor that, if charged to 1 C, has a potential difference of 1 V.

Faraday – the electric charge required to liberate a gram equivalent of a substance; 1 Faraday = 9,6485 coulomb/mole.

Fermat's principle – an electromagnetic wave takes a path that involves the least time when propagating between two points.

First Law of Motion – every object remains at rest or in a state of uniform straight-line motion unless acted on by an unbalanced force.

Fluid – matter that can flow or be poured; the individual molecules of fluid can move, rolling over or by one another.

Focus – the point to which rays that are initially parallel to the axis of a lens or mirror converge or from which they appear to diverge.

Force – a push or pull which tends to change the state of rest or uniform motion, the direction of motion or the shape and size of a body; a vector quantity; the SI unit is a Newton, denoted by N; one N is the force which when acting on a body of mass 1 kg produces an acceleration of 1 m/s^2.

Force of gravitation – the force with which two objects attract by their masses; acts even if the two objects are not connected; an action-at-a-distance force.

Fracture strain – an adjustment to stress in which materials crack or break because of the stress.

Fraunhofer lines – the dark lines in the spectrum of the sun or a star.

Freefall – the motion of a body falling to Earth with no other force except the force of gravity acting on it; all free-falling bodies are weightless.

Freezing point – the temperature at which a phase change of liquid to solid takes place; the same temperature as the melting point for a given substance.

Frequency – the number of oscillations completed in 1 second by an oscillating body.

Frequency (of oscillations) – the number of oscillations made by an oscillating body per second.

Frequency (of waves) – the number of waves produced per second.

Friction – the force that resists the motion of one surface relative to another with which it is in contact; caused by the humps and crests of surfaces, even those on a microscopic scale; the area of contact is small, and the consequent high pressure leads to local pressure welding of the surface; in motion, the welds are broken and remade continually.

Fuel rod – long zirconium alloy tubes are containing fissionable material for use in a nuclear reactor.

Fundamental charge – the smallest common charge known; the magnitude of the charge of an electron and a proton, which is 1.60×10^{-19} coulombs.

Fundamental frequency – the lowest frequency (longest wavelength) at which a system vibrates freely and can set up standing waves in an air column or on a string.

Fundamental properties – a property that cannot be defined in more straightforward terms other than to describe how it is measured; the fundamental properties are length, mass, time, and charge.

G

g – a symbol representing the acceleration of an object in free fall due to the force of gravity; its magnitude is 9.80 m/s^2.

Gamma (γ) ray – a high energy photon of short wavelength electromagnetic radiation emitted by decaying nuclei (γ-decay).

Gases – a phase of matter composed of molecules that are relatively far apart moving freely in constant, random motion and have weak cohesive forces acting between them, resulting in the characteristic indefinite shape and indefinite volume of a gas.

Graham's Law of Diffusion – the rate of diffusion of a gas is inversely proportional to the square root of its density.

Gram-atomic weight – the mass in grams of one mole of an element that is numerically equal to its atomic weight.

Gram-formula weight – the mass in grams of one mole of a compound that is numerically equal to its formula weight.

Gram-molecular weight – the gram-formula weight of a molecular compound.

Gravitational constant G – appears in the equation for Newton's Law of Gravitation; numerically, it is equal to the force of gravitation, which acts between two bodies with a mass of 1 kg each separated by a distance of 1 m; the value of G is 6.67×10^{-11} Nm^2/kg^2.

Gravitational potential at a point – the amount of work done against the gravitational forces to move a particle of unit mass from infinity to that point.

Gravitational potential energy – the potential energy possessed by a body by its height from the ground; equals *mgh*.

Gravity – the gravitational attraction at the surface of a planet or other celestial body.

Greenhouse effect – the process of increasing the temperature of the lower parts of the atmosphere through redirecting energy back toward the surface; the absorption and re-emission of infrared radiation by carbon dioxide, water vapor and a few other gases in the atmosphere.

Ground state – the energy state of an atom with its electrons at the lowest energy state possible for that atom.

H

Half-life – the time required for one-half of the unstable nuclei in a radioactive substance to decay into a new element.

Heat – a form of energy that makes a body hot or cold; measured by the temperature-effect, it produces in any material body; the SI unit is the Joule (J).

Heisenberg uncertainty principle – states that there is a fundamental limit to the precision with which certain pairs of physical properties of a particle (i.e., complementary variables) can be known simultaneously (e.g., one cannot measure both the exact momentum and the exact position of a subatomic particle at the same time – the more one is certain of one, the less certain one can be of the other).

Hertz – unit of frequency (Hz); equivalent to one cycle per second.

Hooke's Law – within the elastic limit, stress is directly proportional to strain.

Horsepower – unit of power; 1 hp = 746 Watts.

Humidity – the ratio of water vapor in a sample of air to the volume of the sample.

Huygens' principle – each point on a light wavefront can be regarded as a source of secondary waves, the envelope of these secondary waves determining the position of the wavefront later.

Hypothesis – a tentative explanation of a phenomenon that is compatible with the data and provides a framework for understanding and describing that phenomenon.

I

Ice-point – the melting point of ice under 1 atm pressure; equal to 0 °C or 32 °F.

Ideal gas equation – PV = nRT.

Impulse – equal to the product of the force acting on a body and the time for which it acts; if the force is variable, the impulse is the integral of Fd_t from t_0 to t_1; the impulse of a force acting for a given time interval is equal to change in momentum produced over that interval; $J = m(v - u)$, assuming that the mass m remains constant while the velocity changes from v to u; the SI units are kg m/s.

Impulsive force – the force which acts on a body for a short time but produces a large change in the momentum of the body.

Incandescent – matter emitting visible light because of high temperature (e.g., a light bulb, a flame from any burning source, the Sun).

Incident ray – line representing the direction of motion of incoming light approaching a boundary.

Index of refraction – the ratio of the speed of light in a vacuum to the speed of light in a material.

Inertia – the property of matter that causes it to resist any change in its state of rest or of uniform motion; there are three kinds of inertia: the inertia of rest, the inertia of motion and the inertia of direction; the mass of a body is a measure of its inertia.

Infrasonic – sound waves at a frequency below the range of human hearing (less than 20 Hz).

Insulators – materials that are poor conductors of heat or electricity (e.g., wood or glass); materials with air pockets slow down the movement of heat because the air molecules are far apart.

Intensity – a measure of the energy carried by a wave.

Interference – the redistribution of energy due to the superposition of waves with a phase difference from coherent sources, resulting in alternate light and dark bands.

Intermolecular forces – forces of interaction between molecules.

Internal energy – the sum of the kinetic energy and potential energy of all molecules of an object.

Inverse proportion – the relationship in which the value of one variable increase while the value of a second variable decreases at the same rate (in the same ratio).

Ionization – a process of forming ions from molecules.

Ionized – an atom or a particle that has a net charge because it has gained or lost electrons.

Isobaric process – in which pressure remains constant.

Isochoric process – in which volume remains constant.

Isostasy – a balance or equilibrium between adjacent blocks of Earth's crust.

Isothermal process – in which temperature remains constant.

Isotope – atoms of the same element with the same atomic number (i.e., number of protons) but with a different mass number (i.e., number of neutrons).

J

Joule – the unit used to measure work and energy; can also be used to measure heat; 1 J = 1N·m.

Joule's Law of Heating – states that the heat produced when a current (I) flows through a resistor (R) for a given time (t) is given by $Q = I^2Rt$.

K

Kelvin scale of temperature – the ice-point (the lower fixed point) is taken as 273.15 K, and the steam-point (the upper fixed point) is taken as 373.15 K; the interval between these two points is divided into 100 equal parts; each division is equal to 1K.

Kelvin's statement of Second Law of Thermodynamics – it is impossible that, at the end of a cycle of changes, heat has been extracted from a reservoir, and an equal amount of work has been produced without producing some other effect.

Kepler's Laws of Planetary Motion – the three laws describing the motion of the planets.

Kepler's First Law – in planetary motion, each planet moves in an elliptical orbit, with the Sun located at one focus.

Kepler's Second Law – a radius vector between the Sun and a planet moves over equal areas of the ellipse during equal time intervals.

Kepler's Third Law – the square of the period of an orbit is directly proportional to the cube of the radius of the major axis of the orbit.

Kilocalorie – the amount of energy required to raise the temperature of 1 kg of water by 1 °C; 1 Kcal = 1,000 calories.

Kilogram – the fundamental unit of mass in the metric system of measurement.

Kinetic energy – energy possessed by a body due to its motion; $KE = ½mv^2$, where m is mass and v is velocity.

L

Laser – a device that produces a coherent stream of light through stimulated emission of radiation.

Latent heat – energy released or absorbed by a body during a constant-temperature phase change.

Latent heat of vaporization – the heat absorbed when one gram of a substance changes from the liquid phase to the gaseous phase; also, the heat released when one gram of gas changes from the gaseous phase to the liquid phase.

Latent heat of fusion – the quantity of heat required to convert one unit mass of a substance from a solid state to a liquid state at its melting point without any change in its temperature; the SI unit is $J\ kg^{-1}$.

Latent heat of sublimation – the quantity of heat required to convert one unit of mass of a substance from a solid state to a gaseous state without any change in its temperature.

Law of Conservation of Energy – states that energy can neither be created nor destroyed but can be transformed from one form to another.

Law of Conservation of Mass – states that mass (including single atoms) can neither be created nor destroyed in a chemical reaction.

Law of Conservation of Matter – matter can neither be created nor destroyed in a chemical reaction.

Law of Conservation of Momentum – the total momentum of a group of interacting objects remains constant in the absence of external forces.

Lenz's Law – the induced current always flows in such a direction that it opposes the cause producing it.

Light-year – the distance that light travels in a vacuum in one year (365.25 days); approximately 9.46×10^{15} m.

Line spectrum – an emission (of light, sound, or other radiation) spectrum consisting of separate isolated lines (discrete frequencies or energies); can be used to identify the elements in a matter of unknown composition.

Lines of force – lines drawn to make an electric field strength map, with each line originating on a positive charge and ending on a negative charge; each line represents a path on which a charge would experience a constant force; having the lines closer indicates a more energetic electric field.

Liquids – a phase of matter composed of molecules that have interactions stronger than those found in gas but not strong enough to keep the molecules near the equilibrium positions of a solid, resulting in the characteristic definite volume but the indefinite shape of a liquid.

Liter – a metric system unit of volume; usually used for liquids.

Longitudinal strain – the ratio of change in the length of a body to its initial length.

Longitudinal waves – the wave in which the particles of the medium oscillate along the direction of propagation of a wave (e.g., sound waves).

Loudness – a subjective interpretation of a sound that is related to the energy of the vibrating source, related to the condition of the transmitting medium and the distance involved.

Luminosity – the total amount of energy radiated into space each second from the surface of a star.

Luminous – an object or objects that produce visible light (e.g., the Sun, stars, light bulbs, burning materials).

Lyman series – a group of lines in the ultraviolet region in the spectrum of hydrogen.

M

Magnetic domain – tiny physical regions in permanent magnets, approximately 0.01 to 1 mm, that have magnetically aligned atoms, giving the domain an overall polarity.

Magnetic field – the region around a magnet where other magnetic objects experience its magnetic force; a model used to describe how magnetic forces on moving charges act at a distance.

Magnetic poles – the ends, or sides, of a magnet about which the force of magnetic attraction seems to be concentrated.

Magnetic quantum number – from the quantum mechanics model of the atom, one of four descriptions of the energy state of an electron wave; describes the energy of an electron orbital as the orbital is oriented in space by an external magnetic field, a kind of energy sub-sublevel.

Magnetic reversal – the changing of polarity of the Earth's magnetic field as the north magnetic pole and the south magnetic pole exchange positions.

Magnetic wave – the spread of magnetization from a small portion of a substance where an abrupt change in the magnetic field has taken place.

Magnification – the ratio of the size of the image to the size of the object.

Magnitude – the size of a measurement of a vector; scalar quantities that consist of a number and unit only.

Malus Law – the intensity of the light transmitted from the analyzer varies directly as the square of the cosine of the angle between the plane of transmission of the analyzer and the polarizer.

Maser – microwave amplification by stimulated emission of radiation.

Mass – the quantity of matter contained in a body; the SI unit is the kg; remains the same everywhere; a measure of inertia, which means resistance to a change of motion.

Mass defect – the difference between the sum of the masses of the individual nucleons forming a nucleus and the mass of that nucleus.

Mass number – the sum of the number of protons and neutrons in a nucleus; used to identify isotopes (e.g., Uranium-238).

Matter – anything that occupies space and has mass.

Mean life – the average time during which a system, such as an atom or a nucleus, exists in a specified form.

Mechanical energy – the sum of the potential energy and the kinetic energy of a body; energy associated with the position of a body.

Mechanical wave – those waves that need a material medium for their propagation (e.g., sound waves and water waves); also elastic waves.

Megahertz – unit of frequency; equal to 106 Hertz.

Melting point – the temperature at which a phase change of solid to liquid takes place.

Metal – matter having the physical properties of conductivity, malleability, ductility, and luster.

Meter – the fundamental metric unit of length.

MeV – unit of energy; equal to 1.6×10^{-13} joules.

Millibar – a measure of atmospheric pressure equivalent to 1,000 dynes per cm^2.

Miscible fluids – fluids that can mix in any proportion.

Mixture – matter made of unlike parts that have a variable composition and can be separated into their parts by physical means.

Model – a mental or physical representation of something that cannot be observed directly; used as an ai to understand.

Modulus of elasticity – the ratio of stress to the strain produced in a body.

Modulus of rigidity – the ratio of tangential stress to the shear strain produced in a body.

Mole – the amount of a substance that contains Avogadro's number of atoms, ions, molecules, or any other chemical unit; 6.02×10^{23} atoms, ions, or other chemical units.

Momentum – a measure of the quantity of motion in a body; the product of the mass and the velocity of a body; SI units are kg·m /s.

Monochromatic light – consisting of a single wavelength.

N

Natural frequency – the frequency of oscillation of an elastic object in the absence of external forces; depends on the size, composition, and shape of the object.

Negative electric charge – one of the two types of electric charge; repels other negative charges and attracts positive charges.

Negative ion – atom or particle that has a surplus or imbalance of electrons and a negative charge.

Net force – the resulting force after all vector forces have been added; if a net force is zero, all the vector forces have canceled, and there is not an unbalanced force.

Newton (N) – a unit of force defined as kg·m/s^2; 1 Newton is needed to accelerate a 1 kg mass by 1 m/s^2.

Newton's First Law of Motion – a body continues in a state of rest or of uniform motion in a straight line unless it is acted upon by an external (unbalanced) force.

Newton's Law of Gravitation – the gravitational force of attraction acting between any two particles is directly proportional to the product of their masses and inversely proportional to the square of the distance between them; the force of attraction acts along the line joining the two particles; real bodies having spherical symmetry act as point masses with their mass assumed to be concentrated at their center of mass.

Newton's Second Law of Motion – the rate of change of momentum is equal to the force applied; the force acting on a body is directly proportional to the product of its mass and acceleration produced by force in the body.

Newton's Third Law of Motion – for every action, there is an equal and opposite reaction; the action and the reaction act on two different bodies simultaneously.

Noise – sounds made up of groups of waves of random frequency and intensity.

Non-uniform acceleration – when the velocity of a body increases by unequal amounts in equal intervals of time.

Non-uniform speed – when a body travels unequal distances in equal intervals of time.

Non-uniform velocity – when a body covers unequal distances in equal intervals of time in a direction, or when it covers equal distances in equal intervals but changes its direction.

Normal – a line perpendicular to the surface of a boundary.

Nuclear energy – the form of energy from reactions involving the nucleus.

Nuclear fission – the splitting of a heavy nucleus into more stable, lighter nuclei with an accompanying release of energy.

Nuclear force – one of four fundamental forces; a strong force of attraction that operates over short distances between subatomic particles; overcomes the electric repulsion of protons in a nucleus and binds the nucleus together.

Nuclear fusion – a nuclear reaction of low mass nuclei fusing to form a more stable and more massive nucleus with an accompanying release of energy.

Nuclear reactor – a steel vessel in which a controlled chain reaction of fissionable materials releases energy.

Nucleons – a collective name for protons and neutrons in the nucleus of an atom.

Nucleus – the central, positively charged, dense portion of an atom; contains protons and neutrons.

O

Ohm – unit of resistance; 1 ohm = 1volt/ampere.

Ohm's Law – states that the current flowing through a conductor is directly proportional to the potential difference across the ends of the conductor.

Open system – a system across whose boundaries both matter and energy can pass.

Optical fiber – a long, thin thread of fused silica; used to transmit light; based on total internal reflection.

Orbital – the region of space around the nucleus of an atom where an electron is likely to be found.

Origin – the only point on a graph where the x and the y variables both have a value of zero at the same time.

Oscillatory motion – the to and fro motion (periodic) of a body about its mean position; also, vibratory motion.

P

Pascal – a unit of pressure equal to the pressure resulting from a force of 1 N acting uniformly over an area of 1 m^2.

Pascal's Law – states that the pressure exerted on a liquid is transmitted equally in all directions.

Paschen series – a group of lines in the infrared region in the spectrum of hydrogen.

Pauli exclusion principle – no two electrons in an atom can have the same four quantum numbers; a maximum of two electrons can occupy a given orbital.

Peltier effect – the evolution or absorption of heat at the junction of two dissimilar metals carrying current.

Period (of a wave) – the time taken by a wave to travel through a distance equal to its wavelength; denoted by T; period of a wave = 1/frequency of the wave.

Period (of an oscillation) – the time taken to complete one oscillation; does not depend upon the mass of the bob and amplitude of oscillation; directly proportional to the square root of the length and inversely proportional to the square root of the acceleration due to gravity.

Periodic wave – a wave in which the particles of the medium oscillate continuously about their mean positions regularly at fixed intervals of time.

Periodic motion – a motion which repeats itself at regular intervals of time.

Permeability – the ability to transmit fluids through openings, small passageways or gaps.

Phase – when the particles in a wave are in the same state of vibration (i.e., in the same position and the same direction of motion).

Phase change – the action of a substance changing from one state of matter to another; always absorbs or releases internal potential energy that is not associated with a temperature change.

Photons – quanta of energy in the light wave; the particle associated with light.

Photoelectric effect – the emission of electrons in some materials when the light of a suitable frequency falls on them.

Physical change – a change of the state of a substance but not in the identity of the substance.

Planck's constant – proportionality constant in the ratio of the energy of vibrating molecules to their frequency of vibration; a value of 6.63×10^{-34} J·s.

Plasma – a phase of matter; a hot highly ionized gas consisting of electrons and atoms that have been stripped of their electrons because of high kinetic energies.

Plasticity – the property of a solid whereby it undergoes a permanent change in shape or size when subjected to a stress.

Plastic strain – an adjustment to stress in which materials become molded or bent out of shape under stress and do not return to their original shape after the stress is released.

Polarized Light – light whose constituent transverse waves are all vibrating in the same plane.

Polaroid – a film that transmits only polarized light.

Polaroid or polarizer – a device that produces polarized light.

Positive electric charge – one of the two types of electric charge; repels other positive charges and attracts negative charges.

Positive ion – atom or particle that has a net positive charge due to an electron or electrons being torn away.

Positron – an elementary particle having the same mass as that of an electron but an equal and positive charge.

Potential Energy – energy possessed by a body by its position or configuration; see the *Gravitational potential energy* and *Elastic potential energy*.

Power – scalar quantity for the rate of doing work; the SI unit is Watt; 1 W = 1 J/s.

Pressure – a measure of force per unit area (e.g., kilograms per square meter (kg/m^2).

Primary coil – part of a transformer; a coil of wire connected to a source of alternating current.

Primary colors – three colors (red, yellow, and blue), which can be combined in various proportions to producė any other color.

Principal quantum number – from the quantum mechanics model of the atom, one of four descriptions of the energy state of an electron wave; describes the main energy level of an electron regarding its most probable distance from the nucleus.

Principle of calorimetry – states that if two bodies of different temperature are in thermal contact, and no heat is allowed to go out or enter into the system, then heat lost by the body with higher temperature is equal to the heat gained by the body of lower temperature (i.e., the heat lost = the heat gained).

Progressive wave – a wave that transfers energy from one part of a medium to another.

Projectile – an object thrown into space either horizontally or at an acute angle and under the action of gravity; the path followed by a projectile is its trajectory; the horizontal distance traveled by a projectile is its range; the time is taken from the moment it is thrown until the moment it hits the ground is its time of flight.

Proof – a measure of ethanol concentration of an alcoholic beverage; double the concentration by volume (e.g., 50% by volume is 100 proof).

Properties – qualities or attributes that, taken together, are usually unique to an object (e.g., color, texture, and size).

Proportionality constant – a constant applied to a proportionality statement that transforms the statement into an equation.

Pulse – a wave of short duration confined to a small portion of the medium at any given time; also a wave pulse.

Q

Quanta – fixed amounts; usually referring to fixed amounts of energy absorbed or emitted by matter.

Quantum limit – the shortest wavelength; present in a continuous x-ray spectrum.

Quantum mechanics – model of the atom based on the wave nature of subatomic particles and the mechanics of electron waves; also wave mechanics.

Quantum numbers – numbers that describe the energy states of an electron; in the Bohr model of the atom, the orbit quantum numbers could be any whole number (e.g., 1, 2, 3, etc.); in the quantum mechanics model of the atom, four quantum numbers are used to describe the energy state of an electron wave (n, m, l, and s).

Quark – one of the hypothetical fundamental particles; has a charge with magnitudes of one-third or two-thirds of the charge on an electron.

R

Rad – a measure of radiation received by a material (radiation-absorbed dose).

Radiant energy – the form of energy that can travel through space (e.g., visible light and other parts of the electromagnetic spectrum).

Radiation – the emission and propagation of waves transmitting energy through space or some medium.

Radioactive decay – the natural, spontaneous disintegration or decomposition of a nucleus.

Radioactive decay constant – a specific constant for a isotope that is the ratio of the rate of nuclear disintegration per unit of time to the total number of radioactive nuclei.

Radioactive decay series – series of decay reactions that begins with one radioactive nucleus that decays to a second nucleus that decays to a third nucleus and so on, until a stable nucleus is reached.

Radioactive Decay Law – the rate of disintegration of a radioactive substance is directly proportional to the number of undecayed nuclei.

Radioactivity – spontaneous emission of particles or energy from an atomic nucleus as it disintegrates.

Rarefaction – a part of a longitudinal wave where the density of the particles of the medium is less tha the typical density.

Real image – an image generated by a lens or mirror that can be projected onto a screen.

Rectilinear motion – the motion of a body in a straight line.

Reflected ray – a line representing the direction of motion of light reflected from a boundary.

Refraction – the bending of a light wave, a sound wave, or another wave from its straight-line path as it travels from one medium to another.

Refractive index – the ratio of the speed of light in a vacuum to that in the medium.

Relative density – (i.e., *specific gravity*) is the ratio of the density (mass of a unit volume) of a substance to the density of given reference material. *Specific gravity* usually means relative density concerning water. The term *relative density* is more common in modern scientific usage.

Relative humidity – the percentage of the amount of water vapor present in a specific volume of the air to the amount of water vapor needed to saturate it.

Resolving power – a quantitative measure of the ability of an optical instrument to produce separable images of different points of an object.

Resonance – when the frequency of an external force matches the natural frequency of the body.

Restoring force – the force which tends to bring an oscillating body back to its mean position whenever it is displaced from the mean position.

Resultant force – a single force, which acts on a body to produce the same effect on it as done by all other forces collectively; see *Balanced forces*.

Reverberation – apparent increase in the volume of sound caused by reflections from the boundary surfaces, usually arriving within 0.1 seconds after the original sound.

Rigid body – an idealized extended body whose size and shape are fixed and remains unaltered when forces are applied.

S

Saturated air – air in which an equilibrium exists between evaporation and condensation; the relative humidity will be 100 percent.

Saturated solution – the apparent limit to dissolving a given solid in a specified amount of water at a given temperature; a state of equilibrium that exists between dissolving solute and solute coming out of solution.

Scalar quantity – a physical quantity described entirely by its magnitude.

Scientific law – a relationship between quantities; usually described by an equation in the physical sciences; describes a broader range of phenomena and is more important than a scientific principle.

Scientific principle – a relationship between quantities concerned with a specific or narrow range of observations and behavior.

Second – the standard unit of time in both the metric and English systems of measurement.

Second Law of Motion – the acceleration of an object is directly proportional to the net force acting on that object and inversely proportional to the mass of the object.

Secondary coil – part of a transformer; a coil of wire in which the voltage of the original alternating current in the primary coil can be stepped up or down by way of electromagnetic induction.

Second's pendulum – a simple pendulum whose period on the surface of the Earth is 2 seconds.

Semiconductors – elements whose electrical conductivity is intermediate between that of a conductor and an insulator.

Shear strain – the ratio of the relative displacements of one plane to its distance from the fixed plane.

Shear stress – the restoring force developed per unit area when deforming force acts tangentially to the surface of a body, producing a change in the shape of the body without any change in volume.

Siemens – the derived SI unit of electrical conductance; equal to the conductance of an element that has a resistance of 1 ohm; also written as ohm^{-1}.

Simple harmonic motion – the vibratory motion that occurs when the restoring force is proportional to the displacement from the mean position and is directed opposite to the displacement.

Simple pendulum – a heavy point mass (a small metallic ball) suspended by a light inextensible string from frictionless rigid support; a simple machine based on the effect of gravity.

Snell's Law – states that the ratio of *sin* i to *sin* r is a constant and is equal to the refractive index of the second medium concerning the first.

Solenoid – a cylindrical coil of wire that becomes electromagnetic when current passes through it.

Solids – a phase of matter with molecules that remain close to fixed equilibrium positions due to strong interactions between the molecules, resulting in the characteristic definite shape and definite volume of a solid.

Sonic boom – sound waves that pile up into a shock wave when a source is traveling at or faster than the speed of sound.

Specific gravity – see *Relative density*.

Specific heat – the amount of heat energy required to increase the temperature of 1 g of a substance by 1 °C; each substance has its specific heat value.

Speed – a scalar quantity for the distance traveled by a body per unit of time; if a body covers the distance in time, then its speed is given by distance/time; SI units are m/s.

Spin quantum number – from the quantum mechanics model of the atom, one of four descriptions of the energy state of an electron wave; describes the spin orientation of an electron relative to an external magnetic field.

Standing waves – the condition where two waves of equal frequency traveling in opposite directions meet and form stationary regions of maximum displacement due to constructive interference and stationary regions of zero displacement due to destructive interference.

State of motion – when a body changes its position concerning a fixed point in its surroundings; the states of rest and motion are relative to the frame of reference.

State of rest – when a body does not change its position concerning a fixed point in its surrounding; the states of rest and motion are relative to the frame of reference.

Steam-point – the temperature of steam over pure boiling water under 1 atm pressure; taken as the upper fixed point (100 °C or 212 °F) for temperature scales.

Stefan-Boltzmann Law – the amount of energy radiated per second per unit area of a perfectly black body, is directly proportional to the fourth power of the absolute temperature of the surface of the body.

Superconductors – some materials in which, under certain conditions, the electrical resistance approaches zero.

Super-cooled – water in the liquid phase when the temperature is below the freezing point.

Supersaturated – containing more than the average saturation amount of a solute at a given temperature.

Surface tension – the property of a liquid due to which its surface behaves like a stretched membrane.

T

Temperature – a numerical measure of the hotness or coldness of a body; according to the molecular model, it is a measure of the average kinetic energy of the molecules of the body; heat flows from a body at a higher temperature to a body at a lower temperature.

Tensional stress – the opposite of compressional stress; occurs when one part of a plate moves away from another part that does not move.

Tesla – SI unit of magnetic flux density; the magnetic flux density of a magnetic flux of 1 Wb through an area of 1 m^2.

Thermal Capacity – the quantity of heat required to raise the temperature of the whole body by one degree (1 K or 1 °C).

Thermal equilibrium – when two bodies in contact are at the same temperature, and there is no flow of heat between them; also, the average temperature of the bodies in thermal equilibrium.

Thermal expansion – the increase in the size of an object when heated.

Thermometer – a device used for the measurement of temperature; the mercury thermometer is commonly used.

Third Law of Motion – whenever two objects interact, the force exerted on one object is equal in size and opposite in direction to the force exerted on the other object; forces always occur in matched pairs that are equal and opposite.

Total internal reflection – a condition where all light is reflected from a boundary between materials; occurs when light travels from a denser to a rarer medium, and the angle of incidence is greater than the critical angle.

Transformation of energy – the conversion of one form of energy into another (e.g., when a body falls, its potential energy is converted to kinetic energy).

Transverse wave – a wave in which the particles of the medium oscillate in a direction perpendicular to the direction of propagation of the wave (e.g., water waves, light waves, radio waves).

Trough – the point of maximum negative displacement on a transverse wave.

U

Ultrasonic – sound waves too high in frequency (above 20,000 Hz) to be heard by the human ear.

Unbalanced forces – when some forces act on a body and the resultant force is not zero.

Uniform acceleration – when the velocity of a body increases by equal amounts in equal intervals of time.

Uniform circular motion – the motion of an object in a circular path with uniform speed; accelerated motion.

Uniform speed – when a body travels equal distances in equal intervals of time.

Uniform velocity – when a body travels along a straight line in a direction and covers equal distances in equal intervals of time.

Universal Law of Gravitation – every object in the universe is attracted to every other object with force directly proportional to the product of their masses and inversely proportional to the square of the distance between the centers of the two masses.

Unpolarized light – light consisting of transverse waves vibrating in all possible random directions.

V

Van der Waals force – general term for weak attractive intermolecular forces.

Vapor – the gaseous state of a substance that is generally in a liquid state.

Vector quantity – a quantity which needs both magnitude and direction to describe it.

Velocity – distance traveled by a body in a direction per unit time; the displacement of the body per unit time; a vector quantity; the SI units are m/s.

Vibration – a back and forth motion that repeats itself.

Virtual image – an image formed when the reflected or refracted light rays appear to meet; this image cannot be projected on a screen.

Volt – a unit of potential difference equivalent to joules/coulomb.

Voltage drop – the difference in electric potential across a resistor or other part of a circuit that consumes power.

W

Watt – SI unit for power; equivalent to joule/s.

Wave – a disturbance or oscillation that moves through a medium.

Wavelength – the distance between the two nearest points on a wave which are in the same phase; the distance between two adjacent crests or two adjacent troughs.

Wave (mechanical) – a periodic disturbance produced in a material medium due to the vibratory motion of the particles of the medium.

Wave mechanics – alternate name for quantum mechanics derived from the wavelike properties of subatomic particles.

Wave motion – the movement of a disturbance from one part of a medium to another involving the transfer of energy but not the transfer of matter.

Wave period – the time required for two successive crests or other successive parts of the wave to pass a given point.

Wave velocity – the distance traveled by a wave in one second; it depends on the nature of the medium through which it passes.

Weight – the force with which a body is attracted towards the center of the Earth; the SI unit is N; the gravitational units are kg·wt and g·wt; the weight of a body is given by *mg*.

Weightlessness – the state when the apparent weight of a body becomes zero; all objects while falling freely under the action of gravity are seemingly weightless.

Wien's Displacement Law – states that for a black body, the product of the wavelength corresponding to its maximum radiance and its absolute temperature is constant.

Work – work is done when a force acting on a body displaces it; Work = Force × Displacement (W = Fd) in the direction of the force; work is a scalar quantity; the SI unit is Joule.

Y

Young's modulus of elasticity – the ratio of normal stress to the longitudinal strain produced in a body.

Z

Zeeman effect – the splitting of the spectral lines in a spectrum when the source is exposed to a magnetic field.

Zeroth Law of Thermodynamics – states that if body A is in thermal equilibrium with body B, and B is also in thermal equilibrium with C, then A is necessarily in thermal equilibrium with C.

Everything you always wanted to know about…

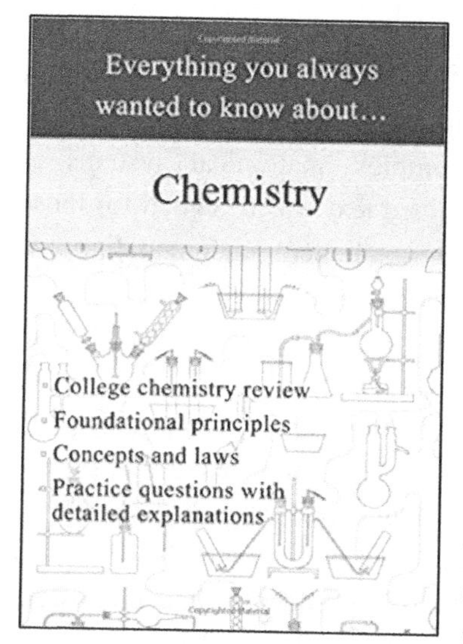

Chemistry

ISBN-13: 978-1947556874

From the foundations of the chemical reactions to the complex mechanisms of atomic particles, this general chemistry guide provides readers with the information necessary to be better equipped to understand these multifaceted chemistry topics. This book is a detailed review of all the fundamental processes and mechanisms affecting general chemistry and physical processes at the atomic level. The content was designed for those who want to develop a better understanding of the electronic structure of elements, principles of chemical bonding, phases of matter, types and mechanisms of chemical reactions, as well as essential principles of solution chemistry and acid-base equilibria. Learn about rate processes in chemical reactions, empirical and molecular formulas, bond dissociation energy for the heats of formation, Gibbs free energy, enthalpy, entropy, oxidation number, the laws of thermodynamics, and electrochemistry.

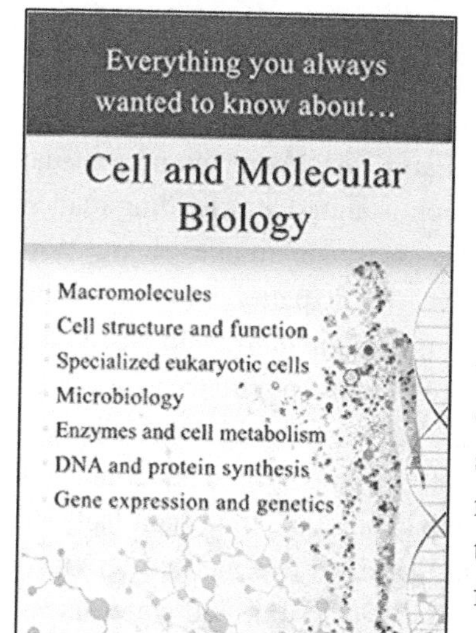

Cell and Molecular Biology

ISBN-13: 978-1947556683

Created by highly qualified science teachers, researchers, and education specialists, this book educates and empowers both the average and the well- informed readers, helping them develop and increase their understanding of biology. From the foundations of a living cell to the complex mechanisms of gene expression, this self-teaching guide provides readers with the information necessary to make them better equipped for navigating these multifaceted biology topics. This book was designed for those who want to develop a better understanding of cell structure and function, cell metabolism, DNA and genetics, as well as the technological and ethical challenges of modern science. The content is focused on an essential review of all the important processes and mechanisms affecting organisms on the cellular and molecular levels. You will learn about macromolecules, enzymes, cell cycle, photosynthesis, the significance of the various DNA mutations and heredity, as well as how different cell processes affect the overall well-being of an organism.

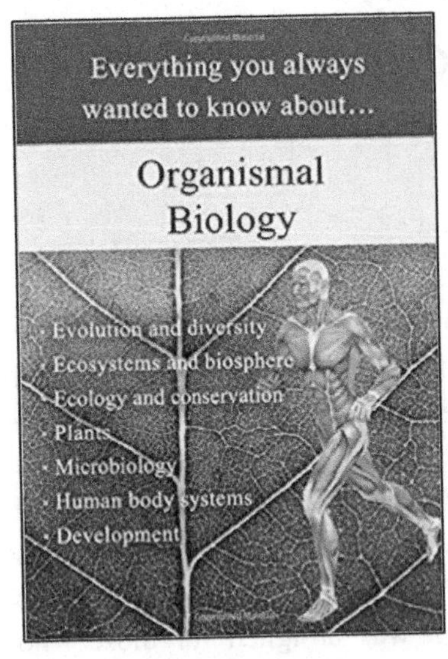

Organismal Biology

ISBN-13: 978-1947556690

Organismal biology is the study of structure, function, ecology and evolution at the organismal level. From the origin of life to the complex anatomical systems of humans, this clearly explained text was designed for those who want to develop a better understanding of ecosystems, diversity and classification, as well as anatomy and physiology. The content is focused on an essential review of all the important events and mechanisms affecting populations and communities. You will learn about evolution, comparative anatomy, plants, microorganisms, as well as human anatomical systems, along with physiological processes affecting the overall organism. Created by highly qualified science teachers, researchers, and education specialists, this book educates and empowers both the average and the well- informed readers, helping them develop and increase their understanding of biology.

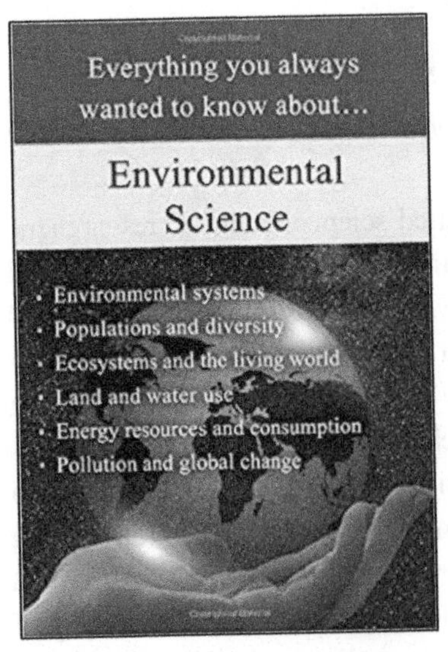

Environmental Science

ISBN-13: 978-1947556645

From the foundations of Earth systems to the present-day climate challenges, this book is aimed at providing readers with the information necessary to make them more engaged and appreciative participants in the global environment. This guide is designed to help develop a better understanding of ecosystems, population dynamics, use of natural resources, as well as the political and social landscape of environmental challenges. The content is focused on an essential review of all the important facts and events shaping the natural world we live in. You will learn about Earth's biochemical cycles, land, and water use, energy resources and their consumption, the significance of the various environmental movements and global initiatives, as well as how different human actions affect the overall balance within ecosystems.

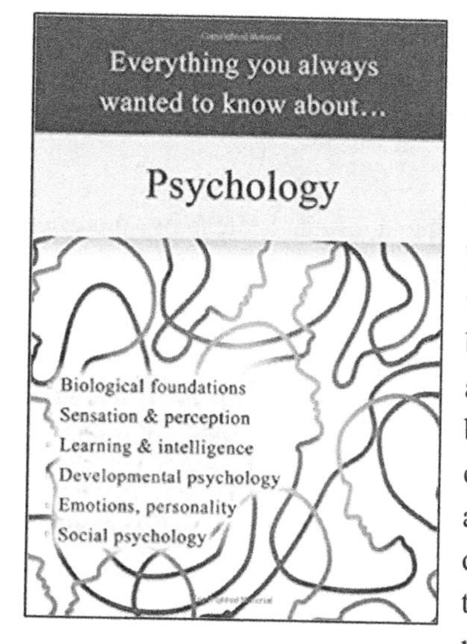

Psychology

ISBN-13: 978-1947556560

This book was designed for those who want to develop a better understanding of the human mind, emotions, feelings and behaviors, as well as the relationships between different historical events. Readers will learn about historically significant psychology researchers, the biological basis of human behavior, basic principles of consciousness and cognition, what drives human emotions and motivations, how early childhood psychological development affects human behavior, as well as develop the ability to compare and interpret theories and scientific methods, and to apply different theoretical frameworks to analyze a given situation. The book also describes all major groups of psychological disorders and covers the foundations of social psychology. From the foundations of human mind theories to the modern neuropsychology challenges, this clearly explained text is a perfect guide for anyone who wants to be knowledgeable about human psychology.

World History

ISBN-13: 9781947556881

Learn the world history starting from the early Neolithic period to the present-day events. This guide is a must-have book for anyone who wants to be knowledgeable about the history of human civilizations. As it goes through the sequence of the major events of the past, it provides readers with the analysis necessary to make them more educated and appreciative participants in the global future. Develop a better understanding of important civilizations, their economic and cultural growth and declines, the political and social challenges they went through, as well as the relationships between different historical events. Learn about historical figures and important events that set the foundations of influential civilizations, the meaning and significance of the historical shifts, as well as how each important historical event shaped its country's cultural heritage and political development.

American History

ISBN-13: 978-1947556553

From the founding of the United States of America government to the present-day challenges, this guide is perfect for anyone who wants to be knowledgeable about the history of America and its democracy. As it goes through the sequence of the events of the past, it provides readers with the analysis necessary to make them more engaged and appreciative participants in the American future. This book was designed for those who want to develop a better understanding of America's founding, economic and cultural growth, the political and social challenges it went through, as well as the relationships between different historical events. Learn about historical figures and important events that established the foundations of American government, the meaning and significance of the various social movements, as well as how each important historical event shaped the country's cultural heritage and political development.

European History

ISBN-13: 978-1947556980

Perfect guide for anyone who wants to learn about European history. From the rise of humanism and Renaissance to the 21st century developments, it goes through the sequence of the events of the past, providing readers with the information necessary to make them more engaged and knowledgeable learners of European history. This book was designed to help readers develop a better understanding of what shaped European governments, their economic and cultural underpinnings, political and social challenges they went through, as well as the causal relationships between different historical events. Learn about historical figures and important events that established the foundations of the European governments, the meaning and significance of the social movements, and how each important historical event shaped regional cultural heritage and political development.

American Government

ISBN-13: 978-1947556546

From the founding of the American government to the present-day challenges, this book is a great resource for anyone who wants to learn about the democratic system and the laws that affect the lives of Americans. This book elucidates the complexity of the U.S. political system as it provides readers with the information necessary to make them more engaged and competent participants in the American system of government. It helps develop a better understanding of American political culture, structures, and governmental functions, as well as the relationships between the branches of government and associated institutions. Learn the meaning and significance of the U.S. Constitution and the Bill of Rights, how each branch of government functions, and how they interact with each other. Understand the influence of special interest groups on policy and politics.

Comparative Government and Politics

ISBN-13: 978-1947556584

Why different countries have different forms of government and political institutions? Why some countries exist as democracies and others are authoritarian regimes? Why are some revolutions successful and others fail? This book was designed for those who want to develop a better understanding of political systems and regimes that affect the lives of people around the world, as well as political cultures, structures, governmental functions, and the relationships between the governments and the governed. Readers will learn about major events that shaped how governments function, the different institutions of government and political cultures that exist around the world, how branches of governments interact with each other and the governed, and how these institutions may be affected by the input from the populace. This clearly explained text is a perfect guide for anyone who wants to be knowledgeable about comparative government and politics.

Made in the USA
Middletown, DE
12 August 2023